Applied Chemical Hydrogeology

Alan E. Kehew
Western Michigan University

Upper Saddle River, New Jersey 07458

Library of Congress Cataloging-in-Publication Data

Kehew, Alan E.
    Applied chemical hydrogeology / by Alan E. Kehew
      p. cm.
    Includes bibliographical references and index.
    ISBN 0-13-270927-9 (alk. paper)
    1. Water chemistry.   2. Hydrogeology.   I. Title.

  GB855.K46 2000
  551.49—dc21   00-042749
                                         CIP

Editorial Director: *Paul F. Corey*
Acquisitions Editor: *Patrick Lynch*
Assistant Editor: *Amanda Griffith*
Editorial Assistant: *Sean Hale*
Executive Managing Editor: *Kathleen Schiaparelli*
Assistant Managing Editor: *Beth Sturla*
Production Editor/Composition: *Preparé Inc.*
Marketing Manager: *Christine Henry*
Marketing Assistant: *Erica Clifford*
Art Director: *Jayne Conte*
Cover Design: *Bruce Kenselaar*
Manufacturing Manager: *Trudy Pisciotti*
Assistant Vice President of Production and Manufacturing: *David W. Riccardi*

 © 2001 by Prentice-Hall, Inc.
Upper Saddle River, New Jersey 07458

Printed in the United States of America

10 9 8 7 6 5 4 3 2 1

ISBN 0-13-270927-9

Prentice-Hall International (UK) Limited, *London*
Prentice-Hall of Australia Pty. Limited, *Sydney*
Prentice-Hall Canada, Inc., *Toronto*
Prentice-Hall Hispanoamericana, S.A., *Mexico*
Prentice-Hall of India Private Limited, *New Delhi*
Prentice-Hall of Japan, Inc., *Tokyo*
Pearson Education Asia Pte., Ltd.
Editora Prentice-Hall do Brasil, Ltda., *Rio de Janeiro*

To Kay,
who shows me
the wherefore and why

Brief Contents

Contents

Preface

This text is intended to serve as a practical introduction to hydrogeochemistry, especially for students who plan to work as consultants or regulators in the environmental field. Hopefully, there is enough theory to provide a foundation for understanding and applying the examples and case histories that follow and to give students the tools they need in hydrogeochemistry to begin an applied career in hydrogeology. For those instructors who prefer a more theoretical treatment of aqueous geochemistry, there are several excellent books available.

I had several goals in mind when I began this project more than five years ago. The first was to introduce and tie together the several diverse disciplines that must be brought to bear on hydrogeochemical problems today. This entailed substantial reading to replace my lack of a formal background in essential topics including organic chemistry and microbiology. I leaned heavily on several specialized books in these areas. When I began to study ground water, Freeze and Cherry's book, *Groundwater*, was just coming out. That book expanded my horizons tremendously into areas that were only briefly mentioned in the hydrogeology courses I had taken, among those being the chemical aspects of ground water. If this book is even a fraction as useful in illustrating the interdisciplinary aspects of this field my goals will have been met.

The second objective for writing this book was to bring to the attention of students some of the best examples and case histories from the explosion in hydrogeological literature over the past several decades. I was particularly interested in including examples with practical applications, the types of problems that professional hydrogeologists deal with on a regular basis. The text hopefully will serve as a window into this vast literature that will stimulate students to explore more deeply on their own. To those whose work I did not include because of space limitations or my lack of awareness of it, I apologize. There is so much excellent basic and applied research being done that it is difficult, if not impossible, to keep up with it all.

A number of people have been instrumental in helping me through the countless times I just wanted to give up. Bob McConnin, who took over as geology editor after the merger of Prentice Hall and Macmillan, was a pleasure to work with. Patrick Lynch has proved to be an able replacement. The reviewers, Robert G. Corbett of Illinois State University, Phil Gerla of the University of North Dakota, and Seth Rose of Georgia State University provided invaluable help in pointing out errors, inconsistences, poor writing style, and in suggesting improvements. Several of these reviewers went well beyond the call of duty in their thoroughness. Their efforts are greatly appreciated.

Bonny True, a graduate student at Western Michigan University who took the course recently, read the entire copy-edited text and page proofs. Her careful reading caught many errors that eluded me, and I am very grateful for her help. Special thanks are due to Mick Lynch, who has allowed me to consult at his company, American Hydrogeology Inc. for the past ten years, with total flexibility in scheduling. This gave me an essential exposure to real-world problems that helped me decide what I wanted to include in a book for future hydrogeologists. I owe the greatest debt of gratitude to my wife Kay and my daughter Liz, who have tolerated six years of evenings and weekends when I was working on this book or stressed out because I should have been working on it.

Alan E. Kehew

About the Author

Alan E. Kehew was born in Pittsburgh, Pennsylvania. After graduation from Bucknell University in 1969, he received M.S. and Ph.D. degrees from Montana State University and the University of Idaho, respectively. He spent three years as an environmental geologist with the North Dakota Geological Survey and six years in the Department of Geology and Geological Engineering at the University of North Dakota. For the past fourteen years he has been in the Department of Geosciences at Western Michigan University, where he currently serves as chair. His major research interests have been ground water quality and glacial deposits and processes. He is married and has three daughters.

CHAPTER 1

Chemical Principles

The ability to understand and interpret ground water chemistry is based upon a basic understanding of chemical principles. Accordingly, we will begin by reviewing some of the chemical concepts that are most useful in the study of hydrogeochemistry.

ATOMIC STRUCTURE

Each element in the periodic table owes its properties to the structure of the smallest possible amount of the element, an *atom*. The structure of these particles consists of a central concentration of mass, the *nucleus*, which is composed of positively charged *protons* and *neutrons*, particles of neither positive nor negative charge. The positive charge of the nucleus is balanced by negatively charged *electrons* which inhabit a hierarchy of shells around the nucleus, with each shell containing subshells with specific energy levels. The number of electrons associated with an element is particularly important because it determines how successive shells are filled. The most important shell is the outermost because it is here that electrons can be removed from or added to the atom to form *ions*. When the former occurs, a positively charged *cation* has been created because the protons in the nucleus now outnumber electrons orbiting around the nucleus. The maximum number of electrons that can occupy the outermost shell is eight; cations tend to be formed when the number of electrons in the outer shell is small. By the removal of these electrons, the atom can then revert to a filled outer shell, the most stable configuration. Elements that tend to attract electrons in order to fill their outer shells form negatively charged *anions*. This occurrence is favored when the outer shell needs only one or two electrons to be complete.

The number of protons in an element, which is identical to the number of electrons in the outer shell, is the basis for the arrangement of the elements in the *periodic table* (inside front cover). One forms the table by ordering the elements in rows (periods) according to the number of protons in the nucleus. This quantity is called the *atomic number;* it increases from left to right in the table. The columns, or groups, of the periodic table include elements with similar numbers of electrons in their outer shells. Elements in a particular group have many chemical similarities. The periodic table also gives other useful information about the elements. For example, the *atomic weight* of each element is given on the table. This quantity is a relative weight which is proportional to the very small true weight of the atoms.

TABLE 1–1 Oxidation numbers of common elements.

Element	Most Common Oxidation Numbers (Valences)
Fixed	
O	−II, 0
H	+I, 0
Ca, Mg	+II, 0
K, Na	+I, 0
Variable	
N	+V, +IV, +III, +II, +I, 0, −III
Cl	+VII, +V, +IV, +III, +I, 0, −I
Mn	+VII, +VI, +IV, +III, +II, 0
S	+VI, +IV, +II, 0, −II
Cu	+II, +I, 0
Cr	+VI, +III, 0
C	+IV, +III, +II, +I, 0, −I, −II, −III, −IV

Atomic weights are commonly measured in grams, and the *gram atomic weight* is just that—the weight of an element expressed in grams. You will also remember that the gram atomic weight, or the gram formula weight in the case of a molecule composed of more than one element, is also referred to as a *mole*. A mole of any compound contains Avogadro's number of atoms or molecules, 6.023×10^{23}.

The loss or gain of electrons to complete an outer shell is indicated by the *oxidation number*, or *valence*, of an ion. For example, to form the compound sodium chloride, chlorine gains one electron for a valence of −I to complete its outer shell. Sodium loses one electron to yield a valence of +I, which also results in a filled outer shell. The oxidation numbers of some of the most important elements are given in Table 1–1.

CHEMICAL BONDING

Attractive forces between atoms are caused by interactions between the electrons in the outer shell, the valence electrons. Thus, different types of *chemical bonds* are formed depending on these interactions. The formation of ions is dependent on the amount of energy required to remove an electron from a neutral atom, the *ionization potential*, which is in turn related to the degree of filling of the outer shell. Nonmetals, which tend to have very high ionization potentials, are therefore unlikely to lose electrons to become cations. The low ionization potential of metals explains their occurrence as positive ions. Nonmetals actually attract electrons, a property known as *electron affinity*. As a result of this attraction, they commonly form anions.

When cations such as sodium (Na^+) are brought near anions such as chloride (Cl^-), there is a strong electrostatic attraction between the oppositely charged ions. The

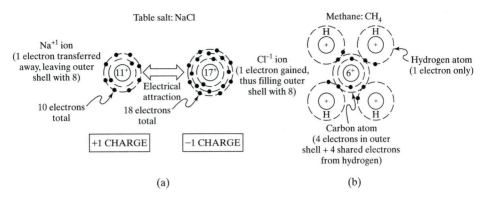

FIGURE 1–1   Formation of ionic bond between sodium and chloride (a) and covalent bond between hydrogen and carbon (b) to form methane.

attraction leads to the formation of an *ionic bond*, one of the two major types of chemical bonds. The resulting ion pair is strongly polar, with definite negative and positive charge centers (Fig. 1–1).

A contrasting situation occurs when two atoms of the same element, hydrogen for example, are brought in close proximity. Here, the single electron possessed by each hydrogen atom is shared equally by both of them to complete the outer shell. A bond of this type is known as a *covalent bond*. Covalent bonding is particularly important in organic compounds. Sharing of electrons can also occur in unlike atoms, for example, between hydrogen and oxygen in water. Electrons from two hydrogen atoms will be shared with one oxygen atom to complete its outer shell. These bonds, however, differ somewhat from the covalent bonds in the hydrogen (H_2) molecule, because hydrogen and oxygen have different affinities for electrons. A relative scale of a property known as *electronegativity* has been set up to measure the degree to which electrons are attracted to an atom (Table 1–2). From Table 1–2, it is apparent that oxygen has a higher electronegativity and therefore attracts electrons more strongly than hydrogen. The electrons are not equally shared in the water molecule, and, as a result, the charge is not equally distributed between the two. This type of bond is a *polar covalent bond* because the molecule has a polar behavior resulting from the unequal charge distribution. The difference in electronegativities between two members of a polar covalent bond is a measure of the relative amount of polarity. The greatest differences in electronegativity are possessed by ions in ionic bonds (Fig. 1–2). Bond types can be predicted by determining the difference in electronegativity between two atoms. A difference of 1.7 units or greater will result in an ionic bond. Polar covalent bonds form with an electronegativity difference between 0.5 and 1.6 units. A difference less than 0.5 units will yield a nonpolar covalent bond. Thus, the bond between carbon and hydrogen ($2.5 - 2.1 = 0.4$) is nonpolar covalent, whereas the bond between oxygen and hydrogen ($3.5 - 2.1 = 1.4$) is polar covalent.

Electronegativity is also useful in assigning oxidation numbers for elements in covalent bonds. The atom with the higher electronegativity is given a negative charge, and

TABLE 1–2   Electronegativity values of the elements.

H 2.1																	He ...
Li 1.0	Be 1.5											B 2.0	C 2.5	N 3.0	O 3.5	F 4.0	Ne ...
Na 0.9	Mg 1.2											Al 1.5	Si 1.8	P 2.1	S 2.5	Cl 3.0	Ar ...
K 0.8	Ca 1.0	Sc 1.3	Ti 1.5	V 1.6	Cr 1.6	Mn 1.5	Fe 1.8	Co 1.8	Ni 1.8	Cu 1.9	Zn 1.6	Ga 1.6	Ge 1.8	As 2.0	Se 2.4	Br 2.8	Kr ...
Rb 0.8	Sr 1.0	Y 1.2	Zr 1.4	Nb 1.6	Mo 1.8	Tc 1.9	Ru 2.2	Rh 2.2	Pd 2.2	Ag 1.9	Cd 1.7	In 1.7	Sn 1.8	Sb 1.9	Te 2.1	I 2.5	Xe ...
Cs 0.7	Ba 0.9	La-Lu 1.1-1.2	Hf 1.3	Ta 1.5	W 1.7	Re 1.9	Os 2.2	Ir 2.2	Pt 2.2	Au 2.4	Hg 1.9	Tl 1.8	Pb 1.8	Bi 1.9	Po 2.0	At 2.2	Rn ...
Fr 0.7	Ra 0.9	Ac-Lr 1.1															

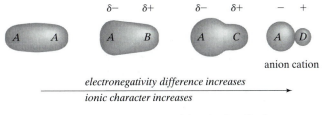

order of electronegativity: $A > B > C > D$

FIGURE 1–2   Charge distribution in bonds between atoms of the same element (left) compared to bonds between elements of increasing negativity difference.

the atom with the lower electronegativity is given a positive charge. For example, carbon is assigned an oxidation number of −IV in methane (CH_4), but in carbon dioxide (CO_2), it is considered to have an oxidation number of +IV. As shown in Table 1–1, many common elements can have several oxidation numbers depending upon the specific compound. In aqueous solutions, molecules with polar covalent bonding can be involved in still another bond type, the *hydrogen bond*. Hydrogen bonding is common between water molecules which, because of the polar covalent bonding between hydrogen and oxygen, have positive and negative charge centers. Adjacent water molecules, therefore, can bond together through hydrogen bonds, although these bonds are considerably weaker than covalent bonds.

AQUEOUS SOLUTIONS AND THE PROPERTIES OF WATER

Throughout this text we will be discussing solutions in which one or more solutes are dissolved in water. All ground waters and surface waters are aqueous solutions of this type. Because dilute aqueous solutions behave much like pure water, it is appropriate to consider the chemical properties of this essential substance. Pure water has a melting point of 0°C and a boiling point of 100°C. Its specific gravity at 4°C, the temperature at which it achieves its maximum density, is 1.000 kg/l.

Atoms of an element with different numbers of neutrons in the nucleus are known as *isotopes*. Hydrogen occurs in three isotopes (Table 1–3): ^1H (protium), ^2H (deuterium), and ^3H (tritium). Although tritium and deuterium constitute only very small percentages of the total reservoir of hydrogen in water, they have very useful hydrologic

TABLE 1–3   Isotopes of hydrogen and oxygen.

Isotope	Relative Abundance (%)	Type
^1H protium	99.984	Stable
^2H deuterium	0.016	Stable
^3H tritium	$0–10^{-15}$	Radioactive; half-life 12.3 years
^{16}O oxygen	99.76	Stable
^{17}O oxygen	0.04	Stable
^{18}O oxygen	0.2	Stable

applications. Deuterium, a stable isotope of hydrogen, makes up about 0.016% of the hydrogen present in water. A number of physical and chemical processes, however, produce small variations in this percentage. The differences resulting from these *fractionation* mechanisms are measurable and become very useful in understanding the processes that have affected a particular water mass. Tritium is a radioactive isotope of hydrogen with a half-life of 12.3 years; and its usefulness lies in determining, or at least constraining, the age of subsurface water. The age refers to the amount of time since recharge of the water. Oxygen occurs in three stable isotopes: ^{16}O, the most abundant, ^{17}O, and ^{18}O. ^{18}O is used along with deuterium as a hydrologic tracer. The applications of these isotopes will be discussed in Chapter 8.

Structure of the Water Molecule

Judging from the electronegativity difference between hydrogen and oxygen (1.4), we can expect the water molecule to be polar covalent. Because this bond approaches the electronegativity difference necessary to form an ionic bond, the polarity of the water molecule is important. In fact, dipolar characteristics are critical in the chemical behavior of water. The opposite poles of individual molecules of water attract each other to form a weak bond known as the *hydrogen bond*. Water molecules are linked together through hydrogen bonds to form aggregates of about 100 molecules at room temperature, a structure that becomes very important in its ability to dissolve various substances. For example, nonpolar molecules, including many organic compounds, find it difficult to fit into these structured aggregates of water molecules. This helps to explain the low solubility in water of most petroleum hydrocarbon compounds that are spilled or leak into the subsurface.

Hydrogen bonding is responsible for many anomalous properties of water, such as its higher than expected boiling point. Water is the dihydride of oxygen, with a molecular weight of 18. Other dihydrides with heavier molecular weights, for example hydrogen sulfide (MW = 34), have lower boiling points and occur as gases at room temperature. Water, because of its hydrogen bonding, has a much higher boiling point and occurs as a liquid at room temperature. Water is also denser than related compounds at any temperature and, unlike many substances, freezes to form a less dense solid. Consider the implications of this property to aquatic life in regions where the temperature drops below the freezing point of water for significant periods of time. Lakes and streams would freeze from the bottom up if ice were denser than water and in winter would freeze solid.

As mentioned earlier, the polarity of the water molecule and the resultant aggregate structure give it its solvent properties. Ionic salts are very soluble because they are composed of positively and negatively charged ions. Water surrounds these ions in a solid and isolates them from their neighboring ions, thus neutralizing the attractive forces that form mineral crystals. These *hydrated* ions are free to move into solution. In their hydrated form, ions are encircled with layers of water molecules. When halite dissolves, it produces sodium and chloride ions according to the reaction

$$NaCl \rightleftharpoons Na^+ + Cl^-.$$

FIGURE 1–3   Polar configuration of water molecule (left), hydrated ions of sodium (middle), and chloride (right).

Sodium forms a hydrated ion with a radius of 0.95 a (Fig. 1–3). Chloride binds its water molecules less tightly, and the resulting hydrated chloride ion has a radius of 1.81 a. In solution, there is still some attraction between the hydrated cations and anions.

CONCENTRATION AND UNIT CONVERSIONS

Now that we have reviewed the properties of water itself, we can turn our attention to the properties of aqueous solutions, of which natural and contaminated ground water are examples. Water is the *solvent* into which other substances, or *solutes,* dissolve to form an aqueous solution. The amount of a particular solute dissolved in water (i.e., its concentration) is an essential characteristic of that particular solution. Several different systems of concentration units are in use.

Mass Concentrations

The most common method of expressing the concentration of solutes in ground water is to use *mass concentrations*, which are the masses of solutes in milligrams (mg) or micrograms (μg) contained in a liter of solution (mg/l or μg/l):

$$\text{mg/l} = \frac{\text{mass of solute (mg)}}{\text{volume of solution (1)}}. \tag{1–1}$$

Most laboratories report concentrations using this system of units. A related system of mass units is parts per million (ppm), which can be defined as

$$\text{ppm} = \frac{\text{mass of solute (mg)}}{\text{mass of solution (kg)}}. \tag{1–2}$$

These two systems of units are related through the density of the solution (ρ), which is the mass per unit volume, by the following expression:

$$\text{conc. in } \frac{\text{mg}}{\text{kg}} \text{ (ppm)} = \text{conc. in } \frac{\text{mg}}{1} \times \frac{1}{\rho}\left(\frac{1}{\text{kg}}\right). \tag{1–3}$$

If the density of the solution approximates that of pure water, $\rho = 1.0$ kg/l, ppm and mg/l will be numerically equal. Because the density of most ground water solutions is low enough to justify this assumption, the two types of units (mg/l and ppm) are commonly used interchangeably.

Sometimes problems arise in the designation of the mass to be used in mass concentration units. For example, we can express the concentration of nitrate (NO_3^-) in two ways. One way is to report the mass concentration of the nitrate ion in mg/l and the other way is to report the mass concentration of nitrate in terms of the mass of nitrogen. To illustrate the difference between these two methods, consider a solution that contains 50 mg NO_3^-/l. This concentration refers to the mass of the nitrate ion, which has a molecular weight of 62 $[14 + (3 \times 16)]$. The alternative would be to report the concentration in terms of the amount of nitrogen, which would be written as mg NO_3^-–N/l. This value would be determined as the concentration of mg NO_3^-/l times the ratio of the atomic weight of nitrogen (14) to the molecular weight of nitrate (62). Thus,

$$50 \, \frac{\text{mg } NO_3^-}{1} \times \frac{14}{62} = 11.3 \, \frac{\text{mg } NO_3^- - \text{N}}{1}.$$

It is important to take note of the form of the concentration unit being used, particularly when a compound with a drinking water standard, such as nitrate, is being reported. The drinking water standard for nitrate is 45 mg NO_3^-/l, or 10 mg NO_3^-–N/l, depending on which method is being used. Both concentrations represent the same amount of nitrate, although you might mistakenly assume that the first concentration was much higher than the second.

Molar Concentrations

It is often necessary to use concentrations based on the molar quantities of solutes. The two types of concentration units available for this purpose are *molarity (M)* and *molality (m)*. Molarity is the number of moles of solute dissolved in 1 liter of solution and molality is the number of moles of solute dissolved in 1000 g of solution. The difference between a 1 molar solution and a 1 molal solution is that the 1 molal solution may have a greater volume than 1 l. However, for most dilute ground water solutions, we can make the same assumption that we made for ppm and mg/l—that they are essentially equal.

Conversion from mass concentrations to molar concentrations can be accomplished through the use of the atomic or molecular weight of the solute. Conversion from mg/l to moles/l is given by

$$\frac{\text{mg}}{1} = \frac{\text{moles}}{1} \times \text{formula wt.} \left(\frac{\text{g}}{\text{mole}} \right) \times \frac{1000 \text{ mg}}{\text{g}}, \qquad (1-4)$$

and the opposite conversion is

$$\frac{\text{moles}}{1} = \frac{\text{mg/l}}{\text{formula wt.} \times 1000}. \qquad (1-5)$$

A form of molar concentration, the *mole fraction*, is used for solid solutions, such as mixtures of calcite and dolomite, and for mixtures of nonaqueous-phase liquids. The mole fraction of substance A in a mixture of A and B would be expressed as

$$\text{mole fraction of } A = \frac{\text{moles of } A}{\text{moles of } A + \text{moles of } B}. \qquad (1-6)$$

Equivalents and Normal Concentrations

The *normality* of a solute is a form of concentration that takes into account the valence of ionic solutes. Because the positive and negative charges in a solution must balance, the sum of the normalities of all ionic solutes in a solution must equal zero. In other words, the sum of the normalities of the positive ions must equal the sum of the normalities of the negative ions. The normality can be determined as the product of the molarity and the charge of the ion. For example, the normality of sulfate in a solution containing 0.003 moles/l of the ion would be 0.006. The latter quantity is also known as the *gram-equivalent weight* of sulfate. This term can be shortened to *equivalents* or *milliequivalents* (milligram-equivalent weight) so that the normal concentration of a solute becomes equivalents/l, or more commonly for typical amounts of solutes in ground water, milliequivalents/l (meq/l).

The concentrations of several analytical parameters in water are expressed as the equivalent weight of calcium carbonate or *mg/l as CaCO₃*. The equivalent weight of calcium carbonate is 50 g/eq because the molecular weight of $CaCO_3$, 100 g/mol, is divided by the ionic charge of calcium (Ca^{2+}) and carbonate (CO_3^{2-}) ions. One water quality parameter reported in such units is *hardness*, an undesirable property of water that produces a scummy residue and scale on plumbing fixtures, sinks, and bathroom tile. Hardness is measured as the amount of polyvalent metallic cations in solution, primarily Ca^{2+} and Mg^{2+}. Hardness can be calculated by multiplying the number of equivalents of calcium and magnesium by 50 or by using the concentrations of calcium and magnesium in mg/l in the following formula:

$$\text{total hardness} = 2.5 \,(\text{conc. of } Ca^{2+} \text{ in mg/l}) + 4.1 \,(\text{conc. of } Mg^{2+} \text{ in mg/l}). \qquad (1\text{--}7)$$

The factors 2.5 and 4.1 are the ratios of the formula weight of calcium carbonate (100) to the weights of calcium (40) and magnesium (24.3). Another parameter that is expressed as the equivalent weight of calcium carbonate, alkalinity, will be discussed in Chapter 3.

COMPOSITION OF GROUND WATER

Solutes in ground water are derived from contact between the water and various solids, liquids, and gases as the ground water makes its way from its recharge area to discharge area. Concentrations of solutes can vary enormously as a function of the mineral content of aquifers through which the ground water flows. Despite this great variety, there are some generalizations that can be made for natural (uncontaminated) ground water. A relatively small number of inorganic solutes are generally present in concentrations greater than 5 mg/l (Table 1–4). These constituents are all ions except for aqueous silica; therefore, the six of them are also referred to as the major ions. The number of minor and trace constituents present in ground water is much greater (Table 1–4). Occasionally, the concentrations of minor constituents can exceed those of some of the major ions. A common example is the high concentration of nitrate in agricultural areas. Iron can be elevated around lakes and wetlands and in contaminant plumes. In addition to the constituents shown in Table 1–4, there is also a wide range of organic solutes that can be present in both natural and contaminated ground water.

TABLE 1–4   Concentration ranges of dissolved inorganic
constituents in ground water. (After Freeze and Cherry, 1979;
Davis and De Wiest, 1966.)

Major constituents (greater than 5 mg/l)

Bicarbonate	Silica
Calcium	Sodium
Chloride	Sulfate
Magnesium	

Minor constituents (0.01–10.0 mg/l)

Boron	Nitrate
Carbonate	Potassium
Fluoride	Strontium
Iron	

Trace constituents (less than 0.1 mg/l)

Aluminum	Molybdenum
Antimony	Nickel
Arsenic	Niobium
Barium	Phosphate
Beryllium	Platinum
Bismuth	Radium
Bromide	Rubidium
Cadmium	Ruthenium
Cerium	Scandium
Cesium	Selenium
Chromium	Silver
Cobalt	Thallium
Copper	Thorium
Gallium	Tin
Germanium	Titanium
Gold	Tungsten
Indium	Uranium
Iodide	Vanadium
Lanthanum	Ytterbium
Lead	Yttrium
Lithium	Zinc
Manganese	Zirconium

CHARGE-BALANCE ERROR

A fundamental property of aqueous solutions is that they are electrically neutral; that is, the total number of equivalents of cations must equal the total number of equivalents of anions. Because the analysis of all the cations and anions that make up a ground water sample involves numerous laboratory and/or field techniques, analytical errors or omissions could result in a situation where the sum of the cations does not equal the sum of the anions. By calculating a quantity known as the *charge-balance error*, we can perform a basic and simple check of the water analysis. The charge-balance error is defined as

$$\text{C.B.E.} = \frac{\Sigma m_c z_c - \Sigma m_a z_a}{\Sigma m_c z_c + \Sigma m_a z_a} \times 100. \qquad (1\text{–}8)$$

The percentage value will have a positive or negative charge, depending upon whether the cations or anions are larger. Freeze and Cherry (1979) recommend that 5% is a reasonable limit for accepting the analysis as valid.

Because of the large number of constituents present in natural waters and the difficulties of achieving analytical perfection, we are not likely to encounter charge-balance errors of zero. Fritz (1994) compiled charge-balance errors from published ground water analyses in the literature and obtained a mean value for more than a thousand analyses of 3.99%, when the absolute values of the charge-balance errors were used. In general, there are two possible explanations for charge-balance errors. One is that a cation or anion of significant concentration in the solution was not measured, and the second is that there have been errors in the analysis. Fritz (1994) considered some of the possible errors for analyses that included all significant ionic species. Random errors would be expected to produce an equal number of positive and negative charge balances, whereas systematic errors would produce a preponderance of either positive or negative errors. One systematic error that Fritz discovered was that positive errors are more common for analyses in which alkalinity, a measure of the bicarbonate ion (HCO_3^-) concentration, was measured in the lab rather than in the field. The reason for this discrepancy is that calcium carbonate precipitated in the samples between the time of sample collection and lab analysis, thus resulting in an erroneously low alkalinity measurement. Calcium and magnesium, the other components of the precipitated carbonate minerals, are measured from a different sample that is acidified at the time of collection so that precipitation cannot take place. Therefore, the analyzed values are representative of their true concentrations in the aquifer. The result is that the analysis of the positive ions is accurate, the analysis of the negative ions is erroneously low, and a positive charge-balance error is recorded. Fritz (1994) discussed other analytical procedures that can lead to charge-balance errors.

The charge-balance error should be calculated for all analyses that include the appropriate cations and anions. This calculation can sometimes point out obvious errors, such as the misplacement of a decimal point in a concentration, or suggest problems with the analysis, requiring investigation of the sampling and analytical procedures.

GRAPHICAL DISPLAYS OF GROUND WATER CHEMISTRY

Graphs are the most useful method of illustrating ground water chemistry for comparison of analyses from different samples. Although many types of graphs have been developed for this purpose, we will focus on bar, Stiff, and Piper plots. Fig. 1–4 shows a bar graph, which is composed of two vertical bars for each water analysis. The bar on the left is for cations and the bar on the right is for anions. Concentrations are plotted in milliequivalents per liter. For a reliable analysis, the cation and anion bars must be the same height. If they do not match, it means the analysis is faulty, incomplete, or both. The Stiff graph (Fig. 1–5) is composed of four horizontal axes on which the cations in milliequivalents per liter are plotted to the left of a vertical axis and the anions are marked

FIGURE 1–4 Four water analyses shown as bar graphs. Left bar of each pair shows cations and right bar shows anions in milliequivalents per liter (from Hem, 1989).

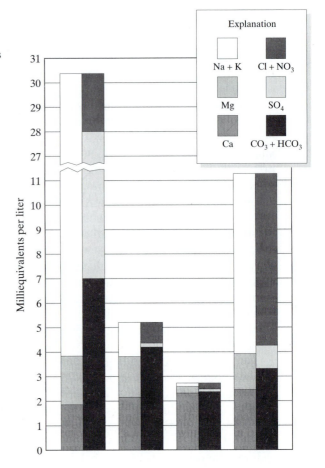

off to the right. When data are plotted in four rows and the points are connected, the result is a polygon whose shape is representative of a particular ground water type. Because different water types produce different shapes, the Stiff graph is a good visual fingerprint of water quality. Its disadvantages are that each analysis requires its own Stiff plot and a limited number of Stiff plots can be shown on a diagram.

The Piper trilinear plot (Fig. 1–6) lacks the shortcoming of the Stiff graph in that many analyses can be plotted. The graph consists of two triangles, one for cations and one for anions, and a centrally located diamond-shape figure. The three axes of the cation triangle are used for concentrations of calcium, magnesium, and sodium + potassium. Values are plotted as percentages of the total of those three groups in milliequivalents per liter. The anion triangle is constructed similarly, with axes for chloride, sulfate, and carbonate + bicarbonate. Thus, each water analysis results in a point on the anion triangle and a point on the cation triangle. These points are then projected diagonally upward, parallel to diagonals on the diagram, to their point of intersection on the diamond. This intersection can be plotted as a single point or as the center of a circle whose

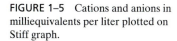

FIGURE 1–5 Cations and anions in milliequivalents per liter plotted on Stiff graph.

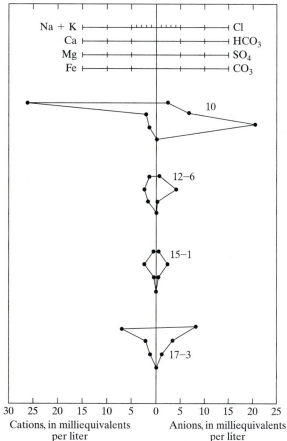

diameter is proportional to the TDS (total dissolved solids) of the analysis. Waters of greatly varying TDS, yet similar relative concentrations of ions, can be differentiated when circles rather than points are plotted on the diamond. In addition, trends of increasing or decreasing dissolved solids concentration, with or without an accompanying change in relative ion concentration, are evident from the diagram. TDS is a useful bulk property of an aqueous solution that is determined by weighing the evaporated residue of a known volume of water. Caution is recommended in interpretation of TDS values because of chemical breakdown of some parameters at the temperature used to evaporate the solution (usually 180°C). For example, bicarbonate ion breaks down to carbonate ion, carbon dioxide, and water at this temperature. For this reason, the value of TDS obtained from the evaporated residue does not always correspond exactly to the value obtained by summing the concentrations of individual species in the solution. Hem (1989) provides more information about TDS determination.

The Piper plot also serves another function—the classification of the water type based on the locations of plotted points in the ion triangles (Fig. 1–7). The classes of

FIGURE 1–6 Water analyses plotted on Piper (trilinear) diagram. Cation percentages in milliequivalents per liter plotted on three axes on left triangle, and anions plotted in same way on right triangle. For example, Ca percentage is calculated as

$$\left[\frac{Ca}{(Ca + Mg + Na + K)} \times 100 \right].$$

Intersection of lines from triangles is plotted as single point on diamond.

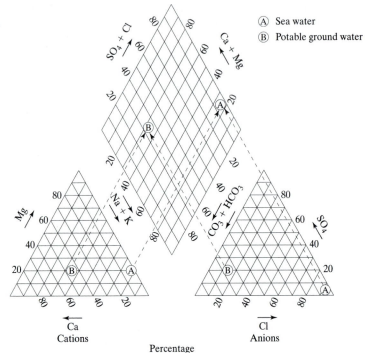

FIGURE 1–7 Classification of hydrochemical facies using the Piper plot. Facies names include combination of cation and anion types, for example, calcium-sulfate type.

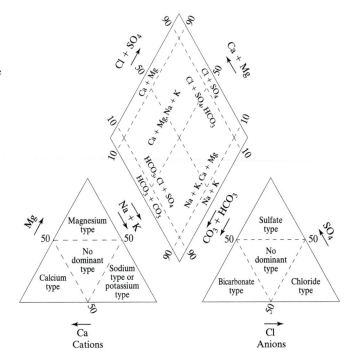

water types are also known as *chemical facies*. Examples of chemical facies names are the calcium-bicarbonate and the sodium-chloride facies. If either the cation or anion composition falls within the region of no dominant type, the facies name would be mixed cation or mixed anion. An example would be a mixed cation-sulfate water. The Piper plot, along with the other types of graphs, is widely used in the literature to illustrate differences, variations, and trends in ground water chemistry. Hypotheses concerning the chemical evolution of ground water are greatly aided by these diagrams, because chemical facies are indicative of specific geochemical processes that take place in aquifers.

PROBLEMS

1. Determine the oxidation number of N in the following chemical species. Name each species.
   a. NO_3^- **b.** NH_4^+ **c.** N_2

2. Determine the oxidation number of S in the following chemical species. Name each species.
   a. SO_4^{2-} **b.** H_2S **c.** HS^- **d.** FeS

3. List the compound or mineral name, electronegativity difference, and bond type (between underlined elements) for each compound.
   a. C$\underline{H}_4$ **b.** $\underline{K}\underline{Cl}$ **c.** $Al(\underline{OH})_3$ **d.** $\underline{CCl}_4$
   e. $\underline{C_6H_6}$ **f.** $Ca\underline{CO}_3$ **g.** $Ca\underline{F}_2$ **h.** $\underline{SiO}_2$

4. Routine analysis of a water sample provides the following concentrations (in mg/l): Ca, 93.9; Mg, 22.9; Na, 19.1; bicarbonate, 344; sulfate, 85.0; chloride, 9.0; pH: 7.20.
   a. Express each concentration in terms of molarity, molality, and normality.
   b. Determine the hardness of the water in mg/l as $CaCO_3$.

5. The following ground water analyses are expressed in mg/l. Plot Stiff and Piper graphs for each analysis. Also calculate the charge balance error for each analysis and state its chemical facies (analyses from Hem, 1989).

	Ca^{2+}	Mg^{2+}	Na^+	K^+	HCO_3^-	SO_4^{2-}	Cl^-	NO_3^-	Fe	pH
1.	48	3.6	2.1		152	3.2	8.0		0.05	7.5
2.	144	55	29		622	60	53	0.3		
3.	394	93	333		157	1150	538	5		
4.	40	22	0.4	1.2	213	4.9	2.0	4.8	0.24	7.4
5.	140	43	21		241	303	38	4.1	0.01	7.4
6.*	3.0	7.4	857	2.4	2080	1.6	71	0.2	0.15	8.3
7.**	40	50	699	16	456	1320	17	1.9	0.4	8.2
8.	353	149	1220	9.8	355	1000	1980	24	0.01	

*$CO_2^{2-} = 57$ **$CO_3^{2-} = 26$

CHAPTER 2

Chemical Equilibrium and Kinetics

As water moves from the ground surface, where it originates as rainfall or snowmelt, to an aquifer, and then toward a discharge area, it chemically reacts with the minerals with which it contacts. These reactions determine the types and amounts of solutes that the ground water will contain at any point. As hydrogeologists, there are several questions that we might have about these chemical processes. First, what are the concentrations that could be produced by contact between a mineral in the aquifer and the ground water moving through it? This information would indicate whether or not a sample that we collect from the aquifer is actively dissolving the mineral in question or whether the mineral could be precipitating out of solution under ambient conditions in the aquifer. The tool that we use for this type of information is the concept of chemical equilibrium, which is based on thermodynamic principles. By knowing the state of equilibrium between the water and minerals within the aquifer, we can predict the type of reactions that are occurring or would be likely to occur. The principles of chemical equilibrium are the foundation for an understanding of the chemical evolution of ground water.

Another basic question is, How fast do reactions take place along the path followed by the ground water? This question is answered by the use of kinetics, which in this case concerns the rate of reaction between ground water and minerals. Answering these questions regarding chemical equilibrium and kinetics is a logical first step in an analysis of ground water chemistry.

THE MEANING OF EQUILIBRIUM

A general reaction in which substances A and B react to form products C and D would be shown as

$$A + B \rightleftharpoons C + D. \tag{2-1}$$

The equation implies that the products can react to form the reactants, that is, the reaction is reversible. In this situation, the forward and reverse reactions will be occurring simultaneously so that the reaction will not go to completion in either direction. Completion would entail the total conversion of A and B to C and D, or vice versa. If we assume that only reactants A and B are present at the beginning of the reaction, the rate of reaction will not be constant as the reaction progresses. In fact, the rate is proportional to the concentrations of A and B according to the expression

$$R_1 = k_1 a_A a_B,$$

(2-2)

where a_A and a_B are the concentrations of A and B, respectively, and k is a constant of proportionality. The concentrations are expressed in a form called activity, which we will describe in more detail later. As this reaction progresses, C and D, initially absent, begin to be produced at an increasing rate. The rate of this reaction is

$$R_2 = k_2 a_C a_D.$$

(2-3)

The change in rates for the two reactions is shown in Fig. 2–1. At some point after the reaction begins, the forward rate [Eq. (2–2)] becomes equal to the reverse rate [Eq. (2–3)], which can be stated as

$$k_2 a_C a_D = k_1 a_A a_B.$$

(2-4)

This equation can be rearranged as

$$\frac{k_1}{k_2} = \frac{a_C a_D}{a_A a_B}.$$

(2-5)

Because k_1 and k_2 are constants, k_1/k_2 is a constant and the equation can be rewritten as

$$K = \frac{a_C a_D}{a_A a_B}.$$

(2-6)

The constant K is known as the *equilibrium constant* of the reaction; and it expresses the constant ratio of the activities of the reactants that is present when the reaction rates have equalized. Equation (2–6) is a statement of the *law of mass action*, which states that, at equilibrium in a reversible reaction, the product of the activities of the products over the product of the activities of the reactants is a constant. In a more general form of the law of mass action, consider the reaction

$$aA + bB \rightleftharpoons cD + dD.$$

(2-7)

In this reaction, the coefficients represent molar quantities of the products and reactants. The corresponding equilibrium expression is written

$$K_{eq} = \frac{a_C^c a_D^d}{a_A^a a_B^b}.$$

(2-8)

The activities are raised to powers of the molar quantities.

FIGURE 2–1   Reaction rates for forward (R_1) and reverse (R_2) directions for reversible reaction $A + B \rightleftharpoons C + D$, assuming only A and B present at beginning. Reaction is at equilibrium when $R_1 = R_2$.

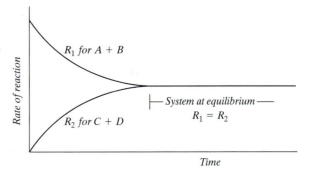

THERMODYNAMICS AND CHEMICAL EQUILIBRIUM

If we had a way of independently determining the equilibrium constant without performing an actual lab experiment in which we mixed reactants and then measured reactants and products, it would give us a powerful tool with which to examine water chemistry. As an example, consider a water sample taken from an aquifer that contains calcite. A reaction could be written for the dissolution of calcite, and the reactants and products could be determined from the water sample. We could then compute the ratio of the products to the reactants and compare this ratio to the equilibrium constant. If the ratio were less than K, it would tell us that the water is not in equilibrium with calcite and that the mineral is still dissolving in this aquifer. If the ratio did equal the equilibrium constant, it would be an indication that the water is in equilibrium with calcite. Calcite would not be expected to dissolve or precipitate under these circumstances.

Fortunately, there is a method for determining equilibrium constants without actually performing an experiment, a method that makes use of known thermodynamic quantities. The most useful relationship involves the parameter known as *Gibbs free energy, G*, which is a measure of a reaction's ability to accomplish "useful" work, that is, energy released by the reaction can be converted to work. This conversion requires that the amount of energy within the system decreases as the reaction proceeds. The concept of Gibbs free energy recognizes that a portion of the energy involved in a chemical reaction is unavailable to do work.

It is actually the change in Gibbs free energy of a reaction, ΔG, that will be of use to us; and it is related to other important quantities by the expression

$$\Delta G = \Delta H - T\Delta S, \tag{2–9}$$

where ΔH is the change in *enthalpy*, or total energy of the reaction, ΔS is the change in *entropy*, and T is the temperature in degrees Kelvin. The Gibbs free energy change occurs at constant temperature and pressure. Entropy refers to the degree of randomness or disorder in the system. The molecules of a gas, for example, which disperse throughout the volume that they occupy, have a higher degree of disorder than molecules of the same substance in a liquid or solid phase. Thus a change from the liquid to the gas phase would entail an increase in entropy.

A useful analogy in understanding free energy in a chemical system is to look at potential energy in a mechanical system. As shown in Fig. 2–2, a cart that is released on a hillside will come to rest at the position of minimum potential energy, or elevation. In a similar way, all spontaneous chemical reactions (those that occur without the addition of energy) proceed in a direction that tends to minimize the Gibbs free energy of the system. This minimum energy level of the reaction is a manifestation of the equilibrium condition that we defined earlier.

Returning again to our general reaction, $aA + bB \rightleftharpoons cC + dD$, consider the energy change that would take place if the reaction began with only A and B or with only C and D. The plots of free energy vs. time for these two conditions would arrive at the same final level of free energy, the minimum level for the reaction (Fig. 2–3). This level would represent equilibrium. A further conclusion that can be drawn from the above discussion is that a spontaneous reaction proceeds in a direction in which Gibbs free energy decreases, so that the free energy change will be $-\Delta G$. According to Equation (2–9),

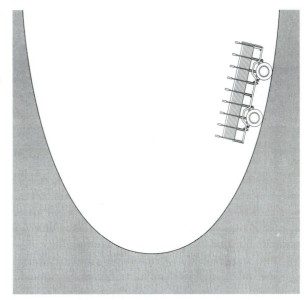

FIGURE 2–2 Mechanical analogy to chemical equilibrium. Cart will come to rest at point of lowest potential energy as chemical reactions will achieve chemical equilibrium at minimum Gibbs free energy.

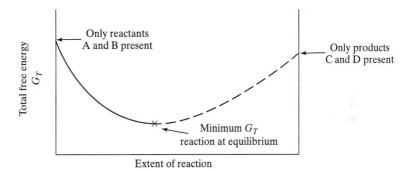

FIGURE 2–3 Energy changes in reaching equilibrium for reaction $aA + bB \rightleftharpoons cC + dD$. Same point is reached if initially only A and B are present or C and D are present.

the free energy can decrease by releasing heat (exothermic), in which ΔH is negative, or by increasing the entropy of the system, in which $T\Delta S$ increases, or by any combination of changes in enthalpy or entropy resulting in a negative ΔG.

Chemical Potential

So far we have discussed only the free energy (or its change) of the entire reaction. To assess the role of each individual species in solution, we must evaluate its contribution to the free energy of the system. This can be done by use of the *chemical potential, μ,* which is defined as the free energy per mole of an individual species. Thus,

$$G_i = n_i\mu_i, \tag{2–10}$$

where n_i is the number of moles of species i and μ_i is the chemical potential of species i. Chemical potential, and therefore free energy, has two components: one that represents the free energy per mole under *standard state* conditions and one that reflects the temperature and concentration of the solution in question. Again, we will use activity as the form of concentration for these relationships. The two components of chemical potential can be expressed as

$$\mu_i = \mu_i^0 + RT \ln a_i, \tag{2-11}$$

in which μ_i^0 is the standard chemical potential (chemical potential in the standard state), R is the universal gas constant $(8.3143 \times 10^{-3}\,\text{kJ/mol})$, a_i is the activity of species i, and T is temperature in degrees K.

Activity

When a salt such as sodium chloride dissolves in water, it forms an *electrolyte* solution that will conduct electricity, because the atoms that dissolve into solution form ions. As illustrated in Chapter 1, the cations and anions become surrounded by water molecules, which tends to neutralize the attractive forces between cations and anions and the repulsive forces between similarly charged ions. Even though these interactive electrical forces are partially neutralized by the surrounding water molecules, some *residual interionic forces of attraction* still remain. These forces prevent the ions from acting as completely independent particles in solution. For example, the addition of sodium chloride to water will depress the freezing point and raise the boiling point, but not quite as much as theoretically should occur. The sodium chloride behaves as if it were not completely ionized, when, in fact, it is. The residual interionic forces are also known to be a function of the concentration of the solute. The more dilute the solution, the greater the spacing becomes between individual ions. As a result, the residual forces are at a minimum; and the solute behaves as if it were totally ionized. At higher concentrations, there is an increase in the residual interionic forces; consequently, the apparent lack of ionization increases. In this situation, there is a difference between the true concentration of solute that can be determined from an analysis and an apparent concentration, which is lower than the true concentration because the electrolyte acts as if it were only partially ionized. This apparent concentration is known as the *activity* of the solute, which differs from the concentration because of ionic interactions in solution.

The relationship between activity and concentration can be defined as

$$\gamma_i = \frac{a_i}{M_i}, \tag{2-12}$$

where a_i = activity, M_i = molar concentration, and γ_i is the *activity coefficient*. Determination of the activity coefficient will be discussed later in this chapter.

Standard States

To correct for ionic interactions in aqueous solutions, the concept of the standard state has been developed. The standard state is one in which the activity equals the concentration. For ionic solutes, the standard state is defined as a hypothetical 1 molar solution in which the activity equals the molarity ($a = M$). A temperature of 25°C and a pressure of 1 atm are also specified for the standard state. In real solutions, the activity approaches the concentration as the solution becomes more and more dilute.

The chemical potential of a solute in solution can now be expressed in terms of the concepts of activity and standard state [Eq. (2–11)]. This equation reflects the fact that there are two components of chemical potential—one value for the species in its standard state and one to account for the temperature and activity of the species in the solution.

The definition of the standard state for pure water and for solids in contact with the solution is that the standard state is the pure substance at the temperature and pressure of the reaction. This means that the activity of water and solids is always 1 because, if the substance is always in its standard state, then $\mu_i = \mu_i^0$, and thus from Eq. 2–11, $RT \ln a_i = 0$, and $a_i = 1$.

Equilibrium Constant

Using the chemical potential of the reactants and products of our reaction, we can now attempt to use thermodynamic data to calculate the equilibrium constant of a reaction. To do this, we will need to calculate the *standard Gibbs free energy change of reaction*, ΔG^0, using the chemical potentials of the reactants and products. Any reaction that occurs must be accompanied by a free-energy change, ΔG. If the reaction is spontaneous, the free-energy change is negative ($-\Delta G$); if the reaction is at equilibrium, the free-energy change is 0 ($\Delta G = 0$); and if energy must be added to the reaction to drive it, the free-energy change is positive ($+\Delta G$). The general expression for calculating the free-energy change of a reaction, ΔG_R is

$$\Delta G_R = \Sigma \Delta G_{\text{products}} - \Sigma \Delta G_{\text{reactants}}. \qquad (2\text{--}13)$$

The free energy change for

$$aA + bB \rightleftharpoons cC + dD \qquad (2\text{--}7)$$

can be expressed in terms of chemical potentials as

$$\Delta G_R = c\mu_C + d\mu_D - a\mu_A - b\mu_B, \qquad (2\text{--}14)$$

where $c, d, a,$ and b are the molar amounts of $C, D, A,$ and B, respectively. Next, Eq. (2–14) must be expanded by substitutions of Eq. (2–11) for each of the reactants and products:

$$\Delta G_R = c\mu_C^0 + cRT \ln a_C + d\mu_D^0 + dRT \ln a_D - a\mu_A^0 - aRT \ln a_A$$
$$- b\mu_B^0 - bRT \ln a_B. \qquad (2\text{--}15)$$

This expression can be simplified by first collecting terms involving standard chemical potentials:

$$\Delta G_R = \left(c\mu_C^0 + d\mu_D^0 - a\mu_A^0 - b\mu_B^0\right)$$
$$+ RT\left(c \ln a_C + d \ln a_D - a \ln a_A - b \ln a_B\right). \qquad (2\text{--}16)$$

The first term of the right-hand side of the equation can be defined as the *standard free energy of reaction* $\left(\Delta G_R^0\right)$, the difference between the free energies (chemical potential times the molar quantity) of the products and the reactants. The second term of Eq. (2–16) can be simplified using the properties of logarithms:

$$\Delta G_R = \Delta G_R^0 + RT \ln \left(\frac{a_C^c \, a_D^d}{a_A^a \, a_B^b} \right). \tag{2–17}$$

The term in parentheses in Eq. (2–17) is identical to the expression for the equilibrium constant that was presented in Eq. (2–6). Thus, when the reaction is known to be at equilibrium,

$$\Delta G_R = \Delta G_R^0 + RT \ln K. \tag{2–18}$$

Another consequence of the equilibrium state is that ΔG_R is equal to zero. If zero is substituted for ΔG_R in Eq. (2–18), that equation can be rearranged to

$$-\Delta G_R^0 = RT \ln K, \tag{2–19}$$

or

$$2.303 \log K = \frac{-\Delta G_R^0}{RT}. \tag{2–20}$$

In this form, the more familiar base 10 logs are used. The equation now can be solved for K as follows:

$$\log K = \frac{-\Delta G_R^0}{2.303RT}. \tag{2–21}$$

And, at 25°C,

$$\log K = \frac{-\Delta G_R^0}{(2.303)(8.3143 \times 10^{-3})(25 + 273)} = \frac{-\Delta G_R^0}{5.708} \, (\text{kJ/mol}). \tag{2–22}$$

Calculation of ΔG_R^0

We now have an equation that can be solved for K if the standard free energy of reaction can be determined. The quantity can be calculated using an approach similar to Eq. (2–13). Thus,

$$\Delta G_R^0 = \Sigma G_{f\text{-products}}^0 - \Sigma G_{f\text{-reactants}}^0, \tag{2–23}$$

where ΔG_f^0 is the *standard free energy of formation* of a particular species. By convention, each element is assigned a ΔG_f^0 value of 0 in the most common form of the element. The standard free energy of formation of the species is the amount of energy used to form the species from the element or elements under standard state conditions. Another convention is that the ΔG_f^0 of H^+ is zero. Values of ΔG_f^0 are listed for many species of interest in hydrogeology in Appendix 1.

Equilibrium constants are given different names depending on the type of reaction considered. A very useful equilibrium constant is the *solubility product* of a mineral. This constant represents the equilibrium condition between the mineral and the solution with which it is in contact. Aquifers contain numerous minerals that can dissolve in contact with ground water moving through the aquifer. By knowing the equilibrium constant, it is possible to determine whether the solution is *undersaturated*, in equilibrium, or *oversaturated* with respect to the mineral. An undersaturated solution is one in which the mineral in question, if present in the aquifer, would be thermodynamically favored to dissolve in the aquifer and an oversaturated solution is one from which the mineral would be expected to precipitate.

Example 2–1

Determine the solubility product of calcite at 25°C.

The first step is to write an equation for the reaction, which in this case is

$$CaCO_3 \rightleftharpoons Ca^{2+} + CO_3^{2-}.$$

Next, the expression for the solubility product is written in the form of Eq. (2–6) using the law of mass action as a function of the activities of the products and reactants:

$$K_{eq} = \frac{a_{Ca^{2+}} a_{CO_3^{2-}}}{a_{CaCO_3}}.$$

Because the activity of $CaCO_3$, a solid, is 1 as we mentioned in the discussion of standard states, the solubility product is equal to the product of the activities of calcium and carbonate ions when the solution is at equilibrium. ΔG_f^0 values are obtained from Appendix 1.

Species	ΔG_f^0 (kJ/mol)
Ca^{2+}	−553.6
CO_3^{2-}	−527.0
$CaCO_3$	−1128.4

$$\Delta G_R^0 = \Sigma \Delta G_{f\text{-products}}^0 - \Sigma \Delta G_{f\text{-reactants}}^0 = (-553.6) + (-527.0) - (-1128.4) = +47.8 \frac{kJ}{mol}$$

From Eq. (2–22),

$$\log K = \frac{-\Delta G_R^0}{5.708} = \frac{-47.8}{5.708} = -8.37, \text{ and } K = 10^{-8.37}.$$

Disequilibrium

Knowledge of the solubility product for minerals gives us a powerful tool with which to examine the equilibrium condition of aquifers with respect to mineral phases. Equations describing the dissolution of minerals similar to the reaction in Example 1 can be written for any solid phase for which free-energy data are available. The reaction is then put into mass law form. If, however, it is not known whether or not the reaction is at equilibrium with the mineral, the product of the products over the product of the reactants is known by various names, including the *activity product (AP)*, the *ion activity product (IAP)*, or the *reaction quotient (Q)*. For the calcite example, since

$$CaCO_3 \rightleftharpoons Ca^{2+} + CO_3^{2-}, \tag{2–24}$$

the ion activity product is determined using the mass law form of the reaction:

$$IAP_{calcite} = a_{Ca^{2+}} a_{CO_3^{2-}}. \tag{2–25}$$

We have omitted the activity of calcite, as in Example 1, because solids have activities of 1. The Gibbs free-energy change for this reaction can be written in a form similar to Eq. (2–18):

$$\Delta G_R = \Delta G_R^0 + RT \ln(\text{IAP}). \qquad (2\text{–}26)$$

When ΔG_R for this reaction is calculated using the activities of calcium and carbonate ions calculated from the water analysis of the ground water sample, the sign will indicate the direction in which Eq. (2–24) is moving toward equilibrium. If ΔG_R is negative, it is an indication that the reaction is proceeding from left to right, that is, that calcite is dissolving to produce calcium and carbonate ions. A positive sign for the free-energy change, on the other hand, informs us that the direction the reaction is occurring is from right to left, which would suggest the precipitation of calcite in the aquifer. This condition must have been in effect in rocks that contain calcite veins or calcite cements precipitated in the void spaces between grains composed of other minerals. If the free-energy change happens to be zero, the solution is in equilibrium with calcite and the mineral should neither be dissolving nor precipitating. All three of the preceding possibilities are common in aquifers.

For the class of reactions that we have been discussing, *precipitation/dissolution* reactions, there is another parameter that is useful in determining the equilibrium state for the water in the aquifer. This quantity is known as the *saturation index*, SI, and it is defined as

$$\text{SI} = \log\left(\frac{\text{IAP}}{K_{\text{SP}}}\right), \qquad (2\text{–}27)$$

where K_{SP} is the solubility product. If the reaction is in equilibrium, the ion activity product is equal to the solubility product, and the term in parentheses will be equal to 1. The log of this term, therefore, will be zero. Saturation indices with negative signs indicate undersaturation because the IAP is less than the K_{SP} and SI values greater than zero indicate oversaturation of the solution. Saturation indices are commonly calculated in studies of the chemical evolution of ground water. Computer models are available to make these calculations rapidly.

Example 2–2

A ground water sample was obtained from Big Spring, near Huntsville, Alabama in 1952 (Hem, 1989). The water had a calcium ion activity ($a_{Ca^{2+}}$) of $10^{-3.04}$ and a carbonate ion activity ($a_{CO_3^{2-}}$) of $10^{-5.56}$. Is the spring water in equilibrium with calcite and, if not, is calcite tending to dissolve or precipitate?

There are several ways to answer this question. The reaction can be written as before:

$$CaCO_3 \rightleftharpoons Ca^{2+} + CO_3^{2-}.$$

First, the free energy change of the reaction will be calculated:

$$\Delta G_R = \Delta G_R^0 + RT \ln(\text{IAP}).$$

Using the ΔG_R^0 value determined in Example 2–1,

$$\Delta G_R = (+47.8) + (8.3143 \times 10^{-3})(298) \ln\left(10^{-3.04} \times 10^{-5.56}\right) = -1.27.$$

The negative sign of this result indicates that the reaction will be proceeding from left to right and calcite should be dissolving in the aquifer. The other way to answer this problem is to calculate the SI:

$$SI = \log\left(\frac{IAP}{K_{SP}}\right) = \log\left(\frac{10^{-3.04}10^{-5.56}}{10^{-8.37}}\right) = -0.23.$$

The negative value of the saturation index yields the same conclusion as the free-energy change: the solution is slightly undersaturated and calcite should dissolve in contact with this water. This example was simplified by assuming the temperature of the spring was 25°C, whereas the actual temperature was somewhat different. Equilibrium constants are a function of temperature and therefore the saturation index for this water would be slightly different. We will consider the effects of temperatures other than 25°C shortly.

A more quantitative indication of the distance of a reaction from equilibrium is to calculate the term $RT \ln\left(\dfrac{Q}{K}\right)$. This value is the free energy used in converting 1 mole of reactants to 1 mole of products at concentrations existing at the time of the reaction. The magnitude of this free-energy change varies with distance from equilibrium. The farther the reaction is from equilibrium, the greater is the free energy released as the reaction proceeds. As the reaction approaches equilibrium, the magnitude of $RT \ln\left(\dfrac{Q}{K}\right)$ decreases exponentially.

Example 2–3

Determine the equilibrium constant for the reaction in which liquid H_2O dissociates to H^+ and OH^- at 25°C. Is this reaction proceeding as written when $a_{H^+} = 10^{-6}$ and $a_{OH^-} = 5 \times 10^{-8}$?
    The equation that describes the reaction is

$$H_2O \rightleftharpoons H^+ + OH^-,$$

and in mass law form,

$$K = a_{H^+} a_{OH^-}.$$

At equilibrium,

$$\log K = \frac{-\Delta G_R^0}{5.708}.$$

The values of ΔG_f^0 are obtained from Appendix 1, as follows:

ΔG_f^0
$H_2O = -237.1$
$H^+ = 0$
$OH^- = -157.3$

$$\Delta G_R^0 = \Sigma \Delta G_{f\text{-products}}^0 - \Sigma \Delta G_{f\text{-reactants}}^0$$

$$\Delta G_R^0 = 0 - 157.3 - (-237.1) = +79.80 \text{ kJ}$$

Because the temperature of the reaction is 25°C,

$$\log K_{eq} = \frac{-\Delta G_R^0}{5.708} = -79.80/5.708 = -13.98,$$

and $K_{eq} = 10^{-13.98}$, which is approximately equal to 10^{-14}. If the reaction is occurring in pure water with no other solutes, then a_{H^+} must equal a_{OH^-} at equilibrium. If this is true, then

$$a_{H^+} = a_{OH^-} = \left(10^{-14}\right)^{1/2} = 10^{-7}.$$

The activity of the hydrogen ion is usually expressed as pH, which is defined as the *negative log of the hydrogen ion activity*. In this case pH = 7. When the a_{H^+} equals the a_{OH^-}, the solution is said to be neutral. This solution will therefore be neutral at a pH of 7. Notice that a pH value of 7 is only neutral for solutions at 25°C, because the equilibrium constant will be different at other temperatures. When the temperature is different, as is the case for many aquifers, the pH of a neutral solution will be slightly different than 7.0.

To answer the second part of the question, we can calculate the free-energy change of the reaction using Eq. (2–26):

$$\begin{aligned}
\Delta G_R &= \Delta G_R^0 + RT \ln(\text{IAP}) \\
&= \Delta G_R^0 + RT \ln(a_{H^+} a_{OH^-}) \\
&= 79.80 + 5.708 \log(10^{-6}\, 5 \times 10^{-8}) \\
&= 79.80 + 5.708 \log(10^{-6}\, 10^{-7.3}) \\
&= 79.80 + 5.708 \log(10^{-13.3}) \\
&= 79.80 - 75.92 \\
&= +3.88 \text{ kJ/mol.}
\end{aligned}$$

The positive value of ΔG_R indicates that the reaction is not at equilibrium and that it is proceeding to the left. The question can also be answered by calculating IAP/K or log IAP/K as in the saturation index:

$$\frac{\text{IAP}}{K} = \frac{(10^{-6})(5 \times 10^{-8})}{10^{-14}} = \frac{5 \times 10^{-14}}{10^{-14}} > 1$$

$$\log\left(\frac{\text{IAP}}{K}\right) = \log\left(\frac{10^{-13.3}}{10^{-14}}\right) = +0.7.$$

Both of these calculations lead to the conclusion that the reaction is proceeding from right to left.

Variation of K with Temperature

At several points in the preceding discussion of chemical equilibrium, we alluded to the fact that the equilibrium constant is a function of temperature. This variation is important because ground water in many areas occurs at temperatures other than 25°C, the temperature at which free-energy data were used to calculate K values. In the case of mineral solubility, since $K = $ IAP at equilibrium, if K increases, the mineral becomes more soluble. Mineral precipitation or dissolution could therefore occur along a ground water flow path for no other reason than a change in the ground water temperature.

Changes in K within a limited temperature range (0–100°C) can be determined using the thermodynamic variables already introduced. There are several methods that can be used to determine these changes. First, recalling Eq. (2–9), we have

$$\Delta G_R^0 = \Delta H_R^0 - T\Delta S_R^0.$$

Because ΔG_R^0 is equal to $-RT \ln K_{eq}$ at equilibrium, this expression can be substituted into the preceding equation:

$$-RT \ln K_{eq} = \Delta H_R^0 - T\Delta S_R^0,$$

which then can be solved for $\ln K_{eq}$:

$$\ln K_{eq} = \frac{\Delta H_R^0 - T\Delta S_R^0}{-RT}. \tag{2–28}$$

This equation can be solved for K at a temperature T other than 25°C by calculating ΔH_R^0 and ΔS_R^0 using data from Appendix 1. Note that this method assumes that the enthalpy and entropy changes are independent of temperature. This assumption is valid within the temperature range specified above. The standard enthalpy and entropy changes are determined as follows:

$$\Delta H_R^0 = \Sigma \Delta H_{f\text{-products}}^0 - \Sigma \Delta H_{f\text{-reactants}}^0; \tag{2–29}$$

$$\Delta S_R^0 = \Sigma \Delta S_{\text{products}}^0 - \Sigma \Delta S_{\text{reactants}}^0. \tag{2–30}$$

Notice that the entropy values in Appendix 1 are given in joules rather than kilojoules.

The second method for determining changes in K_{eq} with changes in temperature involves an equation known as the *van't Hoff equation*, which can be stated as

$$\ln K_{T_2} - \ln K_{T_1} = \frac{\Delta H_R^0}{R}\left(\frac{1}{T_1} - \frac{1}{T_2}\right), \tag{2–31}$$

where K_{T_2} is the equilibrium constant at the desired temperature; K_{T_1} is the value of the constant at the reference temperature, 25°C; ΔH_R^0 is the standard enthalpy change as defined in Eq. (2–29); and T_1 and T_2 are the reference and new temperatures, respectively. If the equilibrium constant at 25°C is known, the value of the constant at the new temperature can be determined using the van't Hoff equation.

Example 2–4

Determine the solubility product of calcite at 40°C.

The solution to this problem will be shown for both methods. The reaction is

$$CaCO_3 \rightleftharpoons Ca^{2+} + CO_3^{2-}.$$

Using Eq. (2–28), we get

$$\ln K_{40} = \frac{\Delta H_R^0 - (313.15)\Delta S_R^0}{-R(313.15)}.$$

Values of ΔH_f^0 and ΔS^0 can be obtained from Appendix 1:

	H_f^0	S^0
$CaCO_3$	−1207.4	91.97
Ca^{2+}	−543.0	−56.2
CO_3^{2-}	−675.2	−50.0

Then,

$$\Delta H_R^0 = \Sigma \Delta H_{f\text{-products}}^0 - \Sigma \Delta H_{f\text{-reactants}}^0 = -543.0 - 675.2 - (-1207.4) = -10.8 \text{ kJ}$$

$$\Delta S^0 = \Sigma \Delta S_{\text{products}}^0 - \Sigma \Delta S_{\text{reactants}}^0 = -56.2 - 50.0 - 91.97 = -198.17 \text{ (J/mol } ^\circ\text{K)} =$$

$$= -0.19817 \text{ (kJ/mol } ^\circ\text{K)}$$

$$\ln K_{40} = \frac{-10.8 - (313.15)(-0.19817)}{(-8.3143 \times 10^{-3})(313.15)} = -19.72, \text{ and}$$

$$K_{40} = 10^{-8.56}.$$

To apply the van't Hoff equation, we will need to use the value of K_{cal} determined in Example 2–1, $10^{-8.37}$:

$$\ln K_{40} - \ln K_{25} = \frac{\Delta H_R^0}{R}\left(\frac{1}{T_{25}} - \frac{1}{T_{40}}\right) = \frac{-10.8}{8.3143 \times 10^{-3}}\left(\frac{1}{298.15} - \frac{1}{313.15}\right)$$

$$= -0.21$$

$$\ln K_{40} = -0.21 + \ln K_{25} = -0.21 - 19.27 = -19.48$$

$$K_{40} = 10^{-8.46}.$$

The results of Example 2–4 indicate that calcite is less soluble at 40°C than it is at 25°C. This explains why the precipitation of carbonate scale deposits can be a problem in hot water heaters using hard water. By the same token, calcite is more soluble at 10°C than it is at 25°C. These shifts in equilibrium with temperature can be significant in natural settings as well. During the spring thaw in colder climates, for example, the temperature of ground water recharge will be near 0°C. If this water equilibrates with carbonates or other minerals in the vadose zone, precipitation could occur when the water reaches the water table where ground water may be slightly warmer. Calcite could also be precipitated in a deep ground water flow system, where temperatures are higher because of the geothermal gradient. This could provide a mechanism to produce a calcite cement in the pore spaces of a rock that might be present at that depth.

Activity–Concentration Relationships

The activity coefficient was introduced earlier as a correction for nonideal solutions, ones in which residual interionic forces of attraction are present and the activity of dissolved species is less than their concentration. All natural waters are electrolyte solutions of varying strength and thus exhibit this nonideal behavior.

The effects of ionic interactions can be estimated by determining the *ionic strength* of the solution, I, which is defined by

$$I = \frac{1}{2} \Sigma M_i z_i^2, \tag{2-32}$$

where M_i is the molarity of species i and z_i is the charge of the ion. For ground water, the calculation of ionic strength must include all major ions:

$$I = \frac{1}{2} \left[M_{Na^+} + 4M_{Ca^{2+}} + 4M_{Mg^{2+}} + M_{HCO_3^-} + M_{Cl^-} + 4M_{SO_4^{2-}} \right]. \tag{2-33}$$

Other ions, such as potassium, nitrate, and iron, should also be included in the calculation if their concentration is significant.

Deviations from ideal behavior are least important for uncharged solutes. For these species, activity coefficients can be calculated as follows:

$$\gamma = 10^{0.11}. \tag{2-34}$$

The resulting values are commonly close enough to 1.0 that they can be assumed to be 1.0 with little error. Activity corrections are important, however, for ionic species, for which activity coefficients can be calculated using the *Debye-Huckel* equation. The Debye-Huckel equation has several forms, depending on the ionic strength of the solution. One of the most common forms of the equation, which is valid up to an ionic strength of 0.1, is

$$\log \gamma_i = \frac{-A z_i^2 \sqrt{I}}{1 + B a_0 \sqrt{I}}. \tag{2-35}$$

The terms A and B are functions of temperature and a_0 is the hydrated radius of the ion. Values for these parameters are shown in Table 2–1. For more concentrated electrolyte solutions, an additional term can be added to the Debye-Huckel equation, or alternative equations can be used (Drever, 1997). In the range of ionic strengths for most ground

TABLE 2–1   Parameters for the Debye-Huckel equation (from Drever, J. I., *The Geochemistry of Natural Waters* 3/e © 1997. Reprinted by permission of Prentice-Hall, Inc., Upper Saddle River, NJ.)

$T(°C)$	A	$B(\times 10^8)$	Ion	$a_0(\times 10^{-8})$
0	0.4883	0.3241	Ca^{2+}	5.0
5	0.4921	0.3249	Mg^{2+}	5.5
10	0.4960	0.3258	Na^+	4.0
15	0.5000	0.3262	K^+, Cl^-	3.5
20	0.5042	0.3273	SO_4^{2-}	5.0
25	0.5085	0.3281	HCO_3^-, CO_3^{2-}	5.4
30	0.5130	0.3290	NH_4^+	2.5
40	0.5221	0.3305	Sr^{2+}, Ba^{2+}	5.0
50	0.5319	0.3321	Fe^{2+}, Mn^{2+}, Li^+	6.0
60	0.5425	0.3338	H^+, Al^{3+}, Fe^{3+}	9.0

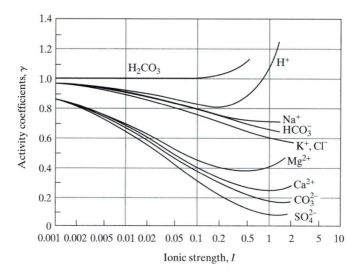

FIGURE 2–4 Activity coefficients (γ) for various species plotted as a function of ionic strength, I (from *Ground Water* by Freeze/Cherry © 1979. Reprinted by permission of Prentice-Hall, Inc., Upper Saddle River, NJ).

waters, γ tends to decrease with increasing ionic strength. This inverse relationship confirms that the more concentrated the solution, the greater the difference between the activity and the concentration of the ion. Figure 2–4 shows the relationship between activity coefficient and ionic strength for several common ions.

Example 2–5 Solubility of gypsum and the effect of activity corrections

Determine the solubility of gypsum in equilibrium with pure water at a temperature of 25°C. The reaction for the dissolution of gypsum is

$$CaSO_4 \cdot 2H_2O \rightleftharpoons Ca^{2+} + SO_4^{2-} + 2H_2O.$$

From this equation, it is clear that when gypsum dissolves in pure water, the molarity of calcium will equal the molarity of sulfate. Also, since 1 mole of gypsum dissolves to produce 1 mole of calcium, by finding the molarity of calcium at equilibrium, we will know the molar quantity of gypsum that has dissolved, and thereby its solubility. The solubility product can be calculated using free energy data according to Eq. (2–21); the resulting value is $10^{-4.41}$. The problem we face is that the ionic strength of the equilibrium solution is unknown, and, therefore, we cannot determine the activity coefficients for calcium and sulfate. The approach that can be taken is to make an initial assumption that the activity coefficients are equal to 1 and then to solve the problem by iteration. With this assumption, we can use the mass law form of the dissolution reaction as follows:

$$K_{gyp} = IAP_{gyp} = 10^{-4.41} = a_{Ca^{2+}} a_{SO_4^{2-}}.$$

Activities can be converted to molarities because $a = \gamma M$. Thus,

$$K_{gyp} = 10^{-4.41} = \left(\gamma_{Ca^{2+}} M_{Ca^{2+}}\right)\left(\gamma_{SO_4^{2-}} M_{SO_4^{2-}}\right).$$

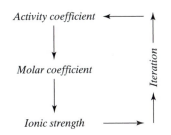

Activity coefficient

Molar coefficient

Ionic strength

Iteration

FIGURE 2–5   Iteration sequence used to determine solubility of gypsum.

TABLE 2–2   Results of iteration calculations for gypsum solubility.

Iteration	γ	$M_{Ca^{2+}}$	I
1	1	6.24×10^{-3}	2.50×10^{-2}
2	0.548	11.40×10^{-3}	4.56×10^{-2}
3	0.476	13.11×10^{-3}	5.24×10^{-2}
4	0.460	13.56×10^{-3}	5.42×10^{-2}
5	0.454	13.75×10^{-3}	5.50×10^{-2}
6	0.452	13.78×10^{-3}	5.51×10^{-2}
7	0.453	13.78×10^{-3}	5.51×10^{-2}

Since both γ values are assumed to be equal to 1 and $M_{Ca^{2+}}$ and $M_{SO_4^{2-}}$ are equal, $M_{Ca^{2+}} = \left(10^{-4.41}\right)^{1/2}$ $= 6.24 \times 10^{-3}$. With this initial estimate of molarity, an initial estimate of ionic strength can be calculated:

$$I = \frac{1}{2}\left(6.24 \times 10^{-3} \times 4 + 6.24 \times 10^{-3} \times 4\right) = 2.50 \times 10^{-2}.$$

Revised activity coefficients can now be determined using the Debye-Huckel equation. The values obtained, $\gamma_{Ca^{2+}} = \gamma_{SO_4^{2-}} = 0.548$, are nearly one-half the original estimate of 1. The corresponding molarity of Ca^{2+} can be found as follows:

$$M_{Ca^{2+}} = \left(\frac{10^{-4.41}}{0.548 \times 0.548}\right)^{1/2} = 11.40 \times 10^{-3}.$$

This value is used to recalculate ionic strength and so on. The iteration process can be continued until values show little or no change between successive values. The iteration sequence is shown in Fig. 2–5 and the results are shown in Table 2–2.

The final molarity of the calcium ion, 13.78×10^{-3}, is more than twice our original estimate with an activity coefficient of 1. Thus the activity correction is very important for this solution, which is concentrated enough to produce relatively strong interionic forces. With the formula weight of gypsum, 172 g/mol, we can now determine its solubility using our calculated molarity:

$$13.78 \times 10^{-3}\ \text{mol/l} \times 172.1\ \text{g/mol} = 2.37\ \text{g/l}.$$

The calcium concentration in the solution can be found as follows:

$$13.78 \times 10^{-3}\ \text{mol/l} \times 40.08\ \text{g/mol} = 0.55\ \text{g/l} = 550\ \text{mg/l}\ Ca^{2+}.$$

Complexes

The solubility of gypsum described in the preceding example neglects one factor, the formation of *complexes*, or *ion pairs*. Complexes are dissolved species that consist of two or more simpler species. In the gypsum example, an important complex is $CaSO_4^0$. It exists in solution as an uncharged molecule of calcium sulfate. Its relevance to the solubility of gypsum is that it is not included in the molarity of Ca^{2+} and therefore must be considered as a separate component of the aqueous solution.

The formation of complexes is represented by an equilibrium constant for a reaction such as

$$Ca^{2+} + SO_4^{2-} \rightleftharpoons CaSO_4^0. \tag{2-36}$$

Note that the reaction is written in the opposite direction from the mineral dissolution reactions that we have considered. The equilibrium constant for Eq. (2–36), which is known as a *stability constant*, is expressed in mass law form as

$$K_{stab} = \frac{a_{CaSO_4^0}}{a_{Ca^{2+}} a_{SO_4^{2-}}}. \tag{2-37}$$

The magnitude of a stability constant is proportional to the strength, or stability, of the complex. The K_{stab} for $CaSO_4^0$ is equal to $10^{2.23}$, and the equilibrium activity of the complex can be determined as follows. Rearranging Eq. (2–37) to solve for $a_{CaSO_4^0}$ yields

$$a_{CaSO_4^0} = K_{stab} a_{Ca^{2+}} a_{SO_4^{2-}}. \tag{2-38}$$

Because $a_{Ca^{2+}} a_{SO_4^{2-}}$ is equal to $10^{-4.41}$ in a solution in equilibrium with gypsum,

$$a_{CaSO_4^0} = 10^{2.23} \times 10^{-4.41} = 10^{-2.18} = 6.61 \times 10^{-3}. \tag{2-39}$$

The activity coefficient for the complex can be assumed to be 1 because it is an uncharged species. The activity of the complex calculated in Eq. (2–39) will therefore be equal to its molarity.

We can now revise the estimate of the solubility of gypsum from Example 2–5 by assuming that $CaSO_4^0$ is the only significant complex in solution and adding its molarity to the molarity of the calcium ion to find a total molarity of calcium:

$$M_{Ca_T} = M_{Ca^{2+}} + M_{CaSO_4^0} = 13.78 \times 10^{-3} + 6.61 \times 10^{-3} = 20.39 \times 10^{-3}. \tag{2-40}$$

This is nearly 50% higher than the value we found without considering the complex! Our estimate of the solubility of gypsum can now be revised again to read

$$20.39 \times 10^{-3} \, mol/l \times 172.1 \, g/mol = 3.51 \, g/l = 3,510 \, mg/l. \tag{2-41}$$

Complexes are obviously important in some solutions. The relationships between complexes, molarities, and activities for selected ions in seawater are shown in Fig. 2–6. Ions such as sulfate and bicarbonate are significantly complexed, whereas chloride and sodium mostly occur as free ions. The degree of complexation of aluminum is nearly total.

KINETICS

The use of equilibrium models in understanding ground water chemistry is only valid when it can be assumed that the reaction has had time to achieve equilibrium. This assumption is based on some estimate of the speed of a reaction, which falls within the realm of *kinetics*. Some reactions, such as those between dissolved species and between

FIGURE 2–6   Importance of complexing and activity corrections for ions in seawater (35 g/l). For each ion, light gray bar is percentage of total concentration of ion that is uncomplexed. Dark gray bar represents activity as a percentage of total concentration. Activity corrections increase with valence of ion (from *Geochemistry, Ground Water and Pollution*. Revised edition, Appelo, C. A. J. & D. Postma, 1996).

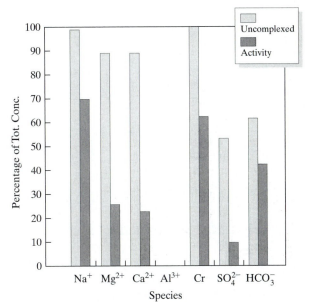

gases and water, take place so rapidly that they can always be assumed to be at equilibrium in ground water flow systems. Others, including many mineral dissolution/precipitation reactions, are much slower. In order to assess the probability that these reactions have reached equilibrium we must know two things. First, we must have some idea of the residence time of the water in the flow system; and second, we need to know how fast the reaction takes place. Rates of generalized classes of reactions are shown in Fig. 2–7, along with residence times for various components of the hydrologic cycle.

The kinetics of many reactions can be described in terms of a *reaction rate* that is proportional to the amount of the reactant raised to a power. If we restrict the type of reaction to those that are irreversible, the reaction rate is the rate involved in converting the reactants to products. In an irreversible reaction, the reaction continues until

FIGURE 2–7   Comparison of residence time of water in system (T_R) with half-times ($T_{1/2}$; amount of time for 1/2 of reactant to be converted to product) of various types of reactions involving ground waters and surface waters (from Langmuir and Mahoney, 1984. Reprinted with the permission of the National Ground Water Association, Copyright, 1984).

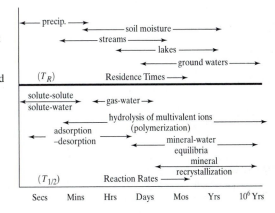

the reactants are totally consumed. Examples of reaction rate equations for reactions involving reactants A, B, and C are as follows, where C represents the concentration of the reactant:

$$\text{Rate} = k \tag{2-42}$$

$$\text{Rate} = kC_A \tag{2-43}$$

$$\text{Rate} = kC_A^2 \tag{2-44}$$

$$\text{Rate} = kC_A C_B \tag{2-45}$$

$$\text{Rate} = kC_A^3 \tag{2-46}$$

$$\text{Rate} = kC_A C_B C_C. \tag{2-47}$$

The *reaction order* is defined as the sum of the exponents for all reactants. Thus, Eq. (2–42) is a *zeroth-order* reaction and Eq. (2–43) is a *first-order* reaction. The change in concentration with time in a zeroth-order reaction is shown in Fig. 2–8. The *rate constant*, k, is the slope of the line. Notice that the rate is not a function of the concentration. The rate constant for a first-order reaction can be determined as the slope of a line plotted on a semi-log graph (Fig. 2–9). The rate for this type of reaction is a linear function of the initial concentration. Equations (2–44) and (2–45) are both second order because the sum of all exponents equals two. Using the same reasoning, Eqs. (2–46) and (2–47) represent third-order reactions.

FIGURE 2–8 Change in concentration with time (left) and reaction rate as a function of concentration (right) for a *zeroth order* reaction.

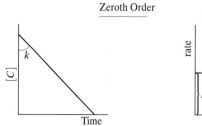

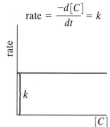

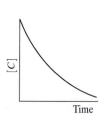

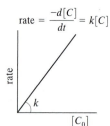

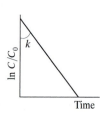

FIGURE 2–9 Plot of concentration vs. time (left), reaction rate as a function of initial concentration (C_0) (middle) and $\ln C/C_0$ vs. time (right) for *first-order* reaction.

First-Order Reactions

To avoid excessive complexity, we will limit our quantitative investigation of reaction kinetics to first-order reactions or to reaction kinetics that can be approximated as first order. These types include many reactions of interest to hydrogeologists.

Radioactive decay is one of the most common examples of first-order kinetics. The rate of decay is directly proportional to the amount of undecayed isotope. That is,

$$\frac{-dC}{dt} = kC, \tag{2-48}$$

where k has the units of reciprocal time. If we rearrange and integrate this equation from time $t = 0$ to $t = t$ and from C_0 to C, it will be set up as follows:

$$-\int_{C_0}^{C} \frac{dC}{C} = k \int_{0}^{t} dt. \tag{2-49}$$

After integration,

$$-\ln\left(\frac{C}{C_0}\right) = kt. \tag{2-50}$$

This can also be expressed as

$$C = C_0 e^{-kt}, \tag{2-51}$$

which illustrates the familiar exponential decay of a radioactive isotope.

The decay of these substances is also expressed as a *half-life* or *half-time*, which is the amount of time necessary for the decay of the original amount to $1/2$ its value. Thus, when $C = \frac{1}{2} C_0, t = t_{1/2}$. The half-life is related to the decay constant by

$$\frac{1}{2} = e^{-kt_{1/2}} \tag{2-52}$$

so that we have

$$2 = e^{kt_{1/2}}, \tag{2-53}$$

$$\ln 2 = kt_{1/2}, \tag{2-54}$$

$$t_{1/2} = \frac{\ln 2}{k} = \frac{0.693}{k}. \tag{2-55}$$

An example of first-order decay is ^{14}C, which has a half-life of 5570 yr and a decay constant of 1.2×10^{-4} yr^{-1}.

Pseudo First-Order Kinetics

The kinetics of second-order reactions can sometimes be approximated as first order if certain simplifying assumptions apply. One example involves the biotransformation of halogenated organic solvents such as trichloroethene. Compounds of this type, which are used as solvents in many manufacturing and degreasing processes, are very common contaminants in aquifers in industrial areas. Even though concentrations are typically in the parts per billion range, the toxicity of these compounds creates an environmental problem at these levels. Microorganisms can break down organic compounds to simpler, and sometimes less hazardous substances, in several ways. If the bacteria can directly utilize the compound as a growth substrate and the concentration is high enough, the microbial population will grow by transforming the compound and the concentration will decline. If, however, the concentration is too low to sustain the growth of a bacterial population, the compound can only be biodegraded if a primary substrate is present. Under these conditions, the halogenated compound is biodegraded as a side effect while the primary substrate is being consumed. This process is called *cometabolism*. The rate expression for such a reaction is

$$\frac{-dC}{dt} = kCX, \tag{2–56}$$

where X is the concentration of bacteria. If the population of bacteria is large, it may be considered to be constant. The reaction rate could then be modified to

$$\frac{-dC}{dt} = k'C, \tag{2–57}$$

in which k' is equal to kX. The rate constant k' is a *pseudo-first-order* constant which has a half-life corresponding to Eqs. (2–52) through (2–55). Half-lives for biodegradation of halogenated solvents are determined in controlled laboratory experiments and rarely adequately describe the breakdown of these compounds in the field.

Variables Affecting Reaction Rates

Numerous difficulties are encountered in the quantitative application of kinetics to ground water flow systems. Temperature, for example, has a significant effect upon reaction rates. A rule of thumb holds that reaction rates double for each 10°C rise in temperature. Another example is the presence or absence of microorganisms that can catalyze the reaction. Langmuir and Mahoney (1984) describe the oxidation of ferrous iron between pH values of 2.2 and 3.5 according to the reaction

$$\text{Fe}^{2+} + 1/4 \, \text{O}_2 + 1/2 \, \text{H}_2\text{O} = \text{FeOH}^{2+}. \tag{2–58}$$

The rate expression in terms of Fe^{2+} is

$$\frac{d(\text{Fe}^{2+})}{dt} = -k \, \text{Fe}^{2+} \text{P}_{\text{O}_2}. \tag{2–59}$$

At 20°C, the value of k is $10^{-3.2}$ day^{-1} atm^{-1}. If the reaction takes place under atmospheric conditions, p_{O_2} is a constant (0.2 atm.) and the reaction is pseudo first order. The half-time of this reaction is 5500 days or 15 years. When the reaction is biologically mediated, which is common for oxidation reactions, the half-time is only 8 minutes, 6 orders of magnitude faster. Above a pH of 4, the half-time for the oxidation of ferrous iron is even smaller, which illustrates the importance of isolating ground water samples from contact with the atmosphere prior to analysis. In a matter of minutes, ferrous iron can oxidize and precipitate, changing the chemical composition of the water from its condition in the aquifer.

Hyperbolic Kinetics

In the preceding discussion, the cometabolism of halogenated solvents was approximated as a first-order process. If instead the contaminant compound can serve as a primary substrate for microorganisms, the reaction rate can be expected to increase as the amount of substrate increases. For example, this relationship applies to spills of petroleum fuels into the subsurface where aerobic bacteria can oxidize the organic contaminants and, in doing so, utilize the energy produced for growth. Reactions of this type are catalyzed by compounds called *enzymes*, which are protein compounds contained in the living organism. When the substrate concentration increases, as would be the case of gasoline leaking from a tank into the subsurface, the reaction rate increases in response to the influx of a readily usable energy source. At first, the reaction rate increases in direct proportion to the increase in substrate concentration. Eventually, however, the enzymes are totally utilized, or *saturated*, and the reaction rate can increase no more. A plot of reaction rate vs. substrate concentration would have the form shown in Fig. 2–10 and would be described by the *Michaelis-Menton* equation,

$$v = \frac{V_{\max}S}{(K_m + S)},$$ (2–60)

in which v is the reaction rate, $V_{\max}$ is the maximum reaction rate, S is the substrate concentration, and K_m is substrate concentration at one-half the maximum reaction rate. To describe the process of biodegradation of oil or gasoline more accurately, a more complicated expression similar to Eq. (2–60) would have to be used to account for both the hydrocarbon being degraded and the oxygen simultaneously consumed.

Mineral Precipitation and Dissolution

One might assume, based on our discussion of equilibrium thermodynamics, that a mineral in contact with a solution that is undersaturated with respect to it will dissolve. Likewise, a solution that is oversaturated with respect to a mineral should precipitate that mineral because a decrease in free energy would result as the solution approaches equilibrium. These expectations, however, do not take into account the effects of kinetics.

FIGURE 2–10   Plot of reaction rate vs. substrate concentration for hyperbolic kinetics.

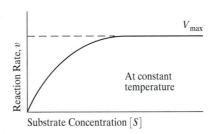

FIGURE 2–11 Plot of free energy changes during precipitation of crystal from oversaturated solution. Free energy must first increase to exceed nucleation barrier prior to decreases in free energy during precipitation. (From Drever, J. I., *The Geochemistry of Natural Waters* 3/e © 1998. Reprinted by permission of Prentice-Hall, Inc., Upper Saddle River, NJ).

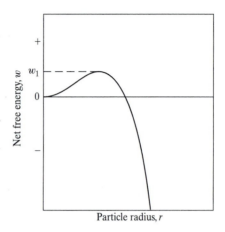

Particle radius, r

Experimental studies have shown that the rates of dissolution of minerals such as calcite vary with the degree of undersaturation (Berner and Morse, 1974). Rates are highest at high degrees of undersaturation and decrease as the solution approaches equilibrium. The experimental dissolution rates of silicate minerals decrease with time after the initiation of the experiment. Several explanations have been proposed for this phenomenon (Drever, 1997). One invokes the formation of a partially leached surface layer which retards the passage of ions to the solution from the fresh mineral beneath the altered layer. Another mechanism for the decrease in rates over time is based upon the difference in dissolution rates between large and small particles. The large surface area of very small mineral particles affects the free-energy relationships such that dissolution is more rapid. Therefore, overall rates decline as the smaller particles are preferentially dissolved, leaving only the larger, less reactive particles to dissolve.

Mineral precipitation is also strongly influenced by kinetics. In order to precipitate from an oversaturated solution, an increase in free energy is initially required prior to a decline in free energy after precipitation begins. This *nucleation barrier* (Fig. 2–11) is related to the energy needed to overcome the high surface area of nuclei of radius, r, forming from the solution. The nucleation barrier is lower if the precipitating mineral is already in contact with the solution so that growth on an existing mineral surface can take place. The degree of saturation is also important in precipitation. The more oversaturated the solution is, the higher the precipitation rate.

The presence of substances in solution other than those involved in the precipitating solid can also affect kinetics. Examples include the inhibition of calcite precipitation by magnesium, phosphate, or organics in the solution. The inhibiting effect of magnesium ions adsorbed to the surface of newly formed calcite explains why aragonite precipitates before calcite in seawater, even though it is thermodynamically less stable in the solution.

PROBLEMS

1. The routine analysis of a water sample provides the following concentrations (as mg/l): Ca^{2+} − 93.9; Mg^{2+} − 22.9; Na^+ − 19.1; HCO_3^- − 344; SO_4^{2-} − 85.0; Cl^- − 9.0 and a pH of 7.20. What is the ionic strength of the water?

2. Write mass law expressions for the following equilibrium expressions.
   a. $CaCO_3 \rightleftharpoons Ca^{2+} + CO_3^{2-}$ **b.** $CO_2(g) \rightleftharpoons CO_2(aq)$ **c.** $Mn^{2+} + Cl^- \rightleftharpoons MnCl^+$

3. For the water sample in problem 1, determine the saturation index for calcite at 25°C given that $a_{CO_3^{2-}} = 0.34 \times 10^{-5}$ and that the equilibrium constant for calcite dissolution is 4.27×10^{-9}. What does the saturation index indicate about the state of saturation with respect to calcite?

4. Will nitrate oxidize sulfide in a water with pH = 8 and T = 25°C? Assume that activity is equal to concentration and the concentrations of NO_3^-, HS^-, SO_4^{2-}, and NH_4^+ are all equal to 10^{-4} M. The reaction is

$$H^+ + NO_3^- + HS^- + H_2O \rightleftharpoons SO_4^{2-} + NH_4^+.$$

5. Calculate the equilibrium constant for gypsum at 50°C using ΔH and ΔS data from Appendix 1: How do the results compare with the results we obtained at 25°C, i.e., is gypsum more or less soluble at 50°C?

6. Find the pH value of neutrality for water at 10°C.

7. If a water is in equilibrium with calcite at 25°C, what is the calcium concentration in mg/l if the CO_3^{2-} is 5 mg/l? Assume activity equals molarity.

8. Calculate the solubility product for the dissolution of carbon dioxide in water at 25°C. If the dissolved CO_2 concentration in a lake at the same temperature is 2.2 mg/l, is the lake in equilibrium with atmospheric CO_2 (partial pressure $= 10^{-3.5}$)? If not, is the gas volatilizing from the lake or dissolving into it?

9. Strontium 90 is a radioactive element which has a half-life of 29 years. How long would a given amount of ^{90}Sr need to be stored to obtain a 99.9% reduction in quantity?

10. Tritium levels in the atmosphere reached their peak of about 1000 T.U. in 1963. Assuming ground water recharge occurring at that time was in equilibrium with atmospheric tritium and that the tritiated water moved along the flow path without dispersion, how long would it take for the tritium level to decay to less than 5 T.U.?

CHAPTER 3

Acid–Base Reactions and the Carbonate System

Acid–base reactions involve the transfer of hydrogen ions between reactants and products. They include many significant reactions occurring in the unsaturated and saturated zones of the subsurface. Particularly important are the reactions involving carbon dioxide, the dissolved species of inorganic carbon, and carbonate minerals. The pH of surface waters and ground waters is controlled by these processes; and in many aquifers, the types and amounts of dissolved solids are determined by reactions of this type.

Acids are defined as compounds that donate hydrogen ions (protons), and bases are compounds that accept protons. The general reaction between acids and bases can be written as

$$\text{Acid}_1 + \text{Base}_2 = \text{Acid}_2 + \text{Base}_1. \tag{3–1}$$

Base_1 is known as the *conjugate base* of Acid_1. The dissociation of the bicarbonate ion [Eq. (3–2)] is an example of an acid–base reaction:

$$\text{HCO}_3^- + \text{H}_2\text{O} \rightleftharpoons \text{H}_3\text{O}^+ + \text{CO}_3^{2-}. \tag{3–2}$$

Here water serves as Base_2 and accepts the proton donated by the bicarbonate ion to become H_3O^+, the hydronium ion. The conjugate base of bicarbonate is the carbonate ion, CO_3^{2-}. The equilibrium constant for this reaction is

$$K_{eq} = \frac{a_{\text{H}_3\text{O}^+} a_{\text{CO}_3^{2-}}}{a_{\text{HCO}_3^-} a_{\text{H}_2\text{O}}}. \tag{3–3}$$

The reaction can be simplified to

$$K_{eq} = \frac{a_{\text{H}^+} a_{\text{CO}_3^{2-}}}{a_{\text{HCO}_3^-}}. \tag{3–4}$$

The hydrogen ion is shown as a free ion in Eq. (3–4), even though it is actually present in solution as the hydronium ion. The constants for Eqs. (3–3) and (3–4) are identical.

Water can also function as an acid as in the following reaction:

$$\text{NH}_3 + \text{H}_2\text{O} \rightleftharpoons \text{NH}_4^+ + \text{OH}^-. \tag{3–5}$$

Ammonia (NH_3) accepts a hydrogen ion from water to become the ammonium ion. As we observed in Chapter 2, pure water will dissociate to some extent, thus behaving as both an acid and a base. In comparison to Example 2–3, this reaction is more properly written as

$$\text{H}_2\text{O} + \text{H}_2\text{O} \rightleftharpoons \text{H}_3\text{O}^+ + \text{OH}^-. \tag{3–6}$$

STRENGTH OF ACIDS AND BASES

At 25°C, K_{eq} for Eq. (3–6) is equal to 10^{-14}, which can be written in a form similar to pH, as $pK_a = 14$. The value of this constant tells us something about the *strength* of water as an acid. Because the activity product of H_3O^+ and OH^- is quite small (10^{-14}), water does not have a strong tendency to dissociate in an aqueous solution. It is therefore a weak acid. Hydrochloric acid, a strong acid, will dissociate completely in an aqueous solution according to the following reaction:

$$HCl = H^+ + Cl^-. \tag{3–7}$$

The pK_a value is undefined for this acid because the activity of the acid molecule, HCl, is zero. Other strong acids include nitric and sulfuric acids. The conjugate base of a strong acid, for example Cl^- in Eq. (3–7), is weak.

Many acids dissociate to a lesser degree in solution. Examples of the pK_a values for these acids are shown in Table 3–1. The conjugate bases of weak acids are strong.

TABLE 3–1   Dissociation constants of some weak acids at 25°C.

Acid	pK_a
Acetic (CH_3COOH)	4.75
Boric (H_3BO_3)	9.2
Carbonic (H_2CO_3)	6.35
Phosphoric (H_3PO_4)	2.1
Hydrosulfuric (H_2S)	7.0
Silicic (H_4SiO_4)	9.71

NATURAL WEAK ACID–STRONG BASE SYSTEMS—THE CARBONATE SYSTEM

Among the weak acids given in Table 3–1 is carbonic acid, one whose properties are of the utmost importance in both surface waters and ground waters. Carbonic acid is the result of the dissolution of carbon dioxide in water, which dissolves until equilibrium is reached according to the reaction

$$CO_2(g) \rightleftharpoons CO_2(aq). \tag{3–8}$$

At equilibrium, the activity of dissolved CO_2 is proportional to the partial pressure of CO_2 in the gas phase in contact with the aqueous phase containing the dissolved CO_2. This is true because the activities of gaseous species in mass law form can be approximated by partial pressures. Thus,

$$K_{CO_2} = \frac{a_{CO_2(aq)}}{p_{CO_2}}. \tag{3–9}$$

In solution, aqueous carbon dioxide reacts with water to form carbonic acid:

$$CO_2(aq) + H_2O \rightleftharpoons H_2CO_3. \tag{3–10}$$

In practice, Eq. (3–8) and Eq. (3–10) are combined to form the following equation:

$$CO_2(g) + H_2O \rightleftharpoons H_2CO_3^*. \tag{3–11}$$

This convention assumes that all the CO_2 in solution, which consists of both $CO_2(aq)$ and H_2CO_3, is present in the form of H_2CO_3, and is designated as $H_2CO_3^*$. In reality, most of the CO_2 is present as dissolved CO_2 rather than H_2CO_3, but the convention of treating it all as carbonic acid is used for simplicity. The equilibrium constant for Eq. (3–11), known as K_{CO_2}, has a value of $10^{-1.46}$ at 25°C (Plummer and Busenberg, 1982) and can be expressed as

$$K_{CO_2} = \frac{a_{H_2CO_3^*}}{p_{CO_2}}. \tag{3–12}$$

The importance of this equation is that the activity of dissolved CO_2 (in both of its forms) is a function of the p_{CO_2} in the associated gas phase. In the atmosphere, p_{CO_2} has a value of $10^{-3.5}$. The p_{CO_2} values calculated for ground water are typically much higher than the atmospheric value, which indicates that ground water solutions have come in contact with a gas phase that has a higher partial pressure of CO_2 than the atmosphere. The asterisk in Eq. (3–12) is commonly omitted in reactions involving dissolved CO_2, even though the term still represents the combination of aqueous CO_2 and H_2CO_3. The reaction between aqueous CO_2 and water is quite slow compared to reactions involving H_2CO_3.

Carbonic acid, once it is formed in an aqueous solution, will begin to dissociate to a degree determined by its K_a. Carbonic is what is known as a *diprotic* acid because it can dissociate twice to yield two hydrogen ions. The first dissociation produces a hydrogen ion and a bicarbonate ion according to the reaction

$$H_2CO_3 \rightleftharpoons H^+ + HCO_3^-. \tag{3–13}$$

The equilibrium constant for this reaction, known as K_1, has a value of $10^{-6.35}$ (Plummer and Busenberg, 1982)

$$K_1 = \frac{a_{H^+} a_{HCO_3^-}}{a_{H_2CO_3}} = 10^{-6.35}. \tag{3–14}$$

This low value is the reason that carbonic acid is considered to be a weak acid, even though it is the slow reaction of $CO_2(aq)$ and water to form H_2CO_3, rather than the dissociation of H_2CO_3 as in Eq. (3–13) that results in a relatively low pK_a. Bicarbonate can also ionize according to the reaction

$$HCO_3^- \rightleftharpoons H^+ + CO_3^{2-}, \tag{3–15}$$

where the equilibrium constant

$$K_2 = \frac{a_{H^+} a_{CO_3^{2-}}}{a_{HCO_3^-}} = 10^{-10.33}. \tag{3–16}$$

Based on Eqs. (3–13) through (3–16), it is clear that when CO_2 dissolves in water, three dissolved species can be produced—H_2CO_3, HCO_3^-, and CO_3^{2-}. The relationships of the activities of these species to one another can be expressed as a function of pH by rearrangement of the K_1 and K_2 expressions. For example,

$$\frac{K_1}{a_{H^+}} = \frac{a_{HCO_3^-}}{a_{H_2CO_3}}. \tag{3–17}$$

When the pH of the solution is equal to 6.35, it is obvious from Eqs. (3–17) and (3–14) that activities of bicarbonate and carbonic acid are equal. If the pH is 7.35, however, the activity of bicarbonate ion is 10 times greater than the activity of H_2CO_3. With an analagous expression of the K_2 equation,

$$\frac{K_2}{a_{H^+}} = \frac{a_{CO_3^{2-}}}{a_{HCO_3^-}}, \tag{3–18}$$

the relative proportions of carbonate ion and bicarbonate ion can be compared. If the pH is 10.33, these activities are equal. When the pH drops to 9.33, the carbonate ion is only 1/10 as abundant as the bicarbonate ion. The pH must be quite high for the activity of carbonate ion to be dominant. But what about the relationship between $a_{CO_3^{2-}}$ and $a_{H_2CO_3}$? If we combine Eq. (3–17) and Eq. (3–18), the ratio between these two species can be expressed as

$$\frac{a_{CO_3^{2-}}}{a_{H_2CO_3}} = \frac{K_1 K_2}{a_{H^+}^2}. \tag{3–19}$$

When the pH is equal to 6.35, the ratio is quite small:

$$\frac{10^{-6.35} 10^{-10.33}}{10^{-12.7}} = 10^{-3.98} = 0.0001.$$

Thus, the activity of carbonic acid (and also bicarbonate, since they are equal at this pH) is 4 orders of magnitude greater than the carbonate ion activity. These relationships can be graphically displayed in a diagram called a *Bjerrum plot* (Fig. 3–1), which shows their activities as a function of pH. The Bjerrum plot assumes that the total concentration of dissolved inorganic carbon—the sum of the activities of carbonic acid, bicarbonate ion, and carbonate ion—is fixed. This condition is indicative of a *closed system*, one in which there is no exchange with its surroundings. Inspection of the Bjerrum plot leads to the observation that in the pH range of most ground waters, viz., 6 to 8, bicarbonate ion is the dominant species. In this range, the activity of the carbonate ion is so

FIGURE 3–1 Bjerrum plot showing activities of inorganic carbon species as a function of pH for a fixed concentration of dissolved inorganic carbon (from *The Geochemistry of Natural Waters* 3/e by J. I. Drever © 1997. Reprinted by permission of Prentice-Hall, Inc. Upper Saddle River, NJ).

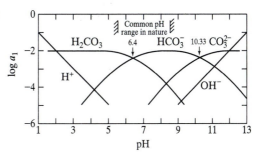

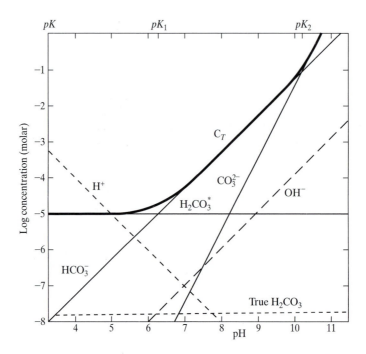

FIGURE 3–2   Plot of log concentrations of inorganic carbon species, H^+ and OH^-, for open-system conditions with a fixed p_{CO_2} of $10^{-3.5}$ atm (from Stumm and Morgan, *Aquatic Chemistry* 3/e, Copyright © 1996, Wiley Interscience. Reprinted by permission of John Wiley & Sons, Inc.).

small that it can be neglected. As the pH rises to 9 and higher, the carbonate activity increases and the carbonic acid activity decreases to a level at which it can be neglected.

A system that is in contact with its surroundings and can exchange constituents such as CO_2 with them is called an *open system*. For reasons that will be discussed later, most shallow ground waters show evidence of open-system conditions. The distribution of species within the carbonate system under open-system conditions is shown in Fig. 3–2, which is drawn for the atmospheric p_{CO_2} value of $10^{-3.5}$ atm. This diagram illustrates that even though the concentrations of HCO_3^- and CO_3^{2-} increase with pH, the general relationships of the Bjerrum plot hold true. In waters with pH values between 6 and 8, bicarbonate is the dominant form of dissolved inorganic carbon. As the pH rises or falls from this range, carbonate ion or carbonic acid, respectively, can become the dominant species. If the partial pressure of CO_2 or the analytical concentration of bicarbonate is known, the concentrations of the other species can be calculated using the K_{CO_2}, K_1, and K_2 equations.

Carbon Dioxide in Ground Water

The reactions involving carbon dioxide in water have been discussed previously in some detail because of their great relevance to ground water chemistry. If the bicarbonate activity of a ground water sample is determined and the *in-situ* pH is known, the p_{CO_2} of the solution can be calculated by combining Eqs. (3–12) and (3–14) as follows:

$$p_{CO_2} = \frac{a_{H^+} a_{HCO_3^-}}{K_1 K_{CO_2}}. \tag{3–20}$$

When p_{CO_2} values are calculated for ground water solutions, they are commonly above the atmospheric p_{CO_2} of $10^{-3.5}$ atm. This can mean either that CO_2 is produced in the aquifer within the ground water flow system or that the water was, prior to its arrival at the water table, in contact with a gas phase containing carbon dioxide at higher levels than in the atmosphere. Although recent work has demonstrated the production of CO_2 at great depths below the water table (Chapelle et al., 1988; McMahon et al., 1990), the presence of high partial pressures of CO_2 in the vadose zone has been known for a long time. Biological production of CO_2 in the vadose zone is sometimes referred to as *soil respiration*. It includes the combination of several processes including root respiration and respiration of soil microbes associated with the decomposition of organic matter in the soil (Fig. 3–3). Root respiration is the process by which live roots absorb oxygen from the soil atmosphere and release carbon dioxide. The decay of organic matter in the soil is mediated by aerobic bacteria, which derive energy from the oxidation of reduced organic carbon. A generic reaction of this type is the equation

$$CH_2O + O_2 \longrightarrow CO_2 + H_2O. \tag{3–21}$$

Although there are many specific compounds present in soil organic compounds, CH_2O represents the reduced carbon in these compounds.

The transfer of carbon dioxide from the soil atmosphere to water infiltrating downward through the vadose zone toward the water table can account for the high p_{CO_2} values calculated for ground water. The complexity of the soil environment, however,

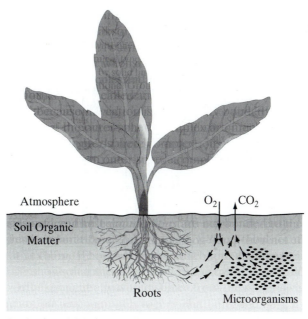

FIGURE 3–3 Production of CO_2 in the soil zone by root respiration and aerobic decomposition of organic matter by microorganisms.

creates great temporal and spatial variability in the production of CO_2 and the interaction between CO_2 and soil water that reaches the water table. The many variables that interact in this system include temperature, soil water content, soil organic matter content, vegetation, soil type, and grain size.

The variation in production of CO_2 is primarily controlled by temperature, moisture content, and amount of organic matter. These variables imply that climate plays a major role in that system. In temperate climates with adequate soil moisture, CO_2 levels in the soil are strongly seasonal (Fig. 3–4). The activity of microorganisms increases with temperature, and CO_2 levels rise accordingly. CO_2 levels begin to increase in the spring, reach their peak in the late summer, and decline throughout the fall and winter (Trainer and Heath, 1976; Reardon et al., 1979; DeJong and Schappert, 1972). The fact that temperature and soil moisture contents generally are inversely related in temperate climates (Fig. 3–5), with soil moisture levels at a minimum when temperatures are highest, suggests that CO_2 production is largely controlled by temperature. In arid climates, however, soil moisture becomes more of a factor. Amundson et al. (1989) studied

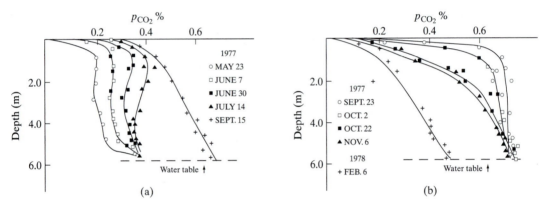

FIGURE 3–4 Increase in p_{CO_2} with temperature during the growing season (a) and decrease during the following fall and winter (b) for a site in Ontario (reprinted from Jour. Hydrol., 43, Reardon, E. J., Allison G. B. and Fritz, P. Seasonal chemical and isotopic variations of soil CO_2 at Trout Creek, Ontario, pp. 355–371. Copyright 1979, with permission from Elsevier Science).

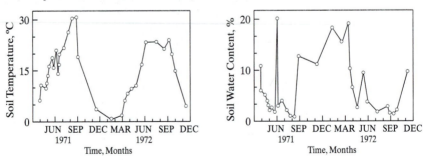

FIGURE 3–5 Inverse relationship between soil moisture and soil temperature in a temperate climatic region (reprinted from Soil Biol. Biochem., 7, Wildung, R. E., Garland, T. R., and Buschbom, R. L., The interdependent effects of soil temperature and water content on soil respiration rate and plant root decomposition in arid grass land soils, pp. 373–378. Copyright 1975, with permission of Elsevier Science).

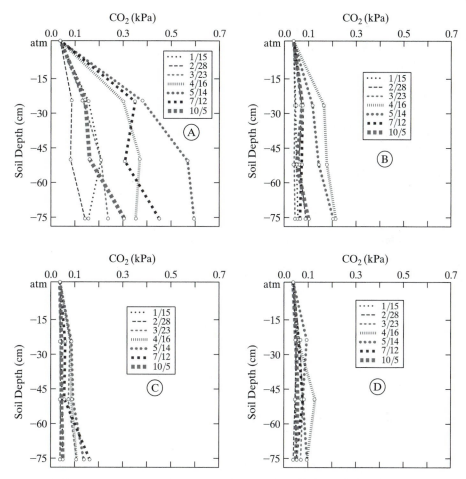

FIGURE 3–6 CO_2 levels in soil plotted against depth for a transect from high (A) to low (D) elevation in the Mojave Desert. Soil moisture contents decrease from high to low elevation. Vegetation types include fir–pine (A), pinyon–juniper (B), blackbrush (C) and creosote (D) (from Amundson et al., 1989. A comparison of soil climate and biological activity along an elevation gradient in the eastern Mojave Desert. *Oceologic*, 80, pp. 395–400. Copyright © Springer-Verlag).

CO_2 production at sites along a transect from high to low elevation in the Mojave Desert. The results (Fig. 3–6) show that CO_2 production is inhibited because of the lack of adequate moisture at lower elevations where temperatures are highest. The higher moisture content at the fir-pine vegetation site (highest elevation) (Fig. 3–7) appears to explain the differences in CO_2 shown in Fig. 3–6.

The role of organic matter content and root respiration in CO_2 production is less apparent. Organic matter accumulates in the soil under a wide range of temperatures, given sufficient humidity (Fig. 3–8). Lower concentrations in desert climates may reinforce the low moisture levels described previously in the low production rates of CO_2. Wildung et al. (1975) concluded that live-root respiration was a much less significant

FIGURE 3–7   Seasonal variation in average soil moisture content at study sites in Mojave Desert shown in Fig. 3–6. Soil moisture contents decrease with decreases in elevation (from Amundson et al., 1989. A comparison of soil climate and biological activity along an elevation gradient in the eastern Mojave Desert. *Oceologic*, 80, pp. 395–400. Copyright © Springer-Verlag).

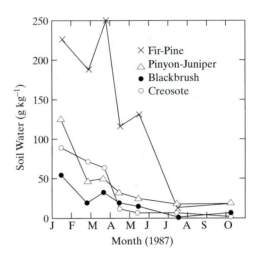

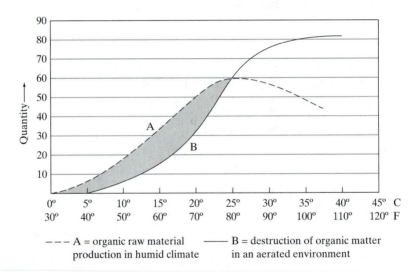

FIGURE 3–8   Production of organic matter (dashed line) in soil and destruction of organic matter (solid line) in a humid climatic region. Organic matter accumulates in soil (gray area) to a mean annual temperature of about 25°C. At higher temperatures destruction exceeds production (from Senstius, 1958).

factor in CO_2 production relative to microbial respiration in humid grassland soils, whereas the studies of Reardon et al. (1979) in a forested Ontario site point to root respiration as the major source of soil carbon dioxide.

The production of CO_2 is only one process in the eventual transfer of CO_2 to ground water. Migration of CO_2 from the root zone is influenced by several factors. In the gas phase of unsaturated soil pores, diffusion to lower partial pressures both upward and downward is an important process. Reardon et al. (1979) and Ring (1995) found high levels of carbon dioxide in soil gas from the root zone down to the water table, although seasonal fluctuations did occur. Even though diffusion may partially explain the downward movement of CO_2, advection by downward-moving vadose water also contributes to the supply of CO_2 reaching the water table. Equilibration of CO_2 and water is so rapid that soil water will quickly reach partial pressures of CO_2 present in the soil gas as it moves downward through the root zone. The timing of ground water recharge from this infiltrating moisture is particularly critical. In temperate climates, most recharge occurs in the spring and fall, that is, before and after the growing season, respectively, because of the consumption of soil water by evapotranspiration in the summer (Trainer and Heath, 1976). Thus recharge takes place when CO_2 levels in the soil are less than their maximum values. Opposing this factor is the greater solubility of CO_2 in water at lower temperatures. Therefore, although the water is cooler and can dissolve greater amounts of CO_2 during the spring recharge event, generally the most significant period of ground water recharge, the available concentration of CO_2 in the soil is near its yearly minimum. Trainer and Heath (1976) compared the bicarbonate ion concentrations of ground waters in carbonate aquifers (an indication of the amount of dissolved inorganic carbon) along a transect from the mid latitudes of North America to the Equator in order to determine the effects of climate on dissolved carbon contents in ground water. Surprisingly, the average bicarbonate values (Fig. 3–9) were relatively constant except for several areas of low bicarbonate in the central and southern United States. Their conclusion was that in these areas, soils in the recharge area were so permeable that CO_2 rapidly escaped to the atmosphere before it could equilibrate with infiltrating water; or recharge took place very rapidly through sinkholes, stream sinks, and other karst pathways and was never in contact with high soil CO_2 partial pressures.

Alkalinity

In any solution, the sum of the normalities of the positively charged species must equal the sum of the normalities of the negatively charged species. This condition is known as *charge balance*. In a pure solution of carbon dioxide dissolved in water, the possible species include H_2O, H^+, OH^-, H_2CO_3 (under the convention established earlier in which this species represents the sum of carbonic acid and aqueous carbon dioxide), HCO_3^-, and CO_3^{2-}. The charge-balance condition or equation for this solution can be expressed as

$$M_{H^+} = M_{HCO_3^-} + 2M_{CO_3^{2-}} + M_{OH^-}. \tag{3–22}$$

It is apparent from this equation that as carbonic acid dissociates to form bicarbonate, the molarity of H^+ must also increase to maintain charge balance. Thus the hydrogen ion activity will be greater than the hydroxide activity and the pH will be lower than 7. Based on the relationships expressed in Figs. 3–1 and 3–2, when the pH is less than 7, the amounts of carbonate ion and hydroxide ion are very small. In fact, they are several orders of magnitude lower than the bicarbonate activity and can be neglected. Under these conditions, the charge-balance equation can be approximated as

$$M_{H^+} = M_{HCO_3^-}. \tag{3–23}$$

In the soil where CO_2 is generated, the reaction between CO_2 and water commonly occurs in the presence of other reactive constituents that can alter the pH and contribute

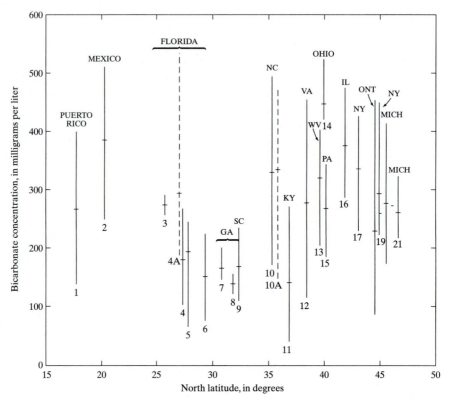

FIGURE 3–9   Concentration of HCO_3^- in ground water at various sites in the northern hemisphere between the mid-latitudes and the equator. Vertical bar shows range of values for each data set and tick mark shows average. Data sets from southern U.S. (Kentucky, South Carolina, Georgia, and Florida) show low HCO_3^- resulting from rapid recharge in karst areas or degassing from very permeable soils in recharge areas (reprinted from Jour. Hydrol., 31, Trainer, F. W. and Heath, R. C., Bicarbonate content of ground water in carbonate rock in eastern North America, Copyright, 1976, with permission from Elsevier Science).

both anions and cations to the solution. For example, if calcite dissolves in the soil, the charge-balance equation would then be

$$2M_{Ca^{2+}} + M_{H^+} = M_{HCO_3^-} + 2M_{CO_3^{2-}} + M_{OH^-}. \qquad (3\text{--}24)$$

If a strong acid were now added to the solution represented by Eq. (3–24), the concentrations of the constituents on the right, as well as H^+, would change as more acid is added because H^+ would react with hydroxide to form water, with carbonate ion to form bicarbonate, and with bicarbonate ion to form carbonic acid as implied by Eqs. (3–13) and (3–15). The concentration of calcium, however, would remain constant. Thus, the terms on the right and H^+ are considered to be nonconservative ions and calcium is classified as a conservative ion. We can group conservative and nonconservative species together as in the equation

$$2M_{Ca^{2+}} = M_{HCO_3^-} + 2M_{CO_3^{2-}} + M_{OH^-} - M_{H^+}. \qquad (3\text{--}25)$$

The term on the right now represents the excess of nonconservative bases over hydrogen, and it is a quantity known as the *total alkalinity*. One formal definition of total alkalinity is that it is the equivalent sum of bases titratable with a strong acid, because the concentrations of the nonconservative bases HCO_3^-, CO_3^{2-} and OH^- will change as they react with H^+ during titration. Perhaps a more practical way to think of total alkalinity is as the acid neutralizing capacity of a solution. The greater the total alkalinity, the more acid the solution could neutralize.

The charge-balance equation for a typical ground water solution could be written as

$$2M_{Ca^{2+}} + 2M_{Mg^{2+}} + M_{Na^+} + M_{K^+} - M_{Cl^-} - 2M_{SO_4^{2-}} = M_{HCO_3^-} + 2M_{CO_3^{2-}} \qquad (3\text{--}26)$$

when the conservative and nonconservative species are grouped together. Hydrogen and hydroxide ions have been omitted because in a typical ground water solution, their molarities are insignificant compared to the other ionic species shown. Of course, carbonate ion would also be so small that it would be neglected, and other ions could be present at high enough concentrations that they might have to be included. The total alkalinity is equal to the right-hand side of Eq. (3–26), or to the molarity of bicarbonate ion.

The measurement of total alkalinity is an important analytical procedure for ground water samples. Because the left-hand side of Eq. (3–26) is a conservative quantity, the sum of the terms on the right-hand side must also be conservative. The only process that could change the alkalinity would be the precipitation of solid phases such as calcite. For this reason, alkalinity is often measured in the field before precipitation can occur.

The actual measurement is made by titration. A known volume of sample is titrated with a strong acid, usually sulfuric acid, until an end point is reached. The end point is the point at which the hydrogen ion concentration equals the bicarbonate ion concentration. At this point, carbonate ion has been converted to bicarbonate ion and most bicarbonate ion has been converted to carbonic acid by reaction with H^+ from the sulfuric acid. The change in pH as acid is added to the sample can be plotted as a titration curve. Fig. 3–10 shows a titration curve for a sodium bicarbonate solution with an initially high pH. The steep section of the curve between pH 9 and 11 is said to be *buffered* because the hydrogen ions react with carbonate ions to form bicarbonate and the pH does not change rapidly as acid is added. After all the carbonate ion has been effectively converted to bicarbonate, the curve flattens out and pH begins to change rapidly with the volume of acid added. Point B is an end point that represents this condition. It is known as the *carbonate alkalinity*. In many ground water solutions, the pH is lower than this value prior to titration and the carbonate alkalinity is zero. As the titration continues, bicarbonate is converted to carbonic acid, and while this is occurring, the solution is again buffered and the curve is steep. The titration is complete at point D where all bicarbonate is effectively converted to carbonic acid. Actually, this represents the point where H^+ is equal to HCO_3^-, which together form an equal amount of carbonic acid. The end point can be determined by a color change of a chemical indicator such as methyl orange or measured more accurately with a pH meter. The pH at the end point is in the range of 4–5.

By knowing the amount and normality of acid added in the titration the alkalinity can be calculated. The convention used to report the total alkalinity is to express it as the equivalent weight of calcium carbonate. This is normally written as mg/l as $CaCO_3$.

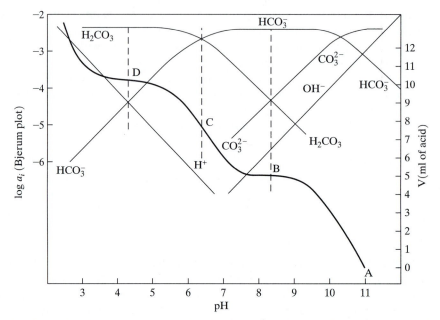

FIGURE 3–10 Titration curve (heavy line) for 5×10^{-3} m Na_2CO_3 solution with amount of acid added on right-hand side of graph. Bjerrum plot for the sum of inorganic carbon species of 5×10^{-3} also included. A is beginning of titration curve, B is carbonate end point, C is region of strong buffering, and D is bicarbonate end point (from *The Geochemistry of Natural Waters*, 3/e by J. I. Drever © 1997. Reprinted by permission of Prentice-Hall, Inc.).

To convert a quantity of acid to this system of units, one must multiply by the equivalent weight of calcium carbonate, which is 50 g/eq because the molecular weight of $CaCO_3$, 100 g/mol, is divided by the ionic charge to obtain the equivalent weight. Because $CaCO_3$ forms ions with a charge of +2 and −2, the equivalent weight is 50 g/eq. This calculation is summarized as follows:

Total alk.(mg/l as $CaCO_3$) =

$$\frac{ml_{acid} \times N_{acid}(eq/l) \times Eq.\ wgt.\ CaCO_3(50\ g/eq) \times 1000\ mg/g}{ml_{acid}}. \quad (3\text{–}27)$$

This procedure is illustrated in Example 3–1.

Example 3–1

In an alkalinity titration, a 100 ml sample is titrated to the methyl orange end point with 2 ml of 0.5 N H_2SO_4. What is the total alkalinity in mg/l as $CaCO_3$, and what is the concentration of HCO_3^- in mg/l?

Solution. The total alkalinity is determined using Equation (3–27):

$$\text{Total alkalinity (mg/l as } CaCO_3) = \frac{2\ ml \times 0.5\ eq/l \times 50\ g/eq \times 1000\ mg/g}{100\ ml}$$

$$= 500\ mg/l\ as\ CaCO_3.$$

The second part of the problem asks for the bicarbonate concentration. Because we assume in most cases that the total alkalinity is contributed by the bicarbonate ion (when the pH is between 6 and 8), Eq. (3–27) can again be used if the equivalent weight of HCO_3^- is used instead of the equivalent weight of $CaCO_3$. Thus,

$$\text{mg/l } HCO_3^- = \frac{2 \text{ ml} \times 0.5 \text{ eq/l} \times 61 \text{ g/eq} \times 1000 \text{ mg/g}}{100 \text{ ml}} = 610 \text{ mg/l}.$$

Notice that the bicarbonate concentration is equal to the total alkalinity times a factor equal to the ratio of the equivalent weights of bicarbonate to calcium carbonate, $61/50 = 1.22$. This relationship can be used to determine the bicarbonate ion concentration whenever the total alkalinity is reported as mg/l as $CaCO_3$.

CARBONATE MINERAL EQUILIBRIA

Among the minerals in the earth's crust that dissolve in ground water solutions to produce the dissolved solids contained in the aqueous phase, the carbonate minerals are perhaps the most abundant and the most important. Although calcite and dolomite are most commonly encountered, others may also occur. We have previously discussed the equilibrium constant of calcite, which is defined as

$$K_{cal} = a_{Ca^{2+}} a_{CO_3^{2-}} = 10^{-8.37} \text{ at } 25°C. \tag{3–28}$$

Aragonite, which is a polymorph of calcite, has an equilibrium constant that is similar in value $\left(K_{arag} = 10^{-8.16} \text{ at } 25°C \right)$ but differs slightly because of the internal structural differences between the two minerals. The larger value of K_{arag} suggests that calcite is the more stable of the two and should precipitate out of solution before aragonite. In seawater, however, this is not the case because calcite precipitation is inhibited by the adsorption of magnesium to the surface of the incipient crystal. Aragonite thus precipitates first but later alters to the more stable calcite. This relationship is a good example of how a system in natural waters behaves contrary to its behavior predicted from calculated equilibrium constants.

The solubility product of dolomite contrasts with that of calcite as a result of its chemical composition. The dissolution of dolomite can be represented by the following reaction:

$$CaMg(CO_3)_2 \rightleftharpoons Ca^{2+} + Mg^{2+} + 2CO_3^{2-}. \tag{3–29}$$

The equilibrium constant for the reaction is

$$K_{dol} = a_{Ca^{2+}} a_{Mg^{2+}} a^2_{CO_3^{2-}}. \tag{3–30}$$

As every introductory geology student knows, the dissolution of dolomite is slower than that of calcite. The possible effects of these kinetic differences will be considered later.

In the presence of pure water, carbonate minerals are only slightly soluble. Under the influence of carbonic acid in nature, however, either at atmospheric levels or at the much greater levels generated by soil respiration, the solubility of the carbonate minerals is greatly enhanced.

The concentrations of species in the equilibrium solution obtained from dissolution of carbonate minerals depends on another factor as well—whether the system is open or closed with respect to CO_2. In a closed system, the CO_2 is used up during the reaction and the solubility is lower than in the open-system case in which CO_2 is constantly replenished as it is consumed in the reaction. For convenience, we will concentrate on open-system conditions.

If we consider the situation in which calcite dissolves in pure water in contact with a gas phase with a fixed p_{CO_2}, the system we are dealing with is the system $H_2O—CO_2—CaCO_3$. The six possible dissolved species in this system for which we want to find concentrations are H^+, OH^-, H_2CO_3, HCO_3^-, CO_3^{2-}, and Ca^{2+}. Six independent equations are necessary to obtain a solution for the unknowns in this problem, five of which are mass action expressions. These equations are

$$K_w = a_{H^+} a_{OH^-} \tag{3–31}$$

$$K_{CO_2} = \frac{a_{H_2CO_3}}{p_{CO_2}} \tag{3–32}$$

$$K_1 = \frac{a_{H^+} a_{HCO_3^-}}{a_{H_2CO_3}} \tag{3–33}$$

$$K_2 = \frac{a_{H^+} a_{CO_3^{2-}}}{a_{HCO_3^-}}, \tag{3–34}$$

and

$$K_{cal} = a_{Ca^{2+}} a_{CO_3^{2-}}. \tag{3–35}$$

The sixth equation is the charge-balance equation for this system:

$$2M_{Ca^{2+}} + M_{H^+} = M_{HCO_3^-} + 2M_{CO_3^{2-}} + M_{OH^-}. \tag{3–36}$$

It is possible to simplify the charge-balance equation by assuming that M_{H^+}, $M_{CO_3^{2-}}$, and M_{OH^-} are small compared to the concentrations of Ca^{2+} and CO_3^{2-}. Thus,

$$2M_{Ca^{2+}} = M_{HCO_3^-}. \tag{3–37}$$

When the p_{CO_2} is known, the concentrations of the other species can be determined as functions of pH and plotted on a log–log plot. For example, from Eq. (3–32), $a_{H_2CO_3} = K_{CO_2} p_{CO_2}$.

At the atmospheric p_{CO_2} value of $10^{-3.5}$,

$$\log a_{H_2CO_3} = \log K_{CO_2} + \log p_{CO_2} = -1.47 + (-3.5) = -4.97. \tag{3–38}$$

This equation can be plotted as a straight line on a log–log plot of concentration vs. pH (Fig. 3–11), indicating that the H_2CO_3 concentration is independent of pH. In a similar fashion, the concentration of HCO_3^- as a function of pH can be derived from the expression

$$a_{HCO_3^-} = \frac{K_1 a_{H_2CO_3}}{a_{H^+}}, \tag{3–39}$$

from which it follows that

$$\log a_{HCO_3^-} = \log K_1 + pH + \log a_{H_2CO_3}$$
$$= -6.35 + pH - 4.97 = -11.32 + pH. \tag{3–40}$$

FIGURE 3–11 Log–log plot of concentration of species in solution in equilibrium with calcite vs. pH at a constant p_{CO_2} of $10^{-3.5}$. If no acid or base added, the concentrations are indicated by vertical arrow (from Stumm and Morgan, *Aquatic Chemistry* 3/e, Copyright © 1996, Wiley Interscience. Reprinted by permission of John Wiley & Sons, Inc.).

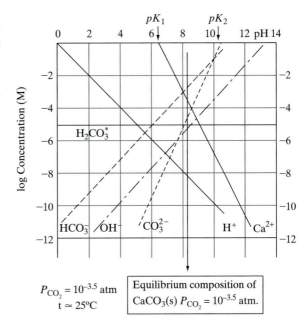

$P_{CO_2} = 10^{-3.5}$ atm
$t \approx 25°C$

Equilibrium composition of $CaCO_3(s)\ P_{CO_2} = 10^{-3.5}$ atm.

This can be plotted as a straight line on Fig. 3–11. Notice that this line crosses the $a_{H_2CO_3}$ line at a pH of 6.35, the value of K_1. The line showing the change in Ca^{2+} concentration has an inverse relationship with pH. The highest calcium concentrations in equilibrium with calcite occur at lower pH values. Figure 3–11 displays the changes in solution composition with the addition of acids or bases. If none are added, the vertical arrow represents the composition of the solution.

Other useful relationships can be obtained when selected constituents are determined as a function of p_{CO_2}. For example, the pH of a solution in equilibrium with calcite can be determined as a function of p_{CO_2}. As a starting point, Eq. (3–35) can be used:

$$K_{cal} = a_{Ca^{2+}}\, a_{CO_3^{2-}}$$

Both terms on the right-hand side can be expressed in terms of pH and p_{CO_2} using substitutions of Eq. (3–32) through (3–36). First from Eq. (3–34),

$$a_{CO_3^{2-}} = \frac{K_2 a_{HCO_3^-}}{a_{H^+}}.$$

Thus,

$$K_{cal} = a_{Ca^{2+}} \frac{K_2 a_{HCO_3^-}}{a_{H^+}}. \tag{3–41}$$

From Eq. (3–33),

$$a_{HCO_3^-} = \frac{K_1 a_{H_2CO_3}}{a_{H^+}}.$$

When this is substituted into Eq. (3–41), along with $a_{H_2CO_3} = p_{CO_2}K_{CO_2}$ from Eq. (3–31), we have

$$K_{cal} = a_{Ca^{2+}}\frac{K_1 K_2 p_{CO_2}K_{CO_2}}{a^2_{H^+}}. \tag{3–42}$$

The $a_{Ca^{2+}}$ term of this equation can be eliminated using the simplified expression for charge balance in this system, $2M_{Ca^{2+}} = M_{HCO_3}$. Because $M = a/\gamma$ for any constituent, the charge balance expression can be converted to activities as follows:

$$\frac{2a_{Ca^{2+}}}{\gamma_{Ca^{2+}}} = \frac{a_{HCO_3^-}}{\gamma_{HCO_3^-}};$$

$$a_{Ca^{2+}} = \frac{a_{HCO_3^-}\gamma_{Ca^{2+}}}{2\gamma_{HCO_3^-}}. \tag{3–43}$$

$a_{HCO_3^-}$ can be replaced in the equation from the K_1 and K_{CO_2} equilibrium expressions (Eqs. 3–33 and 3–32) to yield

$$a_{Ca^{2+}} = \frac{K_1 p_{CO_2}K_{CO_2}\gamma_{Ca^{2+}}}{2\gamma_{HCO_3^-}a_{H^+}},$$

which can then be inserted in Eq. (3–42) to obtain

$$K_{cal} = \frac{K_1 K_{CO_2} p_{CO_2}\gamma_{Ca^{2+}}}{2\gamma_{HCO_3^-}a_{H^+}}\frac{K_2 K_1 p_{CO_2}K_{CO_2}}{a^2_{H^+}}.$$

When this equation is rearranged, we solve for a_{H^+}, to get

$$a^3_{H^+} = \frac{p^2_{CO_2}K^2_1 K^2_{CO_2}\gamma_{Ca^{2+}}K_2}{2K_{cal}\gamma_{HCO_3^-}}. \tag{3–44}$$

A log–log plot of this equation (Fig. 3–12) indicates that pH is inversely related to p_{CO_2} for a solution in equilibrium with calcite. This relationship holds true for ground waters that do not contain excessive amounts of other acids or bases.

The relationships between the concentrations of calcium and bicarbonate vs. p_{CO_2} for solutions in equilibrium with calcite are also very important. Equations (3–32) through (3–36) can be used to derive expressions for these relationships. Contrary to the

FIGURE 3–12   Plot of log p_{CO_2} vs. pH for water in equilibrium with calcite and 25°C (from *The Geochemistry of Natural Waters*, 2/e by J. I. Drever. Copyright © 1988. Reprinted by permission of Prentice-Hall, Inc.).

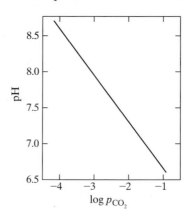

previous derivation, because pH is not a variable of interest, it can be eliminated by dividing Eq. (3–33) by Eq. (3–34). Thus,

$$\frac{K_1}{K_2} = \frac{\dfrac{a_{H^+}\, a_{HCO_3^-}}{a_{H_2CO_3}}}{\dfrac{a_{H^+}\, a_{CO_3^{2-}}}{a_{HCO_3^-}}}$$

$$= \frac{a^2_{HCO_3^-}}{a_{H_2CO_3}\, a_{CO_3^{2-}}},$$

and from Eqs. (3–35) and (3–32),

$$\frac{K_1}{K_2} = \frac{a^2_{HCO_3^-}\, a_{Ca^{2+}}}{K_{cal}\, p_{CO_2}\, K_{CO_2}}. \tag{3–45}$$

The simplified version of the charge balance equation, Eq. (3–37), can be used to eliminate bicarbonate activity from the equation:

$$2M_{Ca^{2+}} = M_{HCO_3^-} = \frac{a_{HCO_3^-}}{\gamma_{HCO_3^-}},$$

$$a_{HCO_3^-} = 2M_{Ca^{2+}}\, \gamma_{HCO_3^-},$$

and

$$a^2_{HCO_3^-} = 4M^2_{Ca^{2+}}\, \gamma^2_{HCO_3^-}.$$

This can be substituted into Eq. (3–45), yielding

$$\frac{K_1}{K_2} = \frac{4M^2_{Ca^{2+}}\, \gamma^2_{HCO_3^-}\, \gamma_{Ca^{2+}}\, M_{Ca^{2+}}}{K_{cal}\, p_{CO_2}\, K_{CO_2}},$$

and

$$M^3_{Ca^{2+}} = \frac{K_1\, K_{cal}\, p_{CO_2}\, K_{CO_2}}{4K_2\, \gamma^2_{HCO_3^-}\, \gamma_{Ca^{2+}}}. \tag{3–46}$$

Plots of this relationship on linear and log–log axes are shown in Fig. 3–13. It is apparent from these plots that there is a nonlinear increase in calcium concentration in equilibrium with calcite with increasing p_{CO_2}, under open-system conditions. The linear plot illustrates an interesting phenomenon that has been observed in field situations. If two solutions lying on the equilibrium curve, points A and B for example, are mixed together in any proportion, the resulting solution will lie on a straight line connecting the two points. This solution will have the ability to dissolve calcite because it lies below the equilibrium line and contains less calcium than required for equilibrium. Mixing of solutions saturated with respect to calcite at different partial pressures of CO_2 can occur in karst terrains. Dissolution caused by undersaturated mixtures of saturated waters has been used as an explanation for the development of caves. We will return to this topic later in the chapter.

A relationship similar to Eq. (3–46) can be developed for the concentration of HCO_3^- in equilibrium with calcite. Langmuir (1971) plotted curves showing the evolution

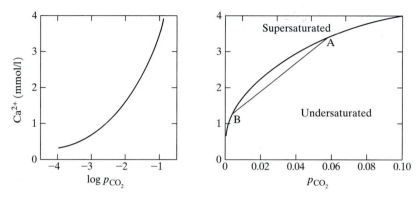

FIGURE 3–13 Plot of calcium concentration vs. log p_{CO_2} (left) and p_{CO_2} on arithmetic scale (right) for pure water in equilibrium with calcite at 25°C (from *The Geochemistry of Natural Waters*, 3/e by J. I. Drever. Copyright © 1997. Reprinted by permission of Prentice-Hall, Inc.).

of HCO_3^- concentrations to saturation with respect to calcite at different partial pressures of CO_2 (Fig. 3–14). Straight lines similar to those shown in Fig. 3–11 result for open-system conditions. In closed systems, however, the increase in HCO_3^- to saturation during reaction with calcite from an initial p_{CO_2} value is nonlinear and much less than the open-system case; and the pH at any point along the curve is higher at equilibrium. These relationships are very relevant for any aquifer in which the mineralogy is dominated by the carbonate minerals. The p_{CO_2} is established in the soil zone as described

FIGURE 3–14 Plot of bicarbonate concentrations approaching equilibrium with calcite and dolomite for three initial p_{CO_2} levels. Straight lines represent constant p_{CO_2} (open system) and curved lines represent consumption of CO_2 during reaction (closed system) (reprinted from Geochim. Cosmochim. Acta, 35, Langmuir, D., The geochemistry of some carbonate ground waters in Central Pennsylvania, pp. 1023–1045, Copyright, 1971, with permission from Elsevier Science).

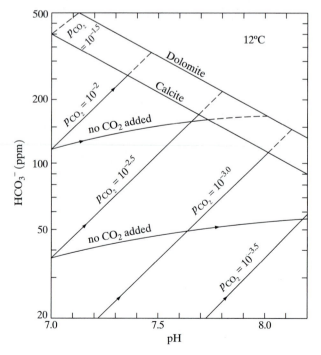

earlier; and as the infiltrating water reacts with calcite, the bicarbonate concentration and pH will follow the straight lines toward equilibrium if open-system conditions prevail. If the system becomes closed to CO_2 at any point in its flow system before equilibrium is reached, the curved paths shown in Fig. 3–14 will be followed.

The Common Ion Effect

The amount of Ca^{2+} in equilibrium with calcite (Fig. 3–13) can change in waters in which either calcium or bicarbonate ion are derived from a more soluble mineral. This phenomenon is called the *common ion effect* because one ion is common to both the more soluble and less soluble mineral. Drever (1997) has demonstrated the effects of adding sodium bicarbonate to water in equilibrium with calcite as a function of p_{CO_2}. The charge balance equation for the resulting solution is

$$M_{Na^+} + 2M_{Ca^{2+}} = M_{HCO_3^-}. \qquad (3\text{–}47)$$

When this is substituted into the derivation leading to Eq. (3–46), Eq. (3–48) is obtained:

$$M_{Ca^{2+}}\left(M_{Na^+} + 2M_{Ca^{2+}}\right)^2 = \frac{p_{CO_2}K_1 K_{cal} K_{CO_2}}{K_2 \gamma_{Ca^{2+}} \gamma_{HCO_3^-}}. \qquad (3\text{–}48)$$

It is apparent from this equation that for any p_{CO_2}, as the sodium concentration increases, the equilibrium calcium concentration decreases. This relationship is shown in Fig. 3–15. If successive amounts of $NaHCO_3$, which is more soluble than calcite, were added to a solution in equilibrium with calcite, the saturation index for calcite would increase and calcite would precipitate. Precipitation would remove calcium and bicarbonate from solution; but because bicarbonate is being added to the solution and calcium is not, the concentration of calcium in equilibrium with calcite would decline as a function of the amount of $NaHCO_3$ added.

A similar circumstance occurs when ground water in equilibrium with calcite encounters gypsum. The more soluble gypsum dissolves, adding Ca^{2+} and SO_4^{2-} to the

FIGURE 3–15 Curves showing calcium concentration in equilibrium with calcite as increasing amounts of $NaHCO_3$ are added to the solution. Addition of common ion (bicarbonate) to solution causes precipitation of calcite and decrease in equilibrium concentration of calcium (from *The Geochemistry of Natural Waters*, 3/e by J. I. Drever. Copyright © 1997. Reprinted by permission of Prentice-Hall, Inc., Upper Saddle River, NJ).

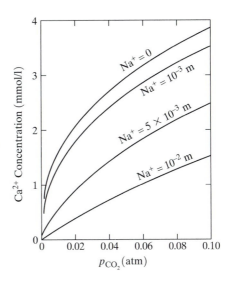

solution. Ca^{2+} is the common ion in this situation. Calcite will become oversaturated and precipitate, thus removing bicarbonate from solution. The amount of calcium in equilibrium with calcite would then rise relative to the amount predicted in Fig. 3–13. According to Appelo and Postma (1993), the solubilities of the two minerals must differ by a factor of three for precipitation to occur by the common ion effect; and, for gypsum and calcite, this criterion is not satisfied unless the temperature is above 25°C.

Incongruent Dissolution of Calcite and Dolomite

When calcite and dolomite are encountered along ground water flow paths, either together or sequentially, the simultaneous dissolution of one mineral and precipitation of the other can occur. This process is called *incongruent dissolution*. The circumstances leading to incongruent dissolution of calcite and dolomite can be considered in light of the values of the equilibrium constants at various temperatures (Freeze and Cherry, 1979). Values of these constants are given in Table 3–2. If a ground water in equilibrium with dolomite at a temperature of 10°C were to come in contact with calcite, what would occur? This question can be answered by rearranging the mass law form of the solubility product of dolomite:

$$K_{dol} = (a_{Ca^{2+}} a_{CO_3^{2-}})(a_{Mg^{2+}} a_{CO_3^{2-}}). \qquad (3-49)$$

Because the activities of calcium and magnesium in equilibrium with dolomite would be equal, the two terms in parentheses would be equal and, furthermore, the quantity $(a_{Ca^{2+}} a_{CO_3^{2-}})$ would be equal to $K^{1/2}_{dol}$. At 10°C, $K^{1/2}_{dol} = 10^{-8.355}$, which is just equal to K_{cal} for this temperature. Thus, calcite would be unable to dissolve since $IAP_{cal} = K_{cal}$. At temperatures other than 10°C, however, $IAP_{cal} \neq K_{cal}$. For example, at 30°C, $IAP_{cal}(K^{1/2}_{dol})$ is less than the value of K_{cal}, which is $10^{-8.510}$. Calcite would dissolve under these conditions. The dissolution of calcite would bring calcium and carbonate ions into solution which, in turn, would oversaturate the solution with respect to dolomite and cause it to precipitate. The simultaneous dissolution of calcite and precipitation of dolomite would be termed *incongruent dissolution of calcite*. The opposite sequence of events would occur at 0°C, where the water in equilibrium with dolomite would have a $K^{1/2}_{dol} = IAP_{cal} = 10^{-8.28}$. This ion activity product is greater than $K_{cal} = 10^{-8.34}$ and

TABLE 3–2   Equilibrium constants for calcite and dolomite in pure water 0–30°C, 1 bar total pressure.

Temp °C	pK_{cal}	pK_{dol}	$pK^{\frac{1}{2}}_{dol}$
0	8.340	16.56	8.280
5	8.345	16.63	8.315
10	8.355	16.71	8.355
15	8.370	16.79	8.395
20	8.385	16.89	8.445
25	8.400	17.00	8.500
30	8.510	17.90	8.950

Source: Freeze and Cherry, 1979.

calcite would therefore precipitate. As calcium and carbonate ions are removed from solution by calcite precipitation, IAP_{dol} would be less than K_{dol}; and dolomite would begin to dissolve incongruently.

Incongruent dissolution reactions can also occur because of the more rapid rate of dissolution of calcite relative to dolomite. If calcite dissolves to equilibrium in an aquifer containing both calcite and dolomite because of its more rapid reaction rate, dolomite would continue to slowly dissolve along the flow path. The solution would be undersaturated with respect to dolomite because of the lack of dissolved magnesium, which is one term in IAP_{dol}. The gradual dissolution of dolomite would induce oversaturation with respect to calcite and, perhaps, precipitation. This would be another example of incongruent dissolution of dolomite.

The Madison Aquifer

An excellent example of the common ion effect and incongruent dissolution occurs in the regional flow system of the Madison aquifer east of the Rocky Mountains. The aquifer is mostly confined except where recharge takes place along the flanks of the mountain ranges (Fig. 3–16). The Madison aquifer is composed of Mississippian

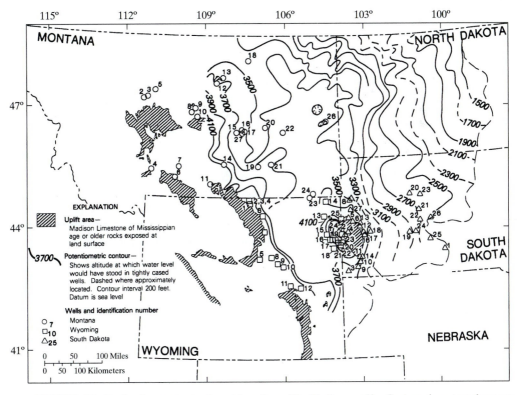

FIGURE 3–16   Predevelopment potentiometric surface of the Madison aquifer. Contours in meters above sea level (from Plummer et al., *Water Resources Research*, v. 26, pp. 1981–2014, 1990. Copyright by the American Geophysical Union).

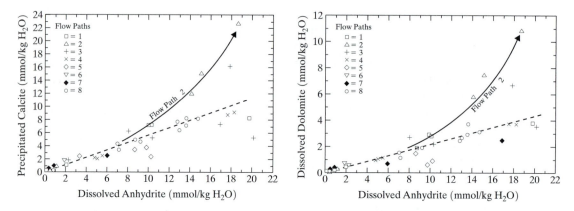

FIGURE 3–17   Precipitation of calcite by the common ion effect as a function of anhydrite dissolution (left) in the Madison aquifer. Right diagram shows incongruent dissolution of dolomite with calcite precipitation. Two trends for various flow paths shown (from Plummer et al., *Water Resources Research*, v. 26, pp. 1981–2014, 1990. Copyright by the American Geophysical Union).

carbonates in which the primary minerals are limestone, dolomite, and anhydrite (Plummer et al., 1990). Along the flow path to the northeast, dissolution of anhydrite causes precipitation of calcite by the common ion effect. The decrease in pH and increase in p_{CO_2} resulting from the precipitation of calcite leads to undersaturation of the solution with respect to dolomite. Dissolution of dolomite is incongruent because of the simultaneous precipitation of calcite. This incongruent reaction is called *dedolomitization*. Fig. 3–17 shows the amounts of calcite precipitated and dolomite dissolved as a function of the amount of anhydrite dissolved. In Fig. 3–18, the calculated saturation indices of calcite, dolomite, and gypsum are shown plotted as a function of the sulfate concentration. Sulfate concentration increases along the flow path until saturation with gypsum occurs and, therefore, serves as a proxy for distance from the recharge area. The general oversaturated condition of calcite throughout the flow system indicates that only a small amount of anhydrite is necessary to produce oversaturation. The saturation index of dolomite is much more variable (Fig. 3–18), with undersaturation present across the range of sulfate concentrations. Undersaturation thus drives the inconguent dissolution of dolomite.

Ionic Strength Effect

Consider again the addition of a salt solution to a solution in equilibrium with calcite (Fig. 3–13). In this case, rather than adding a solution containing a common ion, a sodium chloride solution will be added. Will the concentration of calcium in equilibrium with calcite increase, decrease, or remain the same? Precipitation of calcite is unlikely because the added salt contains neither calcium nor carbonate. The changes that do occur can be examined in light of the calcite mass law expression,

$$K_{cal} = a_{Ca^{2+}} a_{CO_3^{2-}}.$$

FIGURE 3–18   Saturation indices of calcite, dolomite, and gypsum saturation as a function of sulfate concentration for the Madison aquifer. Sulfate concentration increases along the flow path by dissolution of anhydrite, causing oversaturation of calcite. Dolomite saturation index is more variable, with undersaturation occurring throughout flow system (from Plummer et al., *Water Resources Research*, v. 26, pp. 1981–2014, 1990. Copyright by the American Geophysical Union).

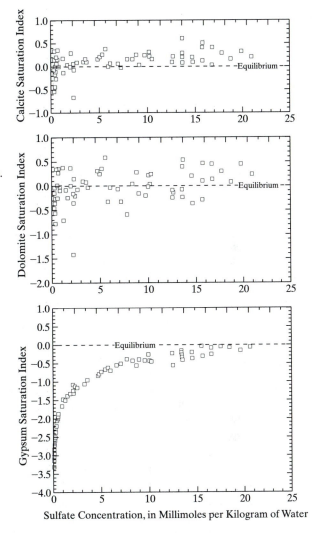

Sulfate Concentration, in Millimoles per Kilogram of Water

If the activities of the species in this equation are converted to molarities using activity coefficients, the expression becomes

$$K_{cal} = (\gamma_{Ca^{2+}} M_{Ca^{2+}})(\gamma_{CO_3^{2-}} M_{CO_3^{2-}}). \qquad (3\text{–}50)$$

The effect of adding sodium chloride to the calcite-saturated solution becomes apparent when the changes in activity coefficients are taken into account. The sodium chloride increases the ionic strength of the solution and, according to Fig. 2–4, activity coefficients decrease as the ionic strength increases. As $\gamma_{Ca^{2+}}$ and $\gamma_{CO_3^{2-}}$ decrease, $M_{Ca^{2+}}$ and $M_{CO_3^{2-}}$ must increase in order to keep the product on the right-hand side of the equation equal to the calcite equilibrium constant. Thus, calcite will dissolve as sodium chloride is added to the solution, which is a consequence of the *ionic strength effect*. Minerals are more soluble in a concentrated solution than in a dilute one.

GROUND WATER CHEMISTRY IN CARBONATE-ROCK AQUIFERS

Carbonate-rock aquifer systems include some of the most prolific and important aquifers for ground water supply. In the United States, aquifers such as the Edwards aquifer of Central Texas and the Floridan aquifer are regionally important aquifers upon which millions of people are dependent for drinking water. An understanding of the hydrogeology and hydrogeochemistry of these aquifers is critical to the management and protection of ground water resources for continued and perhaps expanded usage.

The widespread occurrence of carbonate aquifers results from the deposition of carbonate sediments and rocks in shallow marine waters. When marine sedimentary rocks are present in significant thicknesses, carbonate units are almost certain to be present. Because the North American continent was repeatedly covered by shallow seas, carbonate rocks are common. The physical configuration of these rock units, and consequently their potential as aquifers, is largely a function of their post-depositional tectonic history. Intracratonic basins, including the Williston, Illinois, and Michigan basins, for example, contain thick carbonate units which may or may not contain potable ground water. As these units dip toward the centers of the basins, ground water becomes increasingly saline until brines are encountered near the basin centers. Aquifer potential is limited to the basin margins where fresher water is present. Broad uplifts bring carbonates nearer to the surface, as in the Cincinnati arch, where aquifers can be regionally continuous; and more intense tectonic deformation, such as occurred to form the Appalachians, can localize aquifers in linear bands along the structural strike.

In carbonate rocks, a unique characteristic distinguishes these aquifers from nearly all other types—the solubility of the aquifer framework. This trait leads to the highly variable flow systems observed in carbonate aquifers, flow systems which control water chemistry and its variations. When the carbonate rocks are located at or near land surface and the dissolution of rock actually controls the landscape hydrology, the topography is referred to as *karst*, a term that implies that the surface drainage is partially or totally transferred to the subsurface. The hydrogeology of karst aquifers is particularly complex, both physically and chemically. In this chapter, we will examine aspects of ground water chemistry that are intimately related to near-surface solution of carbonate rocks. In Chapter 9, other aspects of the chemical evolution of ground water in regionally extensive carbonate aquifers will be examined.

Karst Ground Water Flow Systems

Variations in the hydrogeology of karst aquifers have traditionally been attributed to differences in recharge to the aquifer and to flow mechanisms within the aquifer. Recharge to a karst aquifer can occur by direct transfer from the surface to the saturated zone through sinkholes or shafts or by slow infiltration through the vadose zone, not unlike recharge to other aquifer types. Flow varies between intergranular Darcian movement typical of flow in nonkarst systems to turbulent flow in large open conduits either above or below the water table under non-Darcian conditions. Smart and Hobbs (1986) constructed a more comprehensive model of karst hydrogeology encompassing three independent variables: recharge, storage, and flow. Fig. 3–19 illustrates the range of hydrological conditions occurring in karst aquifers. Each variable in the Smart and

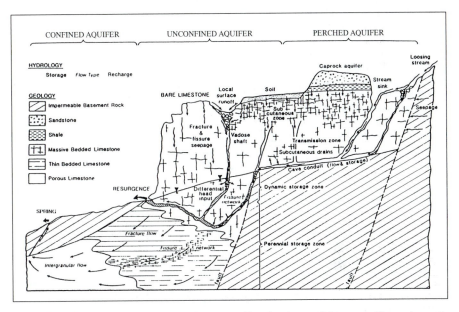

FIGURE 3–19   Conceptual model of carbonate aquifers showing conditions controlling recharge, storage, and flow (from Smart and Hobbs, 1986. Reprinted with the permission of the National Ground Water Association. Copyright © 1986).

Hobbs model functions independently of the other two. Thus, a specific type of recharge does not imply anything about the storage or flow conditions in the aquifer.

Recharge ranges between *concentrated* and *dispersed* end members. The specific processes occurring within this continuum as recognized by Smart and Hobbs (1986) are shown in Fig. 3–20. Stream sinks constitute the most concentrated recharge mechanism. Recharge from the *subcutaneous zone* occupies an intermediate position. The subcutaneous zone is a network of solutionally enlarged fissures and fractures just below the soil zone where CO_2 levels are high. Flow is generally lateral in this zone until a vadose shaft is encountered. Dispersed recharge is derived from infiltration through a soil zone. Recharge of this type would be expected to equilibrate with soil zone CO_2.

Flow occurs between *conduit* and *diffuse* end members (Fig. 3–20). Conduit flow is the hallmark of a karst aquifer. It is best developed in massive limestones with low primary permeability. The development of conduits is associated with the gradual enlargement of widely spaced joints and bedding-plane partings. In the development process, a few select conduits enlarge and form an integrated network by a positive-feedback mechanism. The larger the conduit, the more discharge it can convey; the greater the discharge that moves through the conduit, the more dissolution that can occur. Palmer (1990) and papers contained in White and White (1989) review the hydrogeological aspects of karst aquifers.

The storage variable in the Smart and Hobbs model is more difficult to conceptualize. Types of storage are grouped according to their position in the unsaturated or saturated zones, corresponding to low and high amounts, respectively. Within the unsaturated

FIGURE 3–20 Processes and conditions along the three independent continua of recharge, storage, and flow (from Smart and Hobbs, 1986. Reprinted with the permission of the National Ground Water Association. Copyright © 1986).

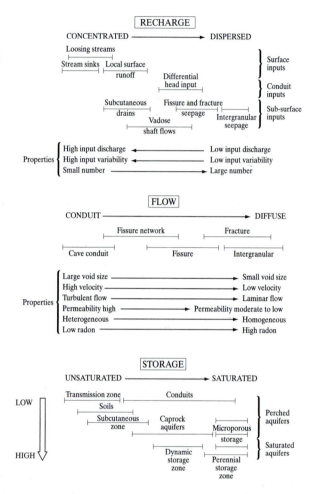

zone, most storage capacity is contained within the subcutaneous network. Saturated storage can be either *dynamic*, if it lies above spring level (the natural discharge point in the aquifer), or *perennial*, if it lies below spring level (Figs. 3–19, 3–20). Fluctuations in spring discharge partially reflect the changes in dynamic storage during the year.

The hydrograph of a karst spring reflects the three independent processes of recharge, storage, and flow in the aquifer. Idealized hydrographs for different combinations of these variables are shown in Fig. 3–21. The shape of each hydrograph is a function of the movement of a precipitation event through the karst aquifer from recharge to discharge at a spring. The combination of concentrated recharge, low storage, and conduit flow produces the least amount of attenuation of the discharge peak as it passes through the aquifer. Thus, the duration of the event is short and the hydrograph is sharply peaked or spiked. When dispersed recharge, high storage, and diffuse flow replace their counterpart end members in the hypothetical model, the peak discharge decreases and the duration of the event, as recorded by the spring discharge, increases.

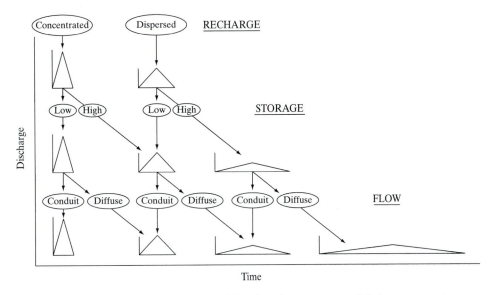

FIGURE 3–21   Hypothetical effects of variations in recharge, storage and discharge upon spring hydrographs (from Smart and Hobbs, 1986. Reprinted with the permission of the National Ground Water Association. Copyright © 1986).

In the field, precise identification of springs that reflect all possible combinations of the three variables is difficult. Figure 3–22 shows three springs in southwest Britain that result from several of the combinations. These hydrographs show weekly discharges during the spring of 1979. The graphs are normalized with respect to the catchment, or recharge area, of the spring. Ogof Ffynnon Ddu has concentrated recharge, low storage, and conduit flow and, therefore, corresponds to the conceptual hydrograph on the left-hand side of the bottom row of Fig. 3–21. St. Dunstan's Well is similar to Ogof Ffynnon Ddu in terms of recharge and flow but has higher storage. As a result, fewer, broader peaks characterize the hydrograph. Banwell Spring has a still less spiky hydrograph because it has dispersed recharge and high storage along with its conduit flow.

Spring hydrographs of the type shown in Fig. 3–22 provide evidence for the source of water discharging at a particular time. Spiky hydrographs imply that the water has had a short residence time in the system and may have been input to the aquifer as concentrated recharge. Broad, attenuated hydrographs suggest that a greater proportion of water derived from long-term storage is discharging from the spring. These differences in the physical hydrogeology of karst aquifers lead to distinct chemical differences in waters within the flow system.

Water Chemistry Variations

Ground water chemistry differences in karst aquifers reflect the complex interactions of recharge, storage, and flow. Residence time within any one of these subsystems is an important factor.

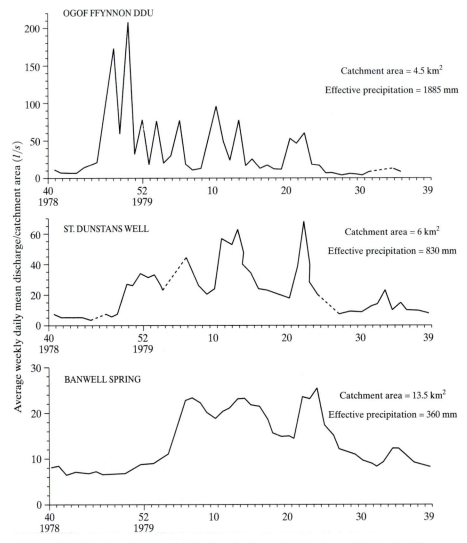

FIGURE 3–22   Annual hydrographs for three hydrographs in southwest Britain with different combinations of recharge, storage, and flow processes (from Smart and Hobbs, 1986. Reprinted with the permission of the National Ground Water Association. Copyright © 1986).

Vadose-zone water-chemistry variations are primarily controlled by the recharge mechanism. Dispersed recharge is derived from infiltration through a soil zone with elevated partial pressures of CO_2. Waters of this type are much more likely to reach equilibrium with carbonate minerals than concentrated recharge transferred rapidly from the surface to the subsurface through large shafts or sinkholes. The dissolution kinetics of calcite help to explain the chemical differences between karst waters of different residence time. Figure 3–23 illustrates that the rate of calcite dissolution decreases rapidly as equilibrium is approached. Ground water derived from concentrated recharge is thus much less likely to achieve saturation with respect to calcite than dispersed recharge.

FIGURE 3–23 Plot of calcite saturation index vs. time under experimental conditions. Saturation index increases very rapidly at first, followed by much slower increase toward saturation (from Hess and White, Chemical Hydrology, in White and White (eds.) *Karst Hydrology: Concepts from the Mammoth Cave Area.* Copyright © 1989 Van Nostrand Reinhold. Reprinted by permission of John Wiley & Sons, Inc. After Herman, 1982).

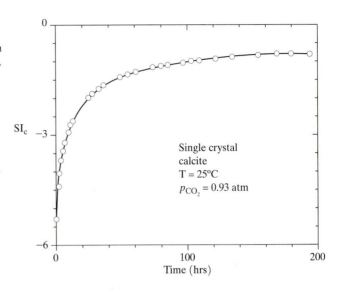

Variations in vadose-zone water chemistry in the Mammoth Cave region of Kentucky were investigated by Hess and White (1989). In part of the area, soluble carbonate rocks lie beneath younger formations; here stratigraphy (Fig. 3–24) controls the vadose ground water chemistry. Cave passages in the vadose zone occur in the Mississippian Girkin and St. Genevieve Limestones, which are overlain by units that contain perched aquifers or confining beds. Ground water discharging from small springs in the

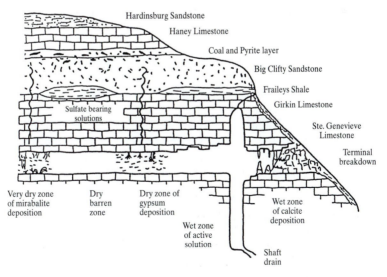

FIGURE 3–24 Cross section of stratigraphy in Mammoth Cave region, Kentucky showing flow paths of vadose water to caves (from Hess and White, Chemical Hydrology, in White and White (eds.), *Karst Hydrology: Concepts from the Mammoth Cave Area.* Copyright © 1989 Van Nostrand Reinhold. Reprinted by permission of John Wiley & Sons, Inc.).

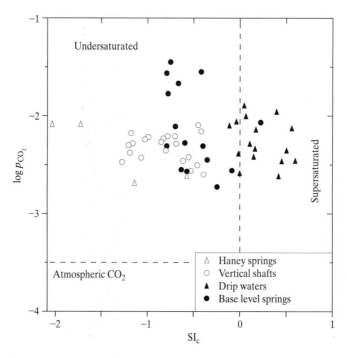

FIGURE 3–25   Plot of calcite saturation index vs. log p_{CO_2} for waters of various origins. Waters in springs with concentrated recharge or conduit flow are most undersaturated with respect to calcite (from Hess and White, Chemical Hydrology, in White and White (eds.), *Karst Hydrology: Concepts from the Mammoth Cave Area.* Copyright © 1989 Van Nostrand Reinhold. Reprinted by permission of John Wiley & Sons, Inc.).

Haney Formation is commonly highly undersaturated with respect to calcite (Fig. 3–25) because of short flow paths and residence time. Vadose seepage is much less abundant lower in the section and is prevented from downward movement by the Fraileys Shale. Where this shale is thin or not present, water that reaches the dry cave passages below precipitates sulfate minerals (gypsum and mirabilite) on cave ceilings. Along the sides of karst valleys where the caprock formations are not present (Fig. 3–24), vadose seepage can enter the cave conduits in several ways. Waters that flow by concentrated recharge into vertical shafts have little contact with CO_2 in the soil zone and are consequently moderately undersaturated (Fig. 3–25). Dispersed recharge also occurs along these valley sides through the soil zone and downward where it emerges as seeps (drip waters) in cave passages. This water is oversaturated with respect to calcite and forms stalactites and stalagmites in the cave passages near the valley walls (Figs. 3–24, 3–25). Ground water that moves through the conduit flow system to regional spring discharge points (base-level springs on Fig. 3–25) remains undersaturated with respect to calcite. The differences in flow paths within the vadose zone explain how undersaturated waters can be dissolving limestone in vertical shafts adjacent to seeps that are precipitating calcite in cave passages.

Flow to a spring discharge point was considered by Shuster and White (1971, 1972) to be the most important variable in determining the chemistry of spring water and its seasonal variations. In fact, seasonal variations were used to classify springs as either conduit or diffuse systems. Shuster and White (1972) used 12 springs in the folded Appalachians of central Pennsylvania in their analysis. Springs were sampled on a two-week interval for a year. Classification of each spring was based on the coefficient of variation (standard deviation divided by mean $\times$ 100) of the hardness. The two models envisioned by Shuster and White are shown in Fig. 3–26. It is likely that their model actually includes an assumption about spring recharge, as well as flow. Conduit springs were assumed to have primarily concentrated recharge whereas diffuse springs were thought to have dispersed recharge. Hardness and temperature are plotted for the study period for representative springs of each type in Fig. 3–27. The conduit spring displays a large seasonal variation of hardness and temperature relative to the diffuse spring, which remains nearly constant throughout the year. Temperature differences reflect differences in residence time in the system. The higher summer temperatures of conduit springs result from the higher temperature of surface water that enters the system as concentrated recharge and then moves rapidly through to the spring.

Other chemical parameters also differ between the two types of springs. The saturation index for calcite is always lower for the conduit springs as compared to the diffuse springs (Fig. 3–28). This difference shows the effect of equilibration with a high soil CO_2 upon the diffuse spring waters. Conduit springs basically discharge surface waters entering the karst flow system in stream sinks. The saturation index curve for the diffuse springs indicates that these waters are more aggressive in the spring and summer than in the winter. To explain these variations, the calculated p_{CO_2} levels must be examined.

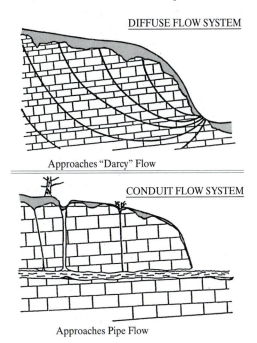

DIFFUSE FLOW SYSTEM

Approaches "Darcy" Flow

CONDUIT FLOW SYSTEM

FIGURE 3–26   Conceptual model of end-member flow systems for springs in the folded Appalachians (from Shuster and White, *Water Resources Research*, v. 8, pp. 1067–1073. Copyright, 1972 by the American Geophysical Union).

Approaches Pipe Flow

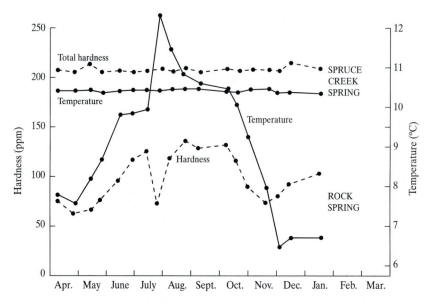

FIGURE 3–27 Plots of temperature and hardness vs. time for a diffuse-flow spring (Spruce Creek Spring) and a conduit-flow spring (Rock Spring) (from Shuster and White, *Water Resources Research*, v. 8, pp. 1067–1073. Copyright, 1972 by the American Geophysical Union).

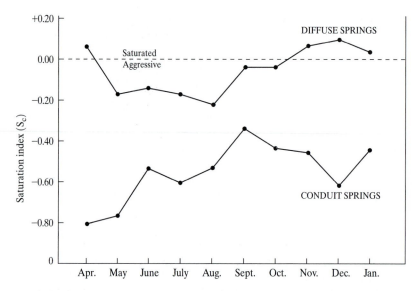

FIGURE 3–28 Saturation indices for calcite for conduit and diffuse springs over time (from Shuster and White, *Water Resources Research*, v. 8, pp. 1067–1073. Copyright, 1972 by the American Geophysical Union).

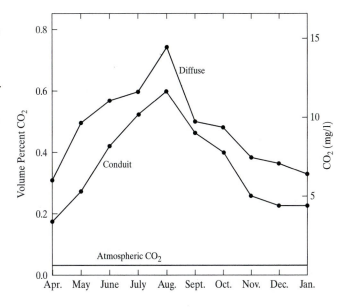

FIGURE 3–29 CO_2 pressures for conduit and diffuse springs vs. time (from Shuster and White, *Water Resources Research*, v. 8, pp. 1067–1073. Copyright, 1972 by the American Geophysical Union).

Fig. 3–29 shows the monthly average p_{CO_2} values for both spring types. Both types show a strong seasonal increase in the summer. The explanation for this increase is the accelerated level of biological activity in the soil during the growing season. The surprising aspect of the curves for the two spring types is that both peak at the same time, even though it might be expected that the higher summer p_{CO_2} levels in the diffuse springs would take longer to propagate through the system from the recharge area to the spring discharge point. Shuster and White concluded that even the diffuse springs tap a relatively shallow portion of the flow system and that there is a rapid exchange of CO_2 between the vadose zone and the saturated zone.

By using a high-frequency, flow-dependent sampling strategy, Ryan and Meiman (1996) were able to explain short-term water-chemistry changes for single storm events at a spring discharge point in the Mammoth Cave area. Analysis of the discharge and water-chemistry hydrographs of this spring in response to a heavy rainfall event enabled Ryan and Meiman to assess the effects of dynamic storage within the flow system. Among the chemical parameters measured, fecal coliform levels and turbidity can be used as indicators of nonpoint source contamination within the watershed. Heavy precipitation events wash these contaminants into the karst flow system to spring discharge points.

The results of Ryan and Meiman's analysis of a September, 1992 storm event are shown in Fig. 3–30. Precipitation and spring discharge were measured during the event, along with water-chemistry parameters sampled at three-hour intervals. The discharge curve indicates that the spring responded with increased discharge almost immediately after the rainfall began. This implies that a continuous, water-filled conduit system rapidly transfers the increase in hydrostatic pressure caused by increases in concentrated recharge in the watershed to the discharge point. Water-chemistry parameters, including turbidity, fecal coliform bacteria, and specific conductance, however, remain at their initial levels for many hours. The water discharging from the spring at higher rates,

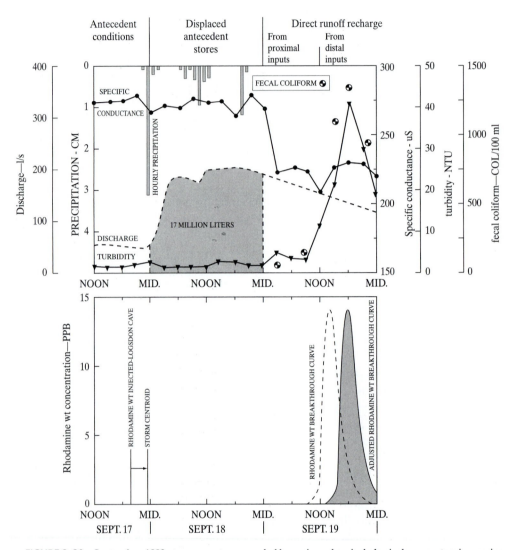

FIGURE 3–30 September, 1992, storm event, as recorded by various chemical physical parameters in a spring in the Mammoth Cave area, Kentucky (top). Water stored in the system was flushed prior to arrival of direct runoff at spring. Dye tracing data from same storm (bottom) showing arrival of pulse of dye from distal input (from Ryan and Meiman, 1996. Reprinted by permission of the National Gound Water Association).

therefore, must represent the flushing of ground water stored in the system prior to the onset of the storm. Twenty-four hours after the beginning of the storm, the specific conductance at the spring began to drop (Fig. 3–30). It is commonly observed in karst flow systems that concentrated recharge moving through a conduit flow system is more dilute because of short residence time than ground water that has been in the flow system longer and has had time to react with the carbonate minerals. The volume of storage flushed (displaced antecedent stores) prior to the drop in specific conductance was calculated as 17 million liters.

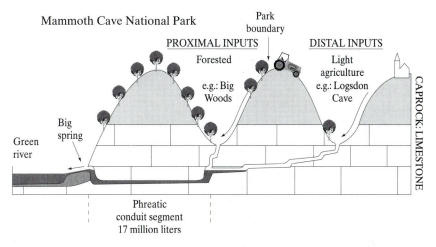

FIGURE 3–31   Conceptual cross section of Mammoth Cave study area showing proximal and distal inputs to spring discharge (from Ryan and Meiman, 1996. Reprinted by permission of the National Ground Water Association).

The fecal coliform bacteria count and turbidity, an indication of suspended sediment carried into the stream sinks, did not begin to increase for an additional 12 hours after the conductivity began to drop. Ryan and Meiman explain this time lag as an indication of the location of the recharge point within the basin. The part of the drainage basin closest to the discharge point lies in Mammoth Cave National Park, where the land cover is forest (Fig. 3–31). Runoff from this area produces a decrease in specific conductance but low levels of turbidity and fecal coliforms. Concentrations of these contaminants rise as recharge from distal parts of the drainage basin outside of the national park reach the discharge point. This hypothesis was confirmed by dye-tracing experiments. A pulse of fluorescent dye introduced to recharge in Logsdon Cave in the distal part of the basin (Fig. 3–31) reached the discharge spring at the same time as the coliform and turbidity increases (Fig. 3–30). The centroid of dye discharge was adjusted slightly to match the rainfall centroid because the dye was injected in anticipation of the storm and the actual beginning of rainfall could not be precisely predicted. Land use is agricultural in the distal part of the watershed in the vicinity of Logsdon Cave, thus explaining the higher turbidity as the result of erosion of soil from crop fields and the coliform bacteria as the result of runoff from livestock facilities.

Although karst springs provide information about ground water chemistry in shallow karst flow systems, ground water sampled from wells that tap deeper portions of the aquifers can be quite different in composition. Langmuir (1971) sampled wells and springs from karst aquifers composed of limestone and dolomite in central Pennsylvania. Most of the springs were associated with limestone rock units and most of the wells were screened in dolomite aquifers. The saturation indicies for calcite and dolomite (Fig. 3–32) differ for these two sample sets. Springs, reflecting their relatively short flow systems, plot farther from saturation than wells. Relative saturation was closely tied to rock type. A majority of the springs were closer to calcite saturation than to dolomite, whereas most wells were closer to saturation with dolomite than to calcite.

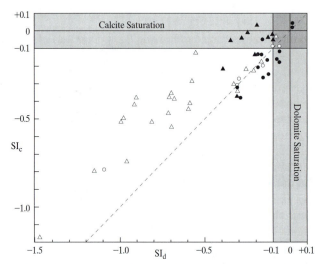

FIGURE 3–32   Saturation indices for calcite and dolomite for spring and well waters in the folded Appalachians of central Pennsylvania. Open symbols represent springs and closed symbols represent wells. Triangles denote limestone source rock and circles denote dolomite source rock. Most springs are undersaturated with respect to both minerals. Most wells are near saturation with respect to calcite but undersaturated with respect to dolomite (reprinted from Geochim. Cosmochim. Acta, 35, Langmuir, D., The geochemistry of some carbonate ground waters in Central Pennsylvania, pp. 1023–1045, Copyright 1971, with permission from Elsevier Science).

Scanlon (1989) recognized three ground water chemical types in the Inner Bluegrass karst region of Kentucky. The Ca-Mg-HCO$_3$ type was most common (Fig. 3–33) and displayed the greatest amount of seasonal variation. The Na-HCO$_3$ type probably developed as a result of ion exchange with interbedded shale units in the carbonate aquifers. No apparent regional trends could explain the differences in water chemistry. Both the Na-HCO$_3$ and the Na-Cl types probably develop in localized zones of shaly, low conductivity rock, in which the residence time of ground water is much greater than in the rocks containing Ca-Mg-HCO$_3$ type water. This study clearly documents the role of storage and residence time on ground water chemical evolution.

Upon emergence from karst aquifers, ground water re-equilibrates with the atmosphere through degassing of CO$_2$. Lorah and Herman (1988) studied the water chemistry changes along Falling Spring Creek, which begins at a cave spring. Downstream samples show a progressive loss of CO$_2$ (Fig. 3–34). The calcite saturation index, which varied seasonally between slightly negative and slightly positive values (Fig. 3–34), gradually increased in response. Recalling Figs. 3–1 and 3–13; as the p_{CO_2} decreases, the pH will rise. A rise in pH should increase the activity of CO$_3^{2-}$ and consequently the saturation index for calcite. Although SI$_{cal}$ rises well above 0 along Falling Spring Creek, a critical level of oversaturation is apparently necessary to initiate precipitation. This threshold is crossed as the stream flows over a waterfall about 1 km downstream from the cave (Fig. 3–34). The agitation and turbulence in the water as the stream goes over the falls induces rapid degassing of CO$_2$. A ubiquitous coating of travertine (freshwater calcite) on rocks, twigs, and anything else in the stream bed at the waterfall provides evidence for the importance of increased turbulence and CO$_2$ degassing in raising the saturation index above the critical point necessary for precipitation.

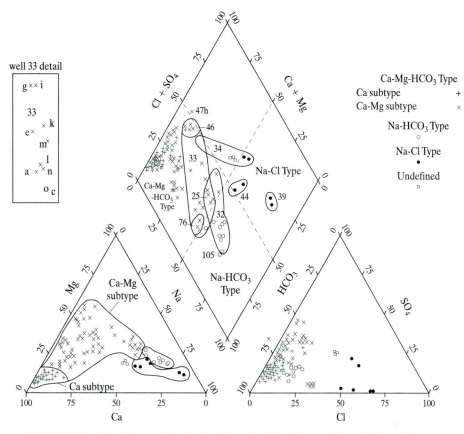

FIGURE 3–33 Piper diagram showing chemistry of wells and springs from the Inner Bluegrass karst area of Kentucky. Water chemical types shown by different symbols. Temporal variations of individual wells shown enclosed by solid lines on diamond-shape figure (from Scanlon, 1989. Reprinted by permission of the National Ground Water Association).

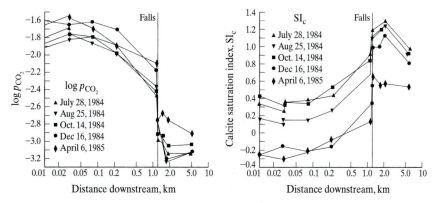

FIGURE 3–34 Partial pressure of CO_2 (left) and calcite saturation index (right) plotted against distance downstream from cave spring in Falling Spring Creek, Virginia. Waterfall has major effect on degassing of CO_2 and calcite precipitation (from Lorah and Herman, *Water Resources Research*, v. 24, pp. 1541–1552. Copyright by the American Geophysical Union).

PROBLEMS

1. Carbon dioxide is in contact with water at a partial pressure of 0.01 atm. How much will dissolve in water at 25°C at equilibrium when the final pH is 7, and also when the final pH is 10? Give your answer as a molar concentration which is the sum of all possible dissolved carbonate species. (Assume a = m).

2. A water analysis from a carbonate spring yields the results shown below. Assume that the value of $K_{cal} = 10^{-8.48}$ at 25° and the values for the other constants at 16°C are $K_{CO_2} = 10^{-1.34}$, $K_1 = 10^{-6.42}$, and $K_2 = 10^{-10.43}$.

   a. What is the p_{CO_2} of this water? Does this indicate an open system or closed system with respect to dissolution of calcite?

   b. What is the saturation index with respect to calcite in the water? Would you expect to see travertine around and downstream from the spring? Why or why not?

	mg/l
Ca^{2+}	48
Mg^{2+}	3.6
Na^+	21
Total alkalinity (as $CaCO_3$)	123
SO_4^{2-}	3.2
Cl^-	8.0
pH	7.5
temp	16°C

3. A shallow pool of rainwater forms on a limestone outcrop and persists long enough to reach equilibrium with calcite. If the temperature is 25°C, what is the pH? Assume a = m.

4. If the outcrop mentioned in question 3 is dolomite, what will the pH be? Assume a = m.

5. Develop an equation for a log–log plot for m_{Ca} as a function of pH at a p_{CO_2} of 10^{-2} and a temperature of 25°C, for a water in equilibrium with calcite. Assume a = m.

Mineral Weathering and Mineral Surface Processes

The low temperature reactions between aqueous solutions and minerals in soils and rocks are collectively called chemical weathering. These processes include both congruent and incongruent dissolution reactions that supply solutes to ground water flow systems and produce secondary mineral products that are, in most cases, more stable than the primary minerals from which they are derived. Among the secondary minerals and amorphous solids that result from weathering reactions are the clay minerals, a group of minerals that have important electrostatic properties associated with their surfaces. When clays or other minerals with charged surfaces exist in rocks or soil containing ground water, reactions between the solutes in the aqueous phase and the mineral surfaces take place which alter the chemical composition of both the solid and liquid phases. In addition to the chemical evolution of natural waters, these minerals strongly influence the ways in which contaminants move through earth materials.

The mostly congruent mineral dissolution reactions discussed in Chapter 3 occur in rocks and soils containing carbonate minerals. Carbonate minerals, however, are overshadowed in the crust as a whole by other mineral groups, among which silicate and oxide minerals are especially significant. Accordingly, we will focus on the weathering and surface properties of these minerals in this chapter.

The weathering of silicates and oxides is more complicated than carbonates for several reasons. Most importantly, the solubilities and dissolution reaction rates among these minerals, which determine their *stability* in weathering environments, vary widely. The relative resistance of silicate minerals to weathering is indicated by *Goldich's stability series* (Goldich, 1938), which is the inverse of Bowen's reaction series. Minerals become increasingly resistant to weathering from bottom to top of the stability series (Fig. 4–1). The significance of this relationship is that minerals near the bottom of the series supply disproportionately greater amounts of solutes to ground water than their abundance in the rock.

PRIMARY MINERALS AND WEATHERING PRODUCTS

The primary silicate minerals in near-surface rocks weather at low temperatures in the general order shown in Goldich's stability series. Some of the more common silicates and their chemical formulas are given in Table 4–1. Their basic mineral structures are shown in Figure 4–2. The relative position of minerals in the series can be explained by the

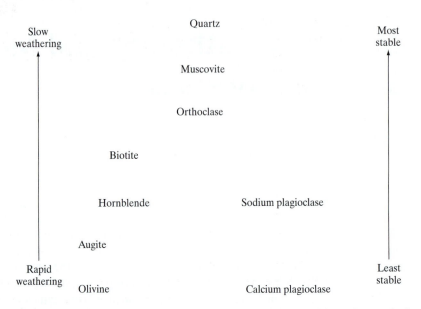

FIGURE 4–1 Goldich's stability series, which shows the relative stability under weathering conditions of primary minerals in igneous rocks.

TABLE 4–1 Names and chemical formulas of primary silicates.

Name	Chemical Formula	Mineral Group
Forsterite	Mg_2SiO_4	Olivine
Fayalite	Fe_2SiO_4	Olivine
Chrysolite	$Mg_{1.8}Fe_{0.2}SiO_4$	Olivine
Enstatite	$MgSiO_3$	Pyroxene
Orthoferrosilite	$FeSiO_3$	Pyroxene
Diopside	$CaMgSi_2O_6$	Pyroxene
Tremolite	$Ca_2Mg_5Si_8O_{22}(OH)_2$	Amphibole
Actinolite	$Ca\ Mg_4FeSi_8O_{22}(OH)_2$	Amphibole
Hornblende	$NaCa_2Mg_5Fe_2AlSi_7O_{22}(OH)$	Amphibole
Muscovite	$K_2[Si_6Al_2]Al_4O_{20}(OH)_4$	Mica
Biotite	$K_2[Si_6Al_2]Mg_4Fe_2O_{20}(OH)_4$	Mica
Phlogopite	$K_2[Si_6Al_2]Mg_6O_{20}(OH)_4$	Mica
Orthoclase	$KAlSi_3O_8$	Feldspar
Albite	$NaAlSi_3O_8$	Feldspar
Anorthite	$CaAl_2Si_2O_8$	Feldspar

molar ratio of silica to oxygen in the mineral. The higher the ratio, the more covalent the structure tends to be and the fewer the number of metal cations are necessary to balance the negative charge imparted to the structure by oxygen. Thus, quartz, with an Si:O ratio of 0.5, is the most stable, or resistant, mineral and olivine, with a ratio of 0.25, is the least stable.

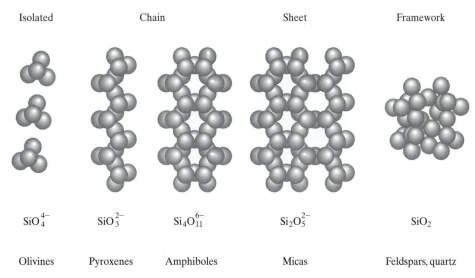

| Isolated | Chain | Sheet | Framework |

SiO_4^{4-} SiO_3^{2-} $Si_4O_{11}^{6-}$ $Si_2O_5^{2-}$ SiO_2

Olivines   Pyroxenes   Amphiboles   Micas   Feldspars, quartz

FIGURE 4–2   Structures of silicate mineral groups. Circles represent oxygen ions at corners of silicon tetrahedra.

The incongruent dissolution reactions of the silicate minerals involve the release to aqueous solution of dissolved silica in the form of silicic acid (H_4SiO_4) and metal cations such as calcium, sodium, magnesium, and potassium. When aluminum and iron are present in the primary mineral, secondary weathering products of low solubility are commonly formed. These include a variety of amorphous and crystalline phases such as the clay minerals and numerous oxides and hydroxides. In an oxidizing environment, iron is oxidized from the Fe^{+2} (ferrous) form to the Fe^{+3} (ferric) form prior to precipitation. The weathering of albite to kaolinite and biotite to kaolinite and amorphous iron oxide is representative of these reactions:

$$2NaAlSi_3O_8 + 2H^+ + 9H_2O \longrightarrow Al_2Si_2O_5(OH)_4 + 2Na^+ + 4H_4SiO_4; \quad (4-1)$$
$$\text{Albite} \qquad\qquad\qquad\qquad \text{Kaolinite}$$

$$2K[Mg_2Fe][AlSi_3]O_{10}(OH)_2 + 10H^+ + 0.5O_2 + 6H_2O \longrightarrow$$
$$\text{Biotite} \qquad\qquad Al_2Si_2O_5(OH)_4 + 2K^+ + 4Mg^{2+} + 2Fe(OH)_3 + 4H_4SiO_4. \quad (4-2)$$
$$\text{Kaolinite}$$

Laboratory studies of dissolution rates have provided support for the arrangement of minerals in Goldich's stability series. Lasaga (1984) normalized these rates for comparison of the lifetimes of crystals of various primary minerals (Table 4–2). The conditions specified for the calculated ages include a crystal radius of 1×10^{-3} m, a pH of 5, and a temperature of 298° K. The age is the amount of time required to totally dissolve a crystal when in continuous contact with a sufficiently undersaturated solution. With the exception of forsterite, the calculated lifetimes of the minerals listed by Lasaga (1984) in Table 4–2 agree exactly with Goldich's stability series.

TABLE 4–2   Mean lifetime of a 1 mm crystal at pH = 5 and T = 298°K.

Mineral	Lifetime	Mineral	Lifetime
Quartz	34 my	Enstatite	8,800 yr
Muscovite	2.7 my	Diopside	6,800 yr
Forsterite	600,000 yr	Nepheline	211 yr
K-feldspar	520,000 yr	Anorthite	112 yr
Albite	80,000 yr		

Source: Lasaga, 1984

Clay Minerals

The clay minerals contain most of the aluminum and some of the silica released from the primary mineral. The structure of this important and highly variable group of minerals is based upon layers composed of silicon in tetrahedral coordination and layers containing metal cations in octahedral coordination with oxygen and hydroxide (Fig. 4–3). Metallic cations of different charge can occupy the octahedral sheet. If a divalent cation

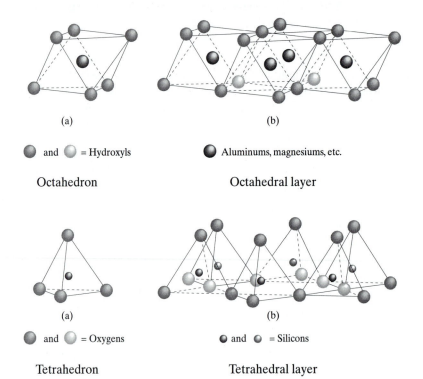

(a)   (b)

● and ○ = Hydroxyls   ● Aluminums, magnesiums, etc.

Octahedron   Octahedral layer

(a)   (b)

● and ○ = Oxygens   ● and ○ = Silicons

Tetrahedron   Tetrahedral layer

FIGURE 4–3   Structures of octahedral and tetrahedral layers in minerals (from Grim, *Clay Mineralogy*, Copyright © 1968. Used by permission of the McGraw-Hill Companies).

such as magnesium occupies the sheet, each potential octahedral location will be filled and the mineral will be classified as *trioctahedral*. Talc [$Mg_3Si_4O_{10}(OH)_2$], although not a clay mineral, has the basic trioctahedral structure. Minerals in which the octahedral holes are filled by ions with a +3 valence such as aluminum are known as *dioctahedral* because only two out of every three locations is occupied. Pyrophyllite [$Al_2Si_4O_{10}(OH)_2$] typifies this configuration. Substitutions of metal cations for silicon in the tetrahedral sheet and among various cations in the octahedral sheet account for the diversity in clay minerals.

The clay minerals can be divided into five groups based upon the *layer type* and the types of substitutions (Table 4–3). The types and amount of substitutions determine the *layer charge* on the mineral. For example, if aluminum (+3) substitutes for silicon (+4) in the tetrahedral sheet, the overall charge will be unbalanced with a deficiency in positive charge. The most simple structural unit is the 1:1 layer of the kaolinite group (Fig. 4–4). Substitutions in either the tetrahedral or octahedral sheets are rare in the kaolinite group; and, as a result, the layer charge is near zero. Three groups have a 2:1 layer type with different substitution patterns (Table 4–3) and different layer charges. The layer charges are balanced by cations that occupy sites between adjacent layers. Calcium and magnesium are common interlayer cations, and potassium can also fill these sites. The layer charge is important in determining how strongly the interlayer cations are held. The smectite group (Fig. 4–5) has the lowest layer charge of the 2:1 minerals and, as a result, the weakest bonds holding the interlayer cations. The very high cation exchange capacity of the smectites is partially the result of the low layer charge. The chlorite group contains an additional layer, making it the most structurally complex of the groups (Fig. 4–6).

TABLE 4–3   Major clay mineral groups.

Group	Layer Type	Layer Charge (x)	Typical Chemical Formula[a]
Kaolinite	1:1	<0.01	$[Si_4]Al_4O_{10}(OH)_8 \cdot nH_2O(n = 0 \text{ or } 4)$
Illite	2:1	1.4–2.2	$M_x[Si_{6.8}Al_{1.2}]Al_3Fe_{0.25}Mg_{0.75}O_{20}(OH)_4$
Vermiculite	2:1	1.2–2.0	$M_x[Si_7Al]Al_3Fe_{0.5}Mg_{0.5}O_{20}(OH)_4$
Smectite[b]	2:1	0.5–1.2	$M_x[Si_8]Al_3Fe_{0.2}Mg_{0.6}O_{20}(OH)_4$
Chlorite	2:1 with hydroxide interlayer	Variable	$(Al(OH)_{2.55})_4 \cdot [Si_{6.8}Al_{1.2}]Al_{3.4}Mg_{0.6}O_{20}(OH)_4$

[a] $n = 0$ is kaolinite and $n = 4$ is halloysite; M = monovalent interlayer cation.
[b] Principally montmorillonite in soils.

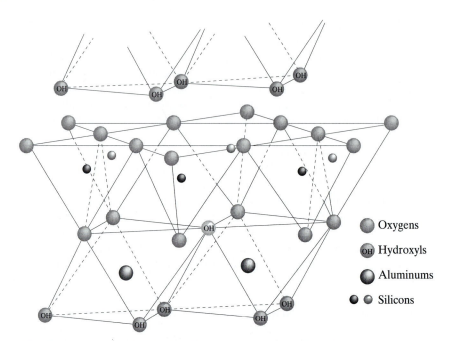

FIGURE 4–4   Structure of the kaolinite layer (from Grim, *Clay Mineralogy*, Copyright © 1968. Used by permission of the McGraw-Hill Companies).

Oxides and Hydroxides

The iron and aluminum that is not incorporated into the clay minerals during the weathering of primary silicates is commonly precipitated by hydrolysis as oxide or hydroxide minerals. Under oxidizing conditions and midrange pH values, these minerals are highly insoluble. The initial precipitates may be amorphous, such as $Fe(OH)_3$; but over time, the corresponding crystalline phases form. The most common examples are goethite [α-FeOOH] and gibbsite [γ-Al(OH)$_3$] (Fig. 4–7). The structure of goethite involves oxygen ions that lie in planes. The ferric iron ions form distorted octahedra with oxygens and hydroxyls. Gibbsite consists of dioctahedral sheets identical to those in the clay minerals. Adjacent sheets are bound together by hydrogen bonds.

DISSOLUTION PROCESSES AND EQUILIBRIA

Surface Complexation and Dissolution

The dissolution of silicates and oxides is more complicated than ionic solids such as halite or gypsum. In these soluble salts, water molecules can easily hydrate surface anions and cations, which are subsequently released to solution as hydrated ions or *solvation complexes*. The covalent bonds between metal ions and oxygen in the silicate and oxide minerals are not so easily broken and the ions are not so easily detached from the mineral surface by hydration.

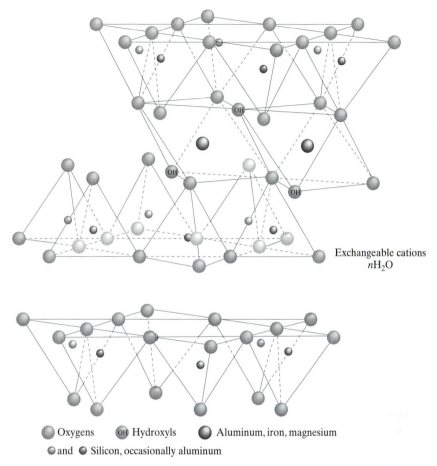

FIGURE 4–5   Structure of smectite minerals (from Grim, *Clay Mineralogy*, Copyright © 1968. Used by permission of the McGraw-Hill Companies).

Silicate and oxide mineral surfaces are dominated by hydroxide ions, which are either part of the mineral structure in minerals such as gibbsite $Al(OH)_3$ or form *surface complexes* with metal ions near the surface of the mineral (Fig. 4–8). Detachment and removal of these near-surface metal ions can occur by either *protonation* or *ligand attack*. Protonation takes place in a two-step process (Sposito, 1989). First, a hydrogen ion from solution bonds to an anion such as OH^- at the mineral surface (Fig. 4–9). The presence of the hydrogen ion polarizes the bonds holding the metal cation to the solid. In the case of gibbsite, $Al(OH)_3$, which is shown in Fig. 4–9, aluminum is then released to solution as the aqueous complex $Al(H_2O)_6^{3+}$.

Ligands, which are the anionic components of complexes, can also promote mineral dissolution. In this process, organic or inorganic ligands can replace a protonated hydroxide such as the one shown in Fig. 4–9. The destabilized aluminum ion then goes into solution as the complex $AlF(H_2O)_5^{2+}$.

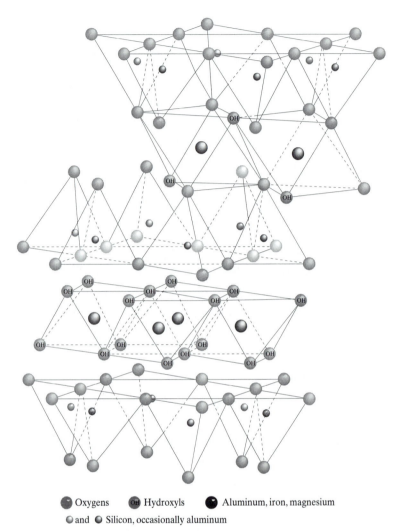

● Oxygens   ⊕ Hydroxyls   ● Aluminum, iron, magnesium

● and ● Silicon, occasionally aluminum

FIGURE 4–6   Structure of chlorite (from Grim, *Clay Mineralogy*, Copyright © 1968. Used by permission of the McGraw-Hill Companies).

The ability of organic acids to promote dissolution of silicate minerals through ligand attack has been demonstrated by many researchers. Humic and fulvic acids, which are complex, high-molecular-weight components of organic soil horizons, probably account for most of the dissolution of silicate minerals and the transport of metal ions such as iron and aluminum within the vadose zone to form soil horizons. Antweiler and Drever (1983) showed by field and lab experiments that the solubility of volcanic ash in Wyoming is greatly enhanced in the presence of humic acid.

Low-molecular-weight organic acids, in particular the dicarboxylic acids such as oxalic acid, also play a major role in the dissolution of silicate minerals through the ligand attack mechanism. Amrhein and Suarez (1988) conducted experimental studies of the dissolution of anorthite ($CaAl_2Si_2O_8$) and found that dissolution involved surface

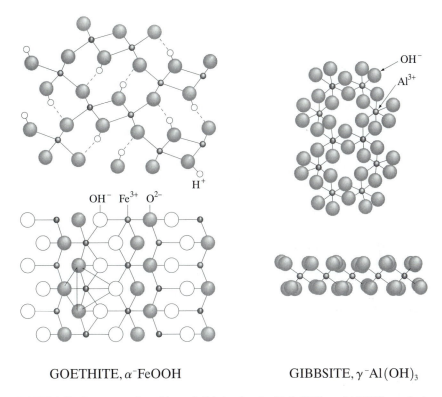

GOETHITE, α-FeOOH   GIBBSITE, γ-Al(OH)$_3$

FIGURE 4–7   Structures of goethite and gibbsite, showing FeO$_3$(OH)$_3$ and Al(OH)$_6$ octahedra in sheets. Dashed lines in goethite structure represent hydrogen bonds between OH and O ions (from *The Surface Chemistry of Soils* by Garrison Sposito. Copyright © 1984 by Garrison Sposito. Used by permission of Oxford University Press, Inc.).

FIGURE 4–8   Surface hydroxide complexes in a metal oxide mineral.

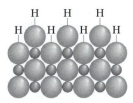

complexes with both hydrogen ions and the organic-acid ligand oxalate (C$_2$O$_4^{2-}$). Protons are more likely to complex and remove surface silicon ions whereas oxalate complexes form with surface aluminum ions. Although the overall dissolution rate increases at lower pH values, the presence of organic acids is very important at the more neutral pH values characteristic of natural waters. The presence of organic acids results in a much higher dissolution rate relative to similar weathering involving only protonation. This effect is more pronounced for calcium-rich plagioclases, which have a higher aluminum content than other feldspars. Welch and Ullman (1993) reached similar conclusions using a variety of low-molecular-weight organic acids and feldspar compositions.

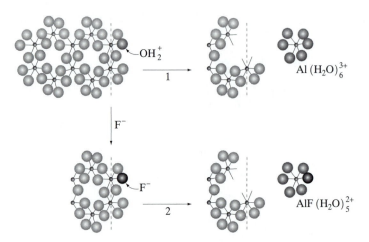

FIGURE 4–9   Dissolution mechanisms for gibbsite. (1) Protonation of a surface OH and detachment of Al^{+3} as a solvation complex. (2) Ligand exchange of OH_2^+ for F^- and detachment of Al^{+3} as the AlF^{+2} complex (from *The Chemistry of Soils* by Garrison Sposito. Copyright © 1989 by Oxford University Press. Used by permission of Oxford Univeristy Press, Inc.).

Dissolution of Quartz

In the absence of organics, and in the midrange of pH values, the solubility of quartz is very low (Drever, 1997). The dissolution reaction below a pH of 9 is

$$SiO_2(\text{quartz}) + 2H_2O = H_4SiO_4(\text{aq}), \tag{4–3}$$

with equilibrium constant

$$K_{eq} = a_{H_4SiO_4}. \tag{4–4}$$

The equilibrium constant shown in Eq. (4–4) has a value of 1×10^{-4} at 25°C. Dissolved silica concentrations are sometimes reported as SiO_2 even though the actual aqueous species is silicic acid as indicated in Eq. (4–3). It is common for ground waters to have silica activities higher than 1×10^{-4} and thus be oversaturated with respect to quartz. Because precipitation of quartz at low temperatures is unlikely, quartz solubility does not control the concentration of silica in ground water. A solid phase that does precipitate at low temperatures, and thus can control silica concentrations, is amorphous silica, SiO_2(amorph). The dissolution reaction is identical to Eq. (4–3), but the equilibrium constant has a value of 2×10^{-3} at 25°C.

At pH values above 9, silicic acid dissociates to form $H_3SiO_4^-$ and $H_2SiO_4^{2-}$, according to the following reactions and equilibrium constants:

$$H_4SiO_4 \rightleftharpoons H_3SiO_4^- + H^+; \tag{4–5}$$

$$K_1 = \frac{a_{H_3SiO_4^-} a_{H^+}}{a_{H_4SiO_4}} = 10^{-9.9}; \tag{4–6}$$

$$H_3SiO_4^- \rightleftharpoons H_2SiO_4^{2-} + H^+; \tag{4–7}$$

$$K_2 = \frac{a_{H_2SiO_4^{2-}} a_{H^+}}{a_{H_3SiO_4^-}} = 10^{-11.7}. \tag{4–8}$$

The total concentration of dissolved silica is the sum of the three dissolved species:

$$(M_{SiO_2})_T = M_{H_4SiO_4} + M_{H_3SiO_4^-} + M_{H_2SiO_4^{2-}}. \tag{4–9}$$

If Eqs. (4–4), (4–6), and (4–8) are rearranged and expressed in terms of $a_{H_4SiO_4}$, and if the activity coefficients are assumed to be 1, they can be substituted in Eq. (4–9). For example, from Eq. (4–6), assuming that $a = M$,

$$M_{H_3SiO_4^-} = \frac{K_1 M_{H_4SiO_4}}{a_{H^+}}. \tag{4–10}$$

Similarly, from Eq. (4–8),

$$M_{H_2SiO_4^{2-}} = \frac{K_2 M_{H_3SiO_4^-}}{a_{H^+}} = \frac{K_2 K_1 M_{H_4SiO_4}}{a_{H^+}^2}. \tag{4–11}$$

Equations (4–10) and (4–11) can then be substituted into Eq. (4–8) to obtain

$$(M_{SiO_2})_T = M_{H_4SiO_4}\left(1 + \frac{K_1}{a_{H^+}} + \frac{K_1 K_2}{a_{H^+}^2}\right). \tag{4–12}$$

The activities of H_4SiO_4 and its dissociation products are plotted as functions of pH in Fig. 4–10. This illustrates the increase in total dissolved silica at high pH values in equilibrium with quartz and amorphous silica that results from the increased concentrations of the dissociation products.

Despite its low solubility in ground water of low dissolved organic carbon and neutral pH, the presence of certain organic acids can considerably enhance the solubility of quartz. Bennett and Siegel (1987) found elevated aqueous silica concentrations in

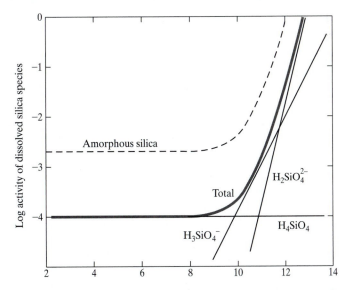

FIGURE 4–10   Activities of dissolved silica species in equilibrium with quartz at 25°C. The heavy line represents the sum of the activities of individual species. Dashed line is corresponding sum for equilibrium with amorphous silica (from *The Geochemistry of Natural Waters* 3/e by Drever © 1997. Reprinted by permission of Prentice-Hall, Inc., Upper Saddle River, NJ).

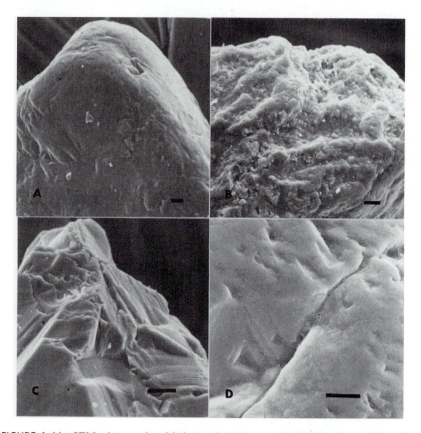

FIGURE 4–11   SEM micrographs of (A) unetched quartz grain from background location, (B) etched quartz grain from contaminated zone, (C) unetched plagioclase grain from background location, and (D) heavily weathered plagioclase with adhering secondary minerals from the contaminated zone. Scale bar=10 μm (from Bennett P. C. and Casey W. H., Organic acids and the dissolution of silicates, *in* E. D. Pittman and M. D. Lewan (eds.), *The Role of Organic Acids in Geological Processes.* Copyright © 1993. Used by permission of Springer-Verlag GmbH & Co.KG).

the contaminant plume produced by the rupture of a crude oil pipeline near Bemidji, Minnesota. In ground water, natural biodegradation of petroleum compounds produces high concentrations of organic acids as degradation products (Cozzarelli et al., 1990). These acids can provide ligands that can enhance the dissolution of aluminosilicate minerals, as previously described. At the Bemidji site, quartz grains collected from the aquifer solids from the core of the plume showed evidence of chemical etching as compared to quartz grains collected from background cores at the site (Fig. 4–11). Bennett and Siegel (1987) concluded that ligand attack involving organic acids was responsible for the enhanced quartz dissolution. Bennett et al. (1988) conducted lab experiments using a variety of organic acids. Their results show that several low-molecular-weight carboxylic acids increased quartz dissolution rates at 25°C and neutral pH (Fig. 4–12). This effect is pH dependent and dissolution decreased at lower pH values. Exposed quartz surfaces, like most silicate minerals in aqueous environments, are characterized by hydroxyl

FIGURE 4–12   Leaching of silica from quartz in 20 mmol/Kg solutions of various acids at pH = 7 compared to leaching of silica from quartz in water at pH = 6.6. Leaching is expressed as micromoles of silica per m² of quartz (reprinted from *Geochim. Cosmochim. Acta*, 52, Bennett, P. C., Melcer, M. E., Siegel, D. I., and Hassett, J. P., pp. 1521–1530, with permission from Elsevier Science).

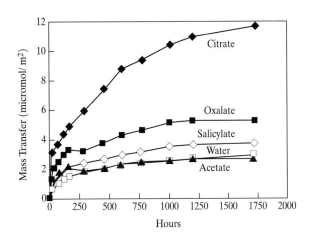

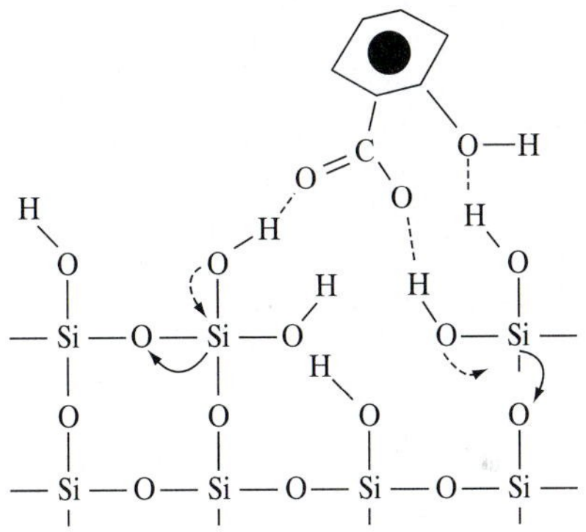

FIGURE 4–13   Dissolution model for hydroxylated silicate mineral. Organic-acid ligand forms complex with surface hydroxide, which polarizes and weakens internal bonds. Complex is then released into solution (reprinted from *Geochim. Cosmochim. Acta*, 52, Bennett, P. C., Melcer, M. E., Siegel, D. I., and Hassett, J. P., pp. 1521–1530, with permission from Elsevier Science).

ions. In the preferred model for the dissolution process (Bennett et al., 1988), bonds between the organic acid ligands and surficial hydroxyls polarize and weaken interior Si-O bonds (Fig. 4–13). The entire organic acid-silica complex can then be released into solution. These aqueous complexes increase the total silica in solution but do not affect the equilibrium between free dissolved silicic acid (H_4SiO_4) and silicate minerals. If the organic ligands are destroyed along the flow path by biodegradation or other processes, the dissolved silica may then become oversaturated with respect to amorphous silica and precipitate out of solution.

Solubility of Aluminum-Bearing Weathering Products

The dissolution reactions involving gibbsite are representative of aluminum-bearing minerals in that they are highly pH dependent. Formation of the Al^{+3} ion occurs by the reaction

$$Al(OH)_3 \rightleftharpoons Al^{+3} + 3OH^- \tag{4–13}$$

with equilibrium constant

$$K_{gibbsite} = a_{Al^{+3}}a^3_{OH^-} = 10^{-32.64}. \tag{4–14}$$

At moderate-to-high pH values, Eq. (4–14) indicates that the activity of Al^{+3} will be extremely small. At higher pH values, however, aluminum forms hydroxy complexes, which can increase the total solubility of gibbsite. These reactions and their stability constants are as follows:

$$Al^{3+} + H_2O \rightleftharpoons Al(OH)^{2+} + H^+; \qquad \log K = -4.99. \qquad (4\text{–}15)$$

$$Al^{3+} + 2H_2O \rightleftharpoons Al(OH)_2^+ + 2H^+; \qquad \log K = -10.13. \qquad (4\text{–}16)$$

$$Al^{3+} + 4H_2O \rightleftharpoons Al(OH)_4^- + 4H^+; \qquad \log K = -22.05. \qquad (4\text{–}17)$$

In solution, the total amount of dissolved aluminum will include the concentrations of all the Al(OH) complexes as well as Al^{+3} according to the relationship

$$(M_{Al})_T = M_{Al^{3+}} + M_{Al(OH)^{2+}} + M_{Al(OH)_2^+} + M_{Al(OH)_4^-}. \qquad (4\text{–}18)$$

Each term on the right-hand side of Eq. (4–18) can be expressed in terms of $M_{Al^{+3}}$ and pH by rearranging Eqs. (4–13) and (4–15) through (4–17). The result will be similar to Eq. (4–12). The concentration of each aqueous species in equilibrium with gibbsite is plotted as a function of pH along with the concentration of total dissolved aluminum, $(M_{Al})_T$ in Fig. 4–14. This diagram shows that at low pH, $(M_{Al})_T$ is relatively high, resulting from the Al^{+3} cation. At high pH, the increasing solubility of the $Al(OH)_4^-$ complex again raises the total solubility to relatively high values. In the near neutral pH range, however, where most natural waters lie, the $(M_{Al})_T$ is at its minimum, indicating a very low solubility of gibbsite.

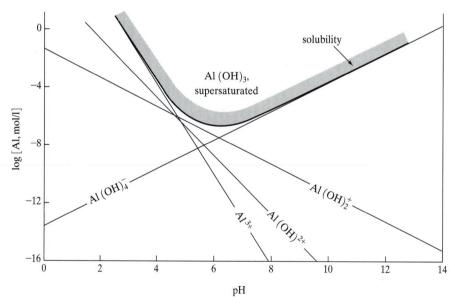

FIGURE 4–14   Activities of dissolved species in equilibrium with gibbsite plotted as a function of pH. Heavy line shows sum of all dissolved species (from Drever, J. I., *The Geochemistry of Natural Waters*, 3/e, © 1997. Reprinted by permission of Prentice-Hall Inc., Upper Saddle River, NJ).

The dissolution reactions for aluminosilicate weathering products such as kaolinite are similar to gibbsite, except that dissolved silica is also produced. The solubility of these minerals in neutral pH waters is very low, suggesting that the precipitation of these weathering products keeps dissolved aluminum concentrations very low, typically on the order of 1 μg/l or less. Because these low levels are difficult to measure accurately, the assumption is commonly made that all aluminum released from the weathering of primary minerals is incorporated into the precipitation of weathering products such as gibbsite and clay minerals.

Incongruent Dissolution and Activity Diagrams

The state of saturation of ground water with respect to a primary silicate mineral could potentially be evaluated in the same manner as a carbonate mineral. For example, the dissolution reaction for K-feldspar could be written as

$$KAlSi_3O_8 + 4H^+ + 4H_2O \longrightarrow K^+ + Al^{3+} + 3H_4SiO_4, \qquad (4-19)$$

with equilibrium constant

$$K_{eq} = \frac{a_{K^+} a_{Al^{3+}} a^3_{H_4SiO_4}}{a^4_{H^+}}. \qquad (4-20)$$

The equilibrium constant could be calculated from free-energy data; and the IAP could be obtained from a water analysis if the aluminum concentration, along with the other constituents, was measured. Finally, a saturation index could be calculated and the solution would be either oversaturated or undersaturated with respect to K-feldspar. The problem with this approach is that dissolved aluminum is not routinely measured for ground water samples; and when samples are analyzed for aluminum, it is frequently below detection limits.

Because of these problems, a different approach is used to evaluate the stability of primary silicate minerals as well as to display the stability relationships along with selected weathering products in graphical form. The diagram used for this purpose is known as an activity, or stability, diagram. The diagram is constructed by assuming that all aluminum is conserved in the weathering product. For example, the incongruent dissolution of K-feldspar to kaolinite would be written

$$2KAlSi_3O_8 + 2H^+ + 9H_2O \longrightarrow Al_2Si_2O_5(OH)_4 + 2K^+ + 4H_4SiO_4 \qquad (4-21)$$

where

$$K_{eq} = \frac{a^2_{K^+} a^4_{H_4SiO_4}}{a^2_H}, \qquad (4-22)$$

or

$$K_{ksp\text{-}kaol} = \frac{a_{K^+} a^2_{H_4SiO_4}}{a_{H^+}}. \qquad (4-23)$$

FIGURE 4–15 Activity diagram for K-feldspar and secondary minerals at 25°C. Boundaries show relative stability of two minerals based on solution concentrations (from Drever, J. I., *The Geochemistry of Natural Waters*, 3/e, © 1997. Reprinted by permission of Prentice-Hall Inc., Upper Saddle River, NJ).

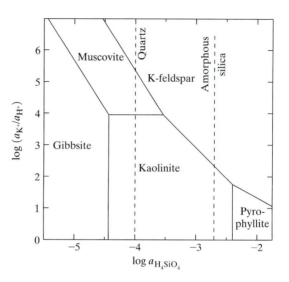

The simplified constant $K_{ksp\text{-}kaol}$ can be calculated from Gibbs free-energy data and a stability boundary between K-feldspar and kaolinite can be plotted on an activity diagram which has $\log a_{K^+}/a_{H^+}$ and $\log a_{H_4SiO_4}$ as its axes (Fig. 4–15). The meaning of this boundary is that when a_{H^+}, a_{K^+} and $a_{H_4SiO_4}$ concentrations are obtained from a water analysis, they can be plotted; and the position of the point relative to the boundary indicates which of the two minerals is more stable in this water. If the point were to plot on the line, the water would be considered to be in equilibrium with both minerals. The boundary does not give us any information as to whether the water is oversaturated or undersaturated with respect to either mineral. That would have to be determined by analyzing for dissolved aluminum and by using Eq. (4–20). The other boundaries on Fig. 4–15 are plotted in exactly the same fashion with the exception of vertical lines for the solubility of quartz and amorphous silica. These boundaries indicate the absolute solubility of these phases based on the concentration of dissolved silica, as indicated in Eqs. (4–3) and (4–4).

The interpretations drawn from an activity diagram are limited by the phases that are included on the diagram. Even if a point plots within the kaolinite field, it could be less stable than another mineral that was not plotted on the diagram. The only conclusion that can be drawn is that kaolinite is the most stable of the minerals included on the diagram.

Boundaries are not drawn between all minerals included on the diagrams. For example, a boundary could be drawn on Fig. 4–15 between pyrophyllite and gibbsite, but this boundary would lie within the kaolinite stability field. This boundary is omitted, however, because within the kaolinite field, both gibbsite and pyrophyllite are unstable with respect to kaolinite.

Activity diagrams for other primary minerals are also used. An activity diagram for anorthite, $CaAl_2Si_2O_8$, along with gibbsite, kaolinite, and Ca-montmorillonite, is shown in Fig. 4–16. These diagrams illustrate that the stable weathering product is primarily a

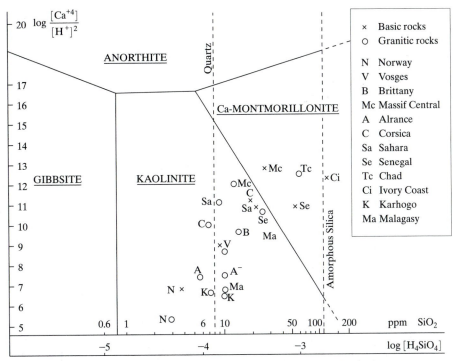

FIGURE 4-16 Activity diagram for anorthite and secondary minerals. Values plotted are surface waters in areas of crystalline rocks (reprinted from *Chemical Geology*, 7, Tardy, Y., Characterization of the principal weathering types by the geochemistry of waters from some European and African crystalline massifs, pp. 253–271, Copyright 1971, with permission from Elsevier Science).

function of the silica concentration. If dissolved silica is low, gibbsite is the stable product. This is consistent with the concentration of oxides such as gibbsite in tropical soils that have been leached of most of their original silica. In the presence of waters with higher silica contents, kaolinite or Ca-montmorillonite become the stable weathering products. The common occurrence of points plotted on activity diagrams to the right of the quartz solubility line supports the conclusion that precipitation of quartz at low temperatures is negligible. By contrast, many waters with high silica concentration cluster along the amorphous silica solubility boundary (Drever, 1997), which indicates the effectiveness of this phase in providing an upper limit to silica concentrations because of the precipitation of amorphous silica.

Ground Water Chemistry in Igneous and Metamorphic Rock Terrains

The initial discussion in this chapter concerning the differing stabilities of various silicate minerals suggests that the chemical composition of ground waters derived from igneous and metamorphic rocks will be determined by the type and abundance of minerals in the rock in which the waters were in contact. This relationship is evident on the bar graphs shown in Fig. 4–17, in which four ground waters are plotted. The presence of dissolved

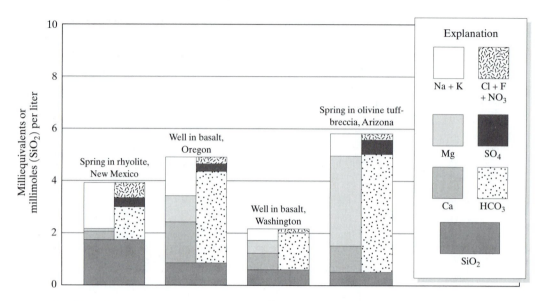

FIGURE 4–17   Composition of ground water from igneous rocks (from Hem, 1989).

silica in each analysis demonstrates the importance of dissolution of silicate minerals to form these waters. Another generalization that can be made from these plots is that bicarbonate is the dominant anion in each analysis. The source of the bicarbonate is soil-zone CO_2, which produces carbonic acid and hydrogen ions that are consumed in weathering reactions with the silicate minerals. The composition of the rhyolite spring, for example, may involve the incongruent dissolution of albite in the presence of CO_2 and the formation of kaolinite, as shown in the following equation:

$$2NaAlSi_3O_8 + 2H_2CO_3 + 9H_2O$$

$$\longrightarrow Al_2Si_2O_5(OH)_4 + 2HCO_3^- + 2Na^+ + 4H_4SiO_4. \qquad (4\text{–}24)$$

The reaction predicts a 2:1 ratio of silica to sodium, compared to a 1:1 ratio shown in Fig. 4–17.

A possible explanation for this discrepancy is the formation of a weathering product with a different silica-to-sodium ratio, such as the alteration of albite to montmorillonite:

$$3NaAlSi_3O_8 + Mg^{2+} + 4H_2O$$

$$\longrightarrow 2Na_{0.5}Al_{1.5}Mg_{0.5}Si_4O_{10}(OH)_2 + 2Na^+ + H_4SiO_4. \qquad (4\text{–}25)$$

Equation (4–25) contains a 1:2 silica-to-sodium ratio. It is possible that the formation of both kaolinite and montmorillonite could produce the observed silica:sodium ratio. Despite the uncertainty in the weathering product, the analysis indicates that dissolution of sodium-rich silicates dominates the weathering process forming this water. The two waters from basalt aquifers shown on Fig. 4–17, despite differences in concentrations of constituents, contain the same proportions of ions. The higher amounts of magnesium and calcium, relative to the rhyolite well, suggest the presence of ferromagnesian minerals and plagioclases more calcic than albite. Both of these characteristics are consistent with the more basic composition of basalt. The cation composition in the spring in olivine tuff-breccia is dominated by magnesium, which is not surprising in light of the relative instability of olivine compared to other silicate minerals.

Because of the predominance of kaolinite as a product of silicate mineral weathering, Garrels (1967) showed the expected relative concentrations produced by the weathering of several common silicate minerals to this clay mineral (Fig. 4–18). These proportions were then applied to waters derived from igneous rocks in which kaolinite was known to be the primary weathering product (Fig. 4–19). The usefulness of the bar

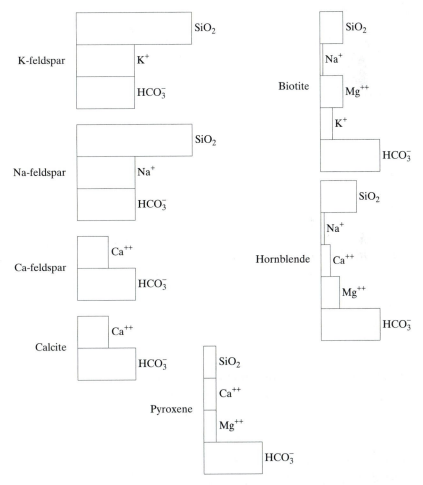

FIGURE 4–18 Calculated proportions of ions in waters resulting from the weathering of various silicate minerals to kaolinite. Compositions are mole ratios of dissolved species to HCO_3^-. Calcite is included to show that water resulting from alteration of anorthite to kaolinite is the same as that resulting from direct solution of calcite. Compositions used for biotite, hornblende, and pyroxene are typical for igneous rocks (from Garrels, 1967).

FIGURE 4–19   Reconstruction of chemical analysis of two waters in which kaolinite is the weathering product. Proportions of ions are obtained from Fig. 4–18. Ab = albite; An = anorthite; Or = K-feldspar; Bi = biotite; Hb = hornblende; Py = pyroxene (from Garrels, 1967).

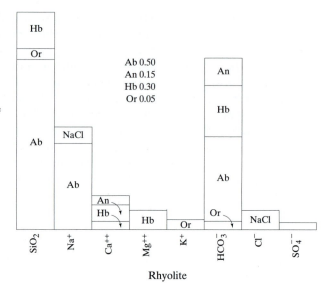

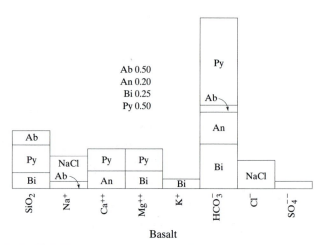

graphs in Fig. 4–19 is that they quantitatively show the weathering reactions that produced the corresponding water composition. The graphs are constructed using the proportions shown in Fig. 4–18. In the rhyolite analysis, for example, the sodium present, except for a small amount used to balance chloride, must come from albite. Since the sodium-to-silica ratio in albite is 1:2 and the sodium-to-bicarbonate ratio is 1:1, a silica concentration equal to twice the amount of sodium and an equal amount of bicarbonate are allocated to albite, as shown by the length of the bar. The relative amounts of silicate minerals weathered can be estimated by the bicarbonate bar, which shows for rhyolite that albite dissolution accounts for 50% of the bicarbonate present. The proportions of hornblende and orthoclase, 0.30 and 0.05 respectively, indicate the relative rates of dissolution of these minerals. Orthoclase is likely to be much more abundant than hornblende in rhyolite, yet the water composition shows that more hornblende than orthoclase has dissolved. This is consistent with the very slow weathering rate of orthoclase.

The basalt weathering reconstruction in Fig. 4–19 displays a lower total silica content and the importance of ferromagnesian minerals in the weathering process. Most of the bicarbonate is produced by the breakdown of pyroxene and biotite and the cation concentrations are a mixture derived from these minerals along with anorthite, the calcium-rich plagioclase end member.

Garrels and Mackenzie (1967) used the ion proportions produced in weathering of primary silicate minerals to kaolinite in a procedure to reconstruct weathering reactions that produced observed spring water concentrations in the Sierra Nevada Mountains. Two types of springs were recognized—shallow ephemeral springs and perennial springs with deeper flow paths. The method used by Garrels and Mackenzie was to write reverse weathering reactions using kaolinite and the molar concentrations of ions in the springs to determine: (1) if primary minerals similar in composition to those observed in the rocks could be produced and (2) if all the dissolved constituents in the spring water could be accounted for. The calculations and concentrations used in this analysis are given in Table 4–4. The first step taken was to subtract the ion concentrations in snow, which melts to recharge the ground water flow systems, from the average ephemeral spring water concentrations. After subtraction, the bicarbonate concentration was increased slightly to maintain charge balance. The remaining concentrations were assumed to represent rock weathering.

Reaction 1 is constructed in the following manner. All the sodium and calcium are assigned to the weathering of plagioclase to produce kaolinite. The sodium and calcium content of plagioclase are determined by normalizing each to a total of 1.0; for sodium this gives $0.110/(0.110 + 0.068) = 0.62$. The proportions of aluminum and silicon can then be calculated to obtain charge balance in the plagioclase. An amount of silica twice that of sodium is used, 0.220 mmol/l, to produce the 2:1 ratio derived from the weathering of albite to kaolinite illustrated in Fig. 4–18; an amount of bicarbonate necessary for charge balance, 0.246 mmol/l, is also used as a reactant. The coefficient of plagioclase in the reaction is the molar amount necessary to produce the dissolved concentrations of sodium and calcium from the specified plagioclase composition; it can be calculated as $0.110/0.62 = 0.177$. The subscripts of aluminum and silicon in the plagioclase formula are established by balancing the charges, and the remaining stoichiometric coefficients in the reaction can then be determined by mass balance. The calculated plagioclase composition resulting from this procedure is similar to the actual composition of plagioclase that is present in the rocks of the study area, which is a good check on the selection of weathering reactions. After reaction 1 is constructed, the masses of dissolved species used are subtracted from the spring water composition.

The reconstruction of reactions 2 and 3 is approached in the same way. All the magnesium in solution, along with appropriate amounts of potassium, bicarbonate, and silica, is used to make $0.022/3 = 0.0073$ mmol of biotite. After these amounts are subtracted, the remaining K^+, HCO_3^- and SiO_2 are reacted with kaolinite to produce K-feldspar according to reaction 3. When these quantities are subtracted, only a small amount of silica remains. The back calculation method therefore accounts for nearly all

TABLE 4-4 Mass balance calculations and weathering reactions for ephemeral spring waters, Sierra Nevada Mountains (from Garrels and Mackenzie, 1967). Concentrations in mmol/l.

Reaction	Na^+	Ca^{2+}	Mg^{2+}	K^{2+}	HCO_3^-	SO_4^{2-}	Cl^-	SiO_2	Products
Initial conc. in springwater	0.134	0.078	0.029	0.028	0.328	0.01	0.014	0.273	
Rock weathering (initial minus snow)	0.110	0.068	0.022	0.020	0.310	0.00	0.00	0.270	
Kaolinite → Plagioclase & subtract masses used	0.00	0.00	0.022	0.020	0.064	0.00	0.00	0.05	$0.177Na_{0.62}Ca_{0.38}$
Plagioclase									
Kaolinite → Biotite & subtract masses used	0.00	0.00	0.00	0.013	0.013	0.00	0.00	0.35	0.0073 biotite
Kaolinite → K-feldspar & subtract masses	0.00	0.00	0.00	0.00	0.00	0.00	0.00	0.02	0.013 K-feldspar

Reaction 1: $0.123 Al_2Si_2O_5(OH)_4 + 0.110 Na^+ + 0.068 Ca^{2+} + 0.246 HCO_3 + 0.220 SiO_2 = 0.177 Na_{0.62}Ca_{0.38}Al_{1.38}Si_{2.62}O_8 + 0.246 CO_2 + 0.367 H_2O$.

Reaction 2: $0.0037 (Al_2Si_2)_5(OH)_4 + 0.0073 K^+ + 0.022 Mg^{2+} + 0.015 SiO_2 + 0.051 HCO_3 = 0.0073 KMg_3AlSi_3(OH)_2 + 0.051 CO_2 + 0.026 H_2O$.

Reaction 3: $0.0065 Al_2Si_2O_5(OH)_4 + 0.013 K^+ + 0.013 HCO_3^- + 0.026 SiO_2 = 0.013 KAlSi_3O_8 + 0.013 CO_2 + 0.0195 H_2O$.

of the solutes present in the ephemeral spring water. Garrels and Mackenzie (1967) conducted a similar analysis of perennial spring waters in the Sierra Nevadas, involving both kaolinite and montmorillonite as weathering products.

Following the work of Garrels and Mackenzie (1967), many investigators have used mass balance calculations to investigate the reactions between ground water and igneous and metamorphic rocks. Geochemical modeling codes are used to perform mass balance calculations and to determine the saturation indices of weathering products, if aluminum analyses are available. Activity diagrams are commonly used to compare the stability of possible weathering products in aquifer systems. These studies have several purposes, one being to relate the ground water chemistry of an area to the rock types present in the aquifer. A second objective in some studies is to develop a quantitative chemical budget of a watershed by accounting for all sources and sinks of solutes. Examples reviewed here illustrate some of the conclusions that have been reached.

Rodgers (1989) compared ground water chemistry in three areas of the northeastern U.S. using a large data base of water analyses. In addition to the differences in bedrock geology between the three areas—New England, New York, and Pennsylvania—the differences between waters in bedrock and glacial-drift aquifers were also examined. A difference between these aquifer types is that glacial-drift aquifers contain a mixture of local and far-traveled lithologies. The generalized bedrock geology of the three regions is shown in Fig. 4–20. The New England area is primarily underlain by

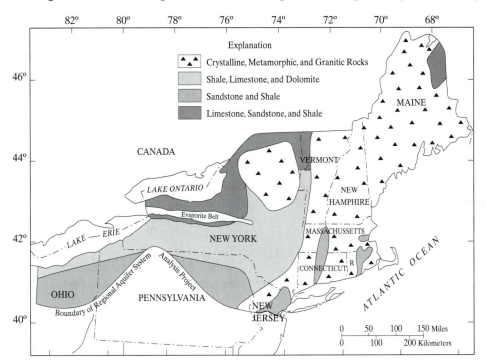

FIGURE 4–20 Geological setting in areas in which ground water chemistry was compared by Rodgers (1989). (Reprinted by permission of the National Ground Water Association, Copyright, 1989.)

silicic igneous and metamorphic rocks, whereas the New York and Pennsylvania study areas are dominated by sedimentary rocks. These are mostly carbonates and evaporites in New York and sandstones and shales in Pennsylvania.

The differences in aquifer lithology between the three areas are reflected in ground water chemistry (Fig. 4–21). Considering total dissolved solids to be a general indicator of reactiveness of the aquifer lithologies, the New England ground waters are the most dilute of the three groups. The New York area, dominated by more reactive carbonate rocks, had the highest TDS values; and the Pennsylvania area was intermediate between New England and New York. Bedrock aquifers had significantly higher TDS values than glacial-drift aquifers in New England and New York but were roughly equal in Pennsylvania. Rodgers (1989) accounted for these differences as the result of the assumed longer residence time of ground water in bedrock aquifers, allowing increased dissolution of the aquifer minerals. Stability diagrams (Fig. 4–22), which were used to investigate silicate weathering reactions, suggest that kaolinite is the dominant secondary product, although the waters evolve toward the kaolinite-smectite boundary. Bedrock waters are more likely to reach and cross this boundary, particularly in the New England study area.

Higher contents of reactive plagioclase and ferromagnesian minerals in basaltic igneous rocks, relative to granitic rocks, yield ground waters with distinctly different chemical compositions. Wood and Low (1986) conducted a comprehensive study of ground water chemistry in the eastern Snake River Plain aquifer system in Idaho. The most significant weathering reactions in the aquifer are the congruent dissolution of olivine and pyroxene and the incongruent weathering of plagioclase to smectite. Volcanic glass probably also dissolves to contribute solutes to the aquifer. Accompanying these dissolution reactions is the precipitation of calcite and microcrystalline silica. Confirmation of these reactions was obtained by analysis of rock mineralogy, mass balance calculations, calculation of mineral saturation indices, and plotting stability diagrams (Fig. 4–23). Wood and Low extended their geochemical analysis of the eastern Snake River Plain aquifer to include a detailed chemical budget of the aquifer system. By considering the concentrations in the Snake River to represent the output from the system as discharge and subtracting all possible inputs of solutes to the system, they were able to demonstrate that about 20% of the solutes leaving the aquifer are produced by the weathering reactions mentioned previously.

Weathering studies consistently show that ferromagnesian minerals weather rapidly and contribute solutes far in excess of their abundance in the rock. In the Filson Creek watershed in Minnesota, Siegel and Pfannkuch (1984) estimated that the release of magnesium from olivine is at least ten times faster than the release of calcium and sodium from plagioclase. The significance of this conclusion is that reactive minerals that constitute only a small proportion of the bulk mineralogy of the aquifer can dominate chemical weathering and ground water chemistry.

Silicate Mineral Weathering and Lake Acidification

In watersheds primarily underlain by igneous and metamorphic rocks, the weathering reactions of silicate minerals have an important relationship to the pH and potential for acidification of surface water bodies in contact with shallow ground water. Acidification of lakes, primarily in igneous and metamorphic rock terrains, has become an

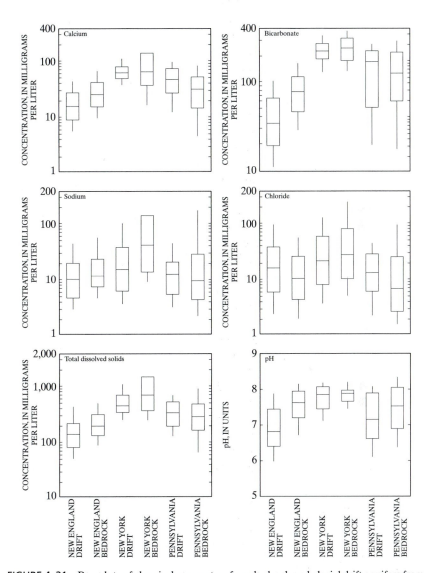

FIGURE 4–21   Box plots of chemical parameters from bedrock and glacial drift aquifers from the three study areas of Rodgers (1989). Horizontal line in box is median; tops and bottoms of boxes represent 75th and 25th percentiles, respectively. Vertical lines extend to 90th and 10th percentiles. (Reprinted by permission of the National Ground Water Association, Copyright 1989.)

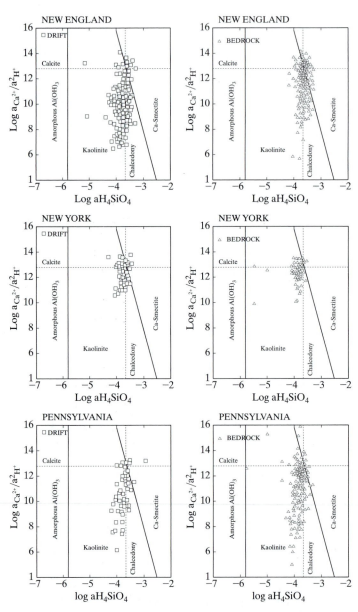

FIGURE 4–22 Activity diagrams for secondary minerals in the three study areas of Rodgers (1989). Most points plot in kaolinite field. Chalcedony (amorphous silica) exerts strong solubility control on waters. (Reprinted by permission of the National Ground Water Association , Copyright 1989.)

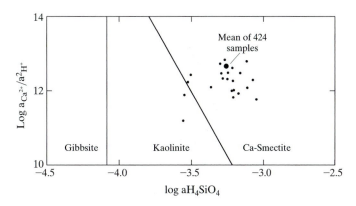

FIGURE 4–23   Stability diagram for secondary minerals in Snake River Plain study. $T = 15°C$ (from Wood and Low, 1986).

increasing problem in the industrialized world. The decrease in pH of precipitation noted over the past few decades is known as acid rain. Acidification of precipitation is caused by the addition of ammonium, sulfur dioxide, and nitrous oxides to the atmosphere by our industrialized society. The following equations conceptually illustrate the transformations of these compounds to acids in the atmosphere:

$$NH_4^+ + 2O_2 \longrightarrow NO_3^- + 2H^+ + H_2O; \tag{4–26}$$

$$SO_{2(aq)} + O_{3(g)} \longrightarrow SO_{3(aq)} + O_{2(g)} \longrightarrow H_2SO_4; \tag{4–27}$$

$$NO \longrightarrow NO_2 \longrightarrow HNO_3. \tag{4–28}$$

Aerosol particles carried in the atmosphere that reach the surface in a process called *dry deposition* also contribute to the acidification problem.

When these acid solutions reach land surface, they can be buffered in lakes or streams that have sufficient alkalinity (Chapter 3), which is also known as *acid neutralizing capacity*. Lakes in carbonate terrains commonly contain alkalinity values sufficient to prevent serious declines in pH. In such waters, alkalinity is balanced by the cations of strong bases, such as calcium, and magnesium. Without the buffering capacity provided by carbonate minerals in the soils and aquifers discharging to lakes and streams in a watershed, surface water bodies are much more susceptible to acidification. An adverse consequence of acidification is that fish cannot survive in the more acidic waters.

Acidified waters may reach the point at which they have no acid neutralizing capacity. With further contributions, they may have measurable *acidity*, which is the capacity to neutralize bases and is measured by titration with a strong base. Acidity can be thought of as a negative alkalinity. In waters of this type, alkalinity is replaced by the anions of acids such as those produced in Eqs. (4–26) through (4–28). Concentrations of lake waters in Switzerland in carbonate terrain compared to lakes in metamorphic rock terrain are shown in Fig. 4–24.

In igneous and metamorphic rock terrains, buffering of acids in precipitation can only take place through the weathering of silicate minerals or through reactions with

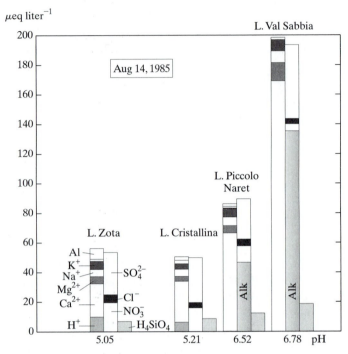

FIGURE 4–24   Water composition of four lakes in Switzerland. Lakes Zota and Cristallina are located in drainage basin of crystalline rocks; the other two lakes are located in drainage basins containing calcite and dolomite (from Stumm and Morgan, *Aquatic Chemistry*, 3/e, Copyright © 1996 by Wiley Interscience. Reprinted by permission of John Wiley & Sons, Inc.).

weathering products (Appelo and Postma, 1993). An example of the latter is the dissolution of gibbsite, yielding dissolved aluminum:

$$Al(OH)_3 + 3H^+ \rightleftharpoons Al^{3+} + 3H_2O. \tag{4–29}$$

Although the pH will rise as a result of this reaction, the release of aluminum can be harmful if the weathering solutions discharge into surface water bodies. The extent to which buffering by silicate mineral weathering can counteract the effects of acid rain is dependent upon the rates of weathering reactions, the abundance of minerals with more rapid weathering rates, and perhaps other factors. April et al. (1986) compared the weathering rates in two watersheds in the Adirondack Mountains, one of which contained a lake with a neutral pH; the pH of the lake in the other basin ranged between 4.5 and 5. Both watersheds are underlain by granitic rocks and till (glacial drift) of mineralogical composition similar to bedrock. The major difference between the two basins was that the glacial-drift deposits in the watershed with the acid lake, Woods Lake, are thinner than in the other watershed, Panther Lake. April et al. (1986) found that hornblende and plagioclase accounted for most of the weathering and, therefore acid neutralization, and that the weathering rates in the Panther Lake watershed are currently much higher than in the Woods Lake watershed. They accounted for this difference by the thicker drift deposits in the Panther Lake watershed and the greater amount of surface runoff (less infiltration and weathering) in the Woods Lake basin. During infiltration through the thicker soils of the Panther Lake watershed, acid rain is buffered by the time it discharges into the lake, thus preventing acidification of the lake. This study illustrates the importance of the ground water contribution to the lake in determining its chemical composition.

A more complete analysis of weathering reactions and acid neutralization must include factors such as ion exchange reactions and the effects of biomass upon element distributions. Drever (1997) and Appelo and Postma (1993) include more in-depth reviews of these topics.

SORPTION AND ION EXCHANGE

During the movement of aqueous solutions (ground water) through rock and soil materials, ionic solutes are brought near and/or in contact with mineral surfaces. The important reactions and processes between solutes and aquifer solids are known by a variety of names. When the process involves mass transfer from the solution to the solid, the process is described as *sorption*. Under this terminology, the species being removed from solution is referred to as the *sorbate* and the solid to which it is sorbed is called the *sorbent*. Sorption can be further clarified based on the type of process through which it is bound to the solid. If the ion or molecule is held primarily at the particle surface as a solvated complex, it is known as *adsorption* (Fig. 4–25). Alternatively, the sorbate may be incorporated into the mineral structure at its surface, a case in which it would be classified as *absorption*. Both subtypes of sorption differ from precipitation because a new crystalline structure is not formed at the surface.

Also of interest to hydrogeologists is the situation in which an ion in solution becomes sorbed to a surface through the release of a similarly charged ion that had occupied the surface position. This quid-pro-quo process is known as *ion exchange*. There are many cases in which ion exchange can significantly change the composition of a solution. In this section, we will first examine the surface properties of solids that promote sorption and ion exchange and follow this by a more detailed discussion of the mechanics of these phenomena.

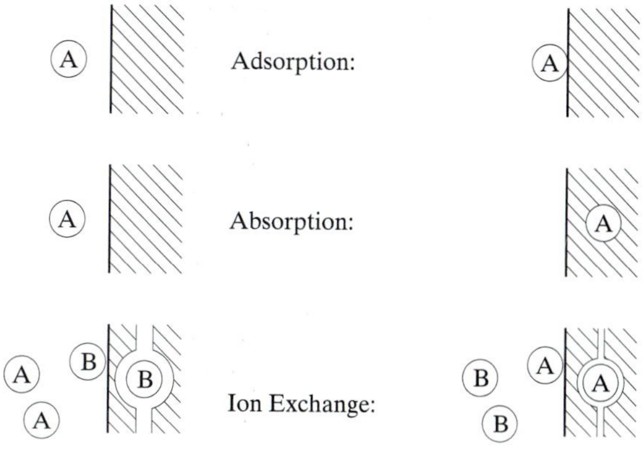

FIGURE 4–25 Processes of adsorption, absorption, and ion exchange.

Acquisition of Surface Charge

The surfaces of certain types of particles in contact with ground water or any electrolyte solution react with the solution because they have developed an electrical charge. The results of these reactions are sorption and ion exchange. Charge can be acquired by a particle surface in two basic ways: (1) isomorphous substitution for a cation in a mineral structure by one of lesser positive charge and (2) reactions involving surface functional groups on the mineral surface and ions in solution. The charge developed by ion substitution is considered to be fixed, whereas the charge due to surface complexation is variable, depending on the pH of the solution.

Ionic substitutions that lead to a fixed charge are significant only in 2:1 clay minerals. In our previous discussion of clay mineral structure, the lack of ion substitutions and, therefore, charge deficiency for the 1:1 clay minerals, such as kaolinite, was mentioned (Table 4–3). Smectites, vermiculite, and other 2:1 minerals, however, are characterized by a distinct deficiency of positive charge resulting from the substitution of cations of lesser charge. The overall negative charge of the surface can be balanced by the sorption of cations in solution in several ways. Coordination of tetrahedra on the outer surface of a 2:1 mineral results in a plane of oxygen atoms known as a *siloxane surface*. Spaced at regular intervals on the siloxane surface are hexagonal *siloxane cavities* (Fig. 4–26), which serve as reactive functional groups for the formation of surface complexes. These complexes, which are analogous to aqueous complexes, can form with either unsolvated or solvated cations. When unsolvated complexes are formed, the cation bonds directly with oxygens in the siloxane cavity through relatively strong ionic or covalent bonds. This type of a surface complex is known as an *inner-sphere complex* (Fig. 4–27). A specific example of an inner-sphere complex is the incorporation of potassium into vermiculite (Fig. 4–28). The potassium ion coordinates with the oxygens in two adjacent siloxane cavities. The diameter of the potassium ion is just the right size for

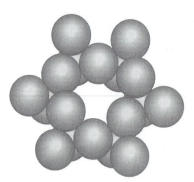

FIGURE 4–26 Siloxane cavity formed by surface oxygens and hydroxyls in metal oxides and silicate minerals (from *The Chemistry of Soils* by Garrison Sposito. Copyright © 1989 by Oxford University Press, Inc. Used by permission of Oxford University Press, Inc.).

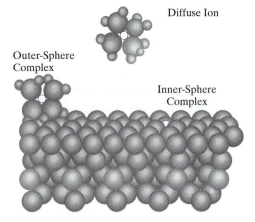

FIGURE 4–27 Interactions and binding of solvated and free cations with siloxane surface (from *The Chemistry of Soils* by Garrison Sposito. Copyright © 1989 by Oxford University Press, Inc. Used by permission of Oxford University Press, Inc.).

FIGURE 4–28   Diagramatic respresentation of inner-sphere complex between K$^+$ and vermiculite (top) and outer-sphere complex between solvated calcium ion and montmorillonite (bottom) (from *Chemistry of the Solid-Water Interface* by Stumm, Copyright © 1992 by John Wiley & Sons, Inc. Reprinted by permission of John Wiley & Sons, Inc.).

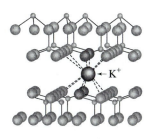

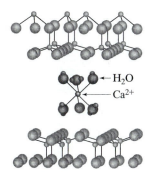

the siloxane cavity and the resulting bond is very strong. The formation of an inner sphere complex is an example of absorption because the sorbed ion is bonded directly to the surface ions of the mineral.

Balance of charge can occur when a solvated cation is held to the siloxane surface by bonds between the water molecules and mineral surface functional groups (Fig. 4–27). The solvated cation forms what is called an *outer-sphere complex*, which differs from an inner-sphere complex because the cation is not directly bonded to the mineral surface. The electrostatic bonds of the outer-sphere complex are much weaker than the inner-sphere complex. As a result, outer-sphere complex ions are readily exchangeable with ions in solution. An example of an outer-sphere complex is the adsorption of calcium or sodium to the surface of montmorillonite (Fig. 4–28) or other 2:1 clay minerals. A final type of adsorption accounts for the remainder of the charge imbalance on a particle surface. This mechanism involves the presence of a diffuse layer of cations concentrated near the mineral surface. These *counter ions* are not bonded to the surface but, because they are more abundant than diffuse anions (*co-ions*), their net effect is to balance the remaining negative charge. The counterions and co-ions are known as the *diffuse double layer*. Figure 4–29 shows that the positive charge provided by the counterions in the diffuse layer drops off exponentially with distance from the mineral surface as the negative charge of the co-ions exponentially increases.

As previously mentioned, only the 2:1 clay minerals have isomorphic substitutions that impart a fixed, or permanent, charge to the mineral surface. Other minerals with

FIGURE 4–29   The diffuse double layer: (a) co-ions and counter ions (black circles) in solution adjacent to mineral surface. (b) σ_p represents net surface charge and σ_D represents diffuse ion charge. (c) charge distribution (concentration of positive and negative ions) with distance from surface (from *Chemistry of the Solid-Water Interface* by Stumm, Copyright © 1992 by John Wiley & Sons, Inc. Reprinted by permission of John Wiley & Sons, Inc.).

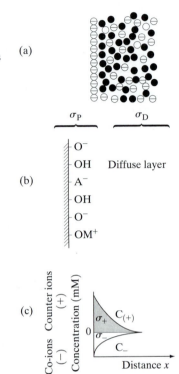

significant surface charges develop their charge by complexation reactions with surface hydroxyls. These charges are not fixed because they depend on interactions with the aqueous solution and are therefore termed *variable charge*. Minerals that exhibit variable charge include the 1:1 clay minerals such as kaolinite and the hydrous metal oxides such as geothite and gibbsite. 2:1 clay minerals can also acquire a variable charge, but it is much lower in magnitude than the fixed charge. The variation in charge is a function of the pH of the solution. At low pH values, the surface hydoxyls react with hydrogen ions in solution to form positively charged surface complexes according to the reaction

$$S{-}OH + H^+ \rightleftharpoons S{-}OH_2^+. \tag{4–30}$$

In Eq. (4–30), S represents a surface metal cation. As the pH rises, the hydrogen ions will return to solution, and surface hydroxyls can even dissociate their original proton:

$$S{-}OH + OH^- \rightleftharpoons S{-}O^- + H_2O. \tag{4–31}$$

Other types of surface complexation reactions involve metal cations and negatively charged ligands of various types. Metal ions in solution compete with the hydrogen ions adsorbed to hydroxyl groups. For example, the adsorption of copper to a hydroxide surface is

$$S{-}OH + Cu^{2+} \rightleftharpoons S{-}OCu^+ + H^+. \tag{4–32}$$

FIGURE 4–30   Inner-sphere complex formed by ligand exchange between protonated hydroxyl and phosphate anion (from *The Chemistry of Soils* by Garrison Sposito. Copyright © 1989 by Oxford University Press, Inc. Used by permission of Oxford University Press, Inc.).

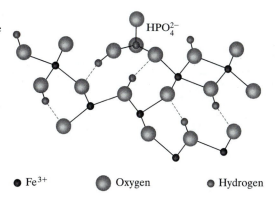

● Fe^{3+} ⬤ Oxygen   ● Hydrogen

Ligand exchange occurs between the protonated hydroxyl and an anion in solution. For example, phosphate forms inner-sphere complexes with goethite (Fig. 4–30) according to the equation

$$Fe-OH_2^+ + HPO_4^{2-} \rightleftharpoons Fe-OPOOOH^- + H_2O. \qquad (4-33)$$

The net surface charge on a particle is therefore fairly complicated—it is the algebraic sum of the fixed charge and all surface complexes, which can carry either negative or positive charges. If this net charge is not zero, it is balanced by the diffuse-ion layer adjacent to the particle. This relationship can be stated as

$$\sigma_P + \sigma_D = 0, \qquad (4-34)$$

where σ_P is the total net surface charge and σ_D is the diffuse ion charge. σ_P, in turn, is equal to the algebraic sum of the fixed charge, σ_0, the net proton charge, σ_H, which is the charge established by the binding or dissociation of hydrogen ions to the surface, σ_{IS} and σ_{OS}, which are the charges due to the presence of inner-sphere and outer-sphere complexes, respectively. Thus,

$$\sigma_P = \sigma_0 + \sigma_H + \sigma_{IS} + \sigma_{OS}. \qquad (4-35)$$

All charge components on the right-hand side of Eq. (4–35) vary as a function of the pH of the solution except σ_0. Hydroxide minerals and their amorphous equivalents, along with kaolinite and other 1:1 clay minerals, develop only a variable charge. At low values of pH, the net charge may be positive; and as the pH rises, the net surface charge (σ_P) may become negative. This relationship implies that there is some pH value at which the net charge is zero. This pH value, at which $\sigma_P = 0$, is a unique property of each mineral, known as the *point of zero charge* (PZC). More specifically, there are several points of zero net charge depending on which of the terms of Eq. (4–35) one considers. For example, the pH value at which $\sigma_H = 0$ is known as the *point of zero net proton charge* (PZNPC). It accounts for the charge produced by the adsorption and dissociation of protons according to Eqs. (4–30) and (4–31). Representative values of PZNPC for various materials are given in Table 4–5.

   The magnitude of surface charge developed on fixed-charge minerals is much greater than that of variable-charge solids. Figure 4–31 compares the negative charge of

TABLE 4–5   Point of zero charge caused by binding or dissociation of protons[a] (Stumm, 1992).

Material	pH_{pznpc}	Material	pH_{pznpc}
α-Al_2O_3	9.1	δ-MnO_2	2.8
α-$Al(OH)_3$	5.0	β-MnO_2	7.2
γ-$AlOOH$	8.2	SiO_2	2.0
CuO	9.5	$ZrSiO_4$	5
Fe_3O_4	6.5	Feldspars	2–2.4
α-$FeOOH$	7.8	Kaolinite	4.6
α-Fe_2O_3	8.5	Montmorillonite	2.5
$Fe(OH)_3$ (amorph)	8.5	Albite	2.0
MgO	12.4	Chrysotile	>10

[a] The values are from different investigators who have used different methods and are not necessarily comparable. They are given here for illustration.

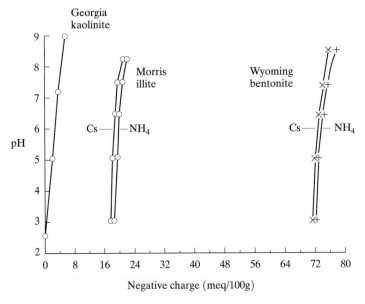

FIGURE 4–31   Fixed and variable surface charges of three clays as a function of pH. Charge determined by sorption of 1.0 mol CsCl or 1.0 mol NH_4Cl (from Langmuir and Mahoney, 1984. Reprinted by permission of the National Ground Water Association).

samples of kaolinite, illite, and montmorillonite (bentonite) as a function of pH. Notice that the net charge of kaolinite is near zero at low pH but becomes slightly negative at higher pH values. This change represents the dissociation of hydrogen ions from surface hydroxyls. Illite and montmorillonite, however, have a substantial negative charge at all pH values. This is their fixed charge. They too show a tendency toward increased negative charge at higher pHs, but the magnitude of the fixed charge is a much greater proportion of the total negative charge than the variable component.

Sorption Isotherms

The ability of a soil to adsorb a solute from solution can be determined in the laboratory through the use of a *batch test*. In a test of this type, a known mass of soil is mixed and allowed to equilibrate with a solution containing a known concentration of the adsorbate of interest. The solution is then separated from the soil and the concentration of the adsorbate is measured. The difference between the initial and final concentrations is the amount that was adsorbed by the soil. Thus the mass adsorbed (S) in $\mu g/g$ or mg/g of dry soil can be calculated as

$$S = \frac{(C_i - C)V}{S_m}, \tag{4–36}$$

where C_i is the initial concentration of the adsorbate in $\mu g/l$ or mg/l, V is the solution volume in ml, and S_m is the soil mass in g. If the test is repeated using different values of adsorbate concentration in the solution, the relationship between the equilibrium concentration and the amount adsorbed can be determined. Graphs of these data are known as *isotherms* because the batch tests are run at constant temperature. The data points on isotherms define nonlinear relationships that can be fit using a variety of equations. Two of the most common are the *Freundlich* and *Langmuir* isotherms. The Freundlich isotherm is defined by the equation

$$S = K\, C^n \tag{4–37}$$

in which K, the partition coefficient, and n are variables. n has a maximum value of 1.0, at which time the plot becomes linear. At n values less than 1.0, the curve is concave with respect to the C axis (Fig. 4–32), indicating that, as solution concentrations increase, the amount of the sorbed species does not increase in proportion to the equilibrium concentration. A linear Freundlich isotherm $(n = 1)$ is a special case in which K is known as the *distribution coefficient, K_d*. This form of the Freundlich isotherm does not fit the adsorption of most inorganic solutes. It is used, however, for the sorption of hydrophobic organics, which will be discussed in Chapter 6.

The Langmuir isotherm fits the case in which sorption of the adsorbate approaches a maximum sorptive capacity, S_{max}, when all available surface sites for sorption of that species are saturated. The equation for the Langmuir isotherm is expressed as

$$S = \frac{S_{max}K\, C}{1 + K\, C}. \tag{4–38}$$

The plot is shown in Fig. 4–32.

Adsorption of Metal Cations

In a solution containing a variety of metal cations that tend to sorb to particle surfaces, there is competition between metals for the available sites. Experiments involving the sorption of metal cations on soils demonstrate that some metals have greater affinities for adsorption. Of several factors that determine this selectivity, *ionic potential*, which is equal to the charge Z of an ion over its ionic radius, r, has a significant effect (Sposito, 1989). Cations with a lower ionic potential tend to release their solvating water molecules more readily so that inner sphere surface complexes can be formed. Among

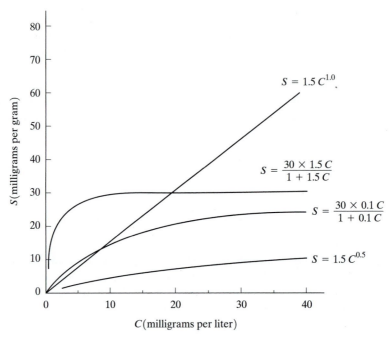

FIGURE 4–32 Examples of Freundlich $\left(S = KC^n \right)$ and Langmuir $\left(S = \dfrac{S_{max}KC}{1 + KC} \right)$ isotherms.

cations of the same valence, selectivity becomes a function of ionic radius only because as the radius gets smaller, the ionic potential, Z/r, increases, and the ion is less likely to be adsorbed. Selectivity sequences are arranged in order of decreasing ionic radius, which results in increasing ionic potential and decreasing affinity or selectivity for adsorption. The following selectivity sequences have been determined (Sposito, 1989):

$$Cs^+ > Rb^+ > K^+ > Na^+ > Li^+;$$

$$Ba^{2+} > Sr^{2+} > Ca^{2+} > Mg^{2+};$$

$$Hg^{2+} > Cd^{2+} > Zn^{2+}.$$

Metals within the transition group differ from those just listed in that electron configuration becomes more important than ionic radius in determining selectivity. The relative affinity of several of these metals is given by

$$Cu^{2+} > Ni^{2+} > Co^{2+} > Fe^{2+} > Mn^{2+}.$$

If the sorbent particles are variable-charge minerals, another relationship develops during sorption of metal cations. For all cations, the fraction sorbed increases with pH because the negative charge of the sorbent also increases as the pH rises. Individual metals, however, have a narrow range of pH values over which the percentage of sorption changes from zero to 100% (Fig. 4–33). The pH ranges of similar metals relate to each other in the same order as the selectivity sequences, with the ion of highest affinity

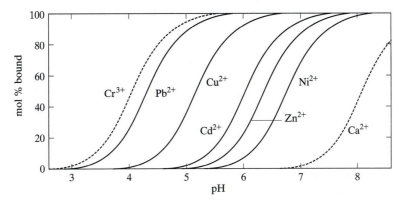

FIGURE 4–33   Surface complexation of dissolved metal ions as a function of pH (from *Aquatic Chemistry* 3/e, by Stumm and Morgan. Copyright © 1996 by Wiley Interscience. Reprinted by permission of John Wiley & Sons, Inc.).

sorbing in the lowest pH range and those of lesser affinity sorbing at progressively higher pH values.

Ion Exchange Reactions

Ions that are adsorbed to soil surfaces by the outer sphere complex and diffuse-ion mechanisms are considered to be readily exchangeable with similarly charged ions in solution. The capacity of a soil for cation exchange, for example, can be determined by displacing the exchangeable surface ions with ions in a standard solution that is brought into contact with the soil. Tests of this type are used to establish the *cation exchange capacity (CEC)* of the soil, which is usually expressed in meq/100 g of soil. Representative values of CEC for various clay minerals are given in Table 4–6.

TABLE 4–6   Cation exchange capacities of minerals and soils (Dragun, 1988).

CEC	meq/100 g
Clay Minerals	
Chlorite	10–40
Illite	10–40
Kaolinite	3–15
Montmorillonite	80–150
Vermiculite	100–150
Oxides and Hydroxides	2–6
Soil Organic Matter	>200
Soils	
Sand	2–7
Sandy loam	2–18
Loam	8–22
Silt loam	9–27
Clay loam	4–32
Clay	5–60

When the exchange reactions involve common, major cations, they are treated as equilibrium processes expressed by the law of mass action. In these reactions, the CEC of the soil is assumed to be fixed, which implies that the pH of the system does not change appreciably during cation exchange. The general form of a cation exchange reaction can be expressed as

$$nA^{m+} + mBX \rightleftharpoons mB^{n+} + nAX, \tag{4-39}$$

in which m and n are the valences of cations A and B, respectively, and X represents the cation that is adsorbed to the soil. The equilibrium coefficient for this reaction is defined in mass action form in the usual manner with one exception. The relationship between concentration and activity for adsorbed species cannot be defined in the same way as is done for aqueous species. One of the conventions used for adsorbed species is to assume the equivalent fraction of the ion is equal to its activity. Thus,

$$K = \frac{a_B^m \, N_A^n}{a_A^n \, N_B^m}. \tag{4-40}$$

The activities of the dissolved species are used in normal fashion, but N refers to the equivalent fraction, defined as

$$N_A = \frac{\text{meq of } A \text{ per 100 g of sediment}}{\text{CEC}}; \tag{4-41}$$

or, if A and B are the only cations on the absorbent,

$$N_A = \frac{\text{meq } A}{\text{meq } A + \text{meq } B}. \tag{4-42}$$

The equilibrium constant in ion exchange reactions is known as the *selectivity coefficient*. If the sediment or aquifer material consists of a variety of exchangers with different selectivity coefficients, a bulk CEC for the sediment can be used in Eq. (4–41). Problems may arise in this approach if these absorbents differ substantially in their exchange properties.

The value of the selectivity coefficient indicates the relative preference of the sorbent for the two cations. As an example of an ion exchange reaction, consider the ion exchange reaction between sodium and calcium, which, according to Eq. (4–39), would be written as follows:

$$2Na^+ + CaX \rightleftharpoons Ca^{2+} + 2NaX; \tag{4-43}$$

$$K_{Na/Ca} = \frac{a_{Ca^{2+}} \, N_{Na^+}^2}{a_{Na^+}^2 \, N_{Ca^{2+}}}. \tag{4-44}$$

If the selectivity coefficient is equal to 1, there is no preference for either ion. The value of $K_{Na/Ca}$ is about 0.4, however, which indicates that the calcium ion is preferred by the sorbent over the sodium ion. Under these conditions, the product of the numerators in Eq. (4–44), $a_{Ca^{2+}}$ and $N_{Na^+}^2$, is less than the product of the denominators, $a_{Na^+}^2$ and $N_{Ca^{2+}}$, which could only be true if more calcium than sodium is adsorbed in the equilibrium system. This example reflects the preference of sorbents for ions of higher charge, as

TABLE 4–7   Values for selectivity coefficients with respect to Na^+ (from Stumm, 1992).

Equation: $Na^+ + 1/i \; I-X \rightleftharpoons Na - X + 1/i \; I^{i+}$ with $K_{Na \setminus I} = \dfrac{[Na-X][I^{i+}]^{1/i}}{[I-X_i]^{1/i}[Na^+]}$

Ion I^+	$K_{Na \setminus I}$	Ion I^{2+}	$K_{Na \setminus I}$	Ion I^{3+}	$K_{Na \setminus I}$
Li^+	1.2 (0.95–1.2)	Mg^{2+}	0.50 (0.4–0.6)	Al^{3+}	0.6 (0.5–0.9)
K^+	0.20 (0.15–0.25)	Ca^{2+}	0.40 (0.3–0.6)	Fe^{3+}	?
NH_4^+	0.25 (0.2–0.3)	Sr^{2+}	0.35 (0.3–0.6)		
Rb^+	0.10	Ba^{2+}	0.35 (0.2–0.5)		
Cs^{2+}	0.08	Mn^{2+}	0.55		
		Fe^{2+}	0.6		
		Co^{2+}	0.6		
		Ni^{2+}	0.5		
		Cu^{2+}	0.5		
		Zn^{2+}	0.4 (0.3–0.6)		
		Cd^{2+}	0.4 (0.3–0.6)		
		Pb^{2+}	0.3		

mentioned earlier. Among ions of the same charge, the preferences follow the selectivity sequences listed above, which are based on the ionic radii of the ions.

Values for a number of selectivity coefficients are listed in Table 4–7. The values are all listed with respect to sodium, that is, in reactions of the form of Eq. (4–43). Coefficients for ion pairs not including sodium can be determined by subtracting the two reactions and dividing their selectivity coefficients as illustrated in Example 4–1.

EXAMPLE 4–1

Determine the selectivity coefficient for the following ion exchange reaction:

$$Ca^{2+} + MgX \rightleftharpoons Mg^{2+} + CaX. \tag{4–45}$$

From Table 4–7, the values of $K_{Na/Mg}$ and $K_{Na/Ca}$ and the corresponding exchange reactions are given as follows:

$$2Na^+ + MgX \rightleftharpoons Mg^{2+} + 2NaX \text{ with } K_{Na/Mg} = 0.5; \tag{4–46}$$

$$2Na^+ + CaX \rightleftharpoons Ca^{2+} + 2NaX \text{ with } K_{Na/Ca} = 0.4. \tag{4–47}$$

By subtracting Eq. (4–47) from Eq. (4–46), Eq. (4–45) is obtained. The value of $K_{Ca/Mg}$ is then equal to

$$\frac{K_{Na/Mg}}{K_{Na/Ca}} = 0.5/0.4 = 1.25.$$

One problem with selectivity coefficients is that they vary with solution properties such as TDS (Stumm and Morgan, 1996). This effect is illustrated in Fig. 4–34, which shows the variation in selectivity of the Ca^{2+}/Na^+ exchange reaction as a function of the concentration of sodium. The dashed line represents a $K_{Na/Ca}$ value of 1, the case in which neither ion is preferred. In dilute solutions ($10^{-3}M$ Na and $10^{-2}M$ Na) calcium is strongly preferred in the exchange because

FIGURE 4–34   Plot of mole fraction of Ca^{2+} bound to the mineral surface vs. the mole fraction of Ca^{2+} in solution in an exchange system. The sodium concentration increases from 10^{-3} to 2 M. In the more dilute solution, calcium is strongly preferred on the solid exchange surface, whereas sodium becomes the preferred ion when the activity of sodium in solution is high. Dashed line indicates no preference. (From *Aquatic Chemistry* 3/e, by Stumm and Morgan. Copyright © 1996 by Wiley Interscience. Reprinted by permission of John Wiley & Sons, Inc.).

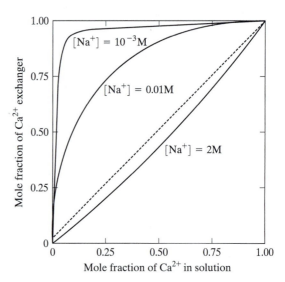

the mole fraction of calcium on the exchanger is much higher than the mole fraction of calcium in solution. As the molarity of sodium increases, the selectivity for calcium decreases, to the point where sodium is preferred over calcium at a sodium molarity of $2M$. These differences can have real significance when comparing clays deposited in seawater versus those deposited in freshwater. Freshwater clays are likely to be dominated by exchangeable calcium, whereas seawater clays are more likely to be dominated by exchangeable sodium. This relationship also explains why water softeners selectively remove calcium but can be stripped of calcium and rejuvenated by flushing with a brine solution with sodium in high concentration.

Ion Exchange in Natural and Contaminated Ground Waters

One of the most common ion exchange reactions occurring in uncontaminated ground waters is the exchange of adsorbed sodium for calcium and magnesium in solution. This natural softening reaction has been recognized in sandstone aquifers in the Atlantic coastal plain (Foster, 1950; Zack and Roberts, 1988; Knobel and Phillips, 1988) and in sedimentary basins in western North America (Thorstenson et al., 1979; Henderson, 1985). The requisite conditions for the reaction include carbonate minerals, high partial pressures of carbon dioxide, and clays with abundant exchangeable sodium. The reaction showing calcium for sodium exchange is

$$CaCO_3 + H_2CO_3 + 2NaX \rightleftharpoons CaX + 2Na^+ + 2HCO_3^-. \qquad (4\text{--}48)$$

In the Fox Hills-Basal Hell Creek aquifer of southwestern North Dakota, Thorstenson et al. (1979) proposed that p_{CO_2} values as high as $10^{-0.7}$ were generated by the oxidation of lignite and related organics in the vadose zone. Under these conditions, calcite dissolution, coupled with calcium for sodium ion exchange on clay minerals, is able to produce water with mean sodium and bicarbonate concentrations of 423 mg/l and 775 mg/l in the recharge area of the confined aquifer. The calcium concentration in the same part of the aquifer is generally less than 5 mg/l.

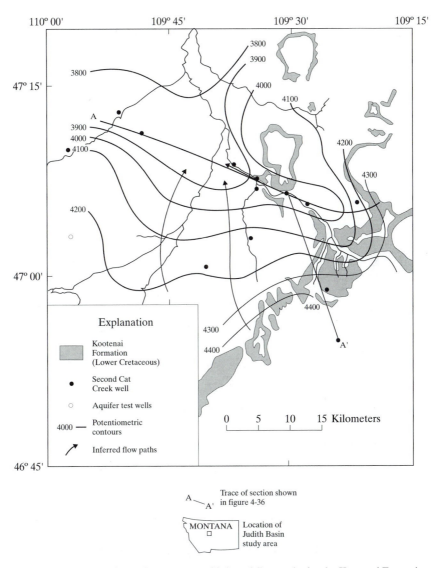

FIGURE 4–35 Potentiometric contours and inferred flow paths for the Kootenai Formation, eastern Judith Basin, Montana. Cross section A–A' shown in Figure 4–26 (from Henderson, 1985).

A similar reaction occurs in aquifers contained within the Lower Cretaceous Kootenai Formation in the Judith Basin of central Montana (Henderson, 1985). Ground water flow in the sandstone aquifer units of the Kootenai is recharged where the formation crops out and trends north-northwest (Fig. 4–35). The aquifer generally dips to the northwest and becomes confined beneath the thick Colorado Shales (Fig. 4–36). Because of chemical variability between aquifer units within the Kootenai Formation, characterization of the water chemistry in the aquifer was restricted to twelve wells completed in the Second Cat Creek Sandstone (Fig. 4–35).

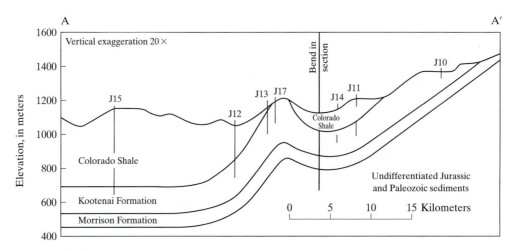

FIGURE 4–36   Cross section of eastern Judith Basin, Montana along line A–A′ in Fig. 4–35. Well locations indicated by vertical lines (from Henderson, 1985).

The chemical evolution of ground water in the Second Cat Creek Sandstone involves a progressive ion exchange reaction between divalent cations and sodium. The net result of this reaction is shown in a plot of $\dfrac{\log(M_{Ca^{2+}} + M_{Mg^{2+}})}{M^2{}_{Na^+}}$ (Fig. 4–37). This parameter changes from positive values near the recharge area, indicating a greater abundance of calcium and magnesium in solution, to a negative value caused by a decrease in calcium and magnesium along with an increase in sodium. Henderson (1985) used geochemical modeling to construct a hypothesis for the evolution of the water. The CO_2 driving the carbonate mineral dissolution was assumed to be derived from adjacent formations because of a lack of evidence for organic matter oxidation within the Second Cat Creek Sandstone. Because the ground water is oversaturated with respect to calcite, Henderson concluded that incongruent dissolution of dolomite was the primary mechanism producing bivalent ions for exchange with sodium adsorbed on the illitic clays in the sandstone. Calcium derived from dolomite dissolution was incorporated into the precipitating calcite as well as exchanging with sodium. The overall reaction is summarized as

$$CaMg(CO_3)_2 + H_2CO_3 + NaX \rightleftharpoons CaCO_3 + 2HCO_3^- + 2Na^+ + MgX. \quad (4\text{–}49)$$

Both geological and hydrogeological properties of aquifers influence ion exchange processes. In the Black Creek aquifer, an Atlantic coastal plain aquifer system, dilute seawater is present in the downgradient portion of the flow system near the coast (Zack and Roberts, 1988). A calcium for sodium ion exchange reaction occurs during flow from the recharge area until the dilute seawater is encountered. There, where the freshwater mixes with water with a much higher sodium-to-calcium ratio, ion exchange reverses and sodium in solution is exchanged for calcium from clay minerals. Ion exchange in the Magothy aquifer of coastal Maryland depends upon the presence or absence of

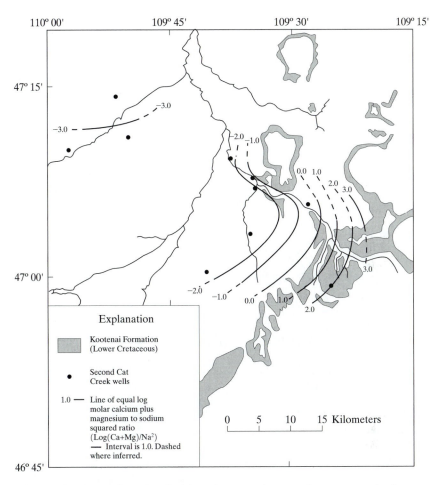

FIGURE 4–37   Log molar ratios of calcium plus magnesium to sodium concentration in waters from Second Cat Creek sandstone of the Kootenai Formation, eastern Judith Basin, Montana. Decreasing ratio indicates ion exchange along flow path (from Henderson, 1985).

overlying confining beds (Knobel and Phillips, 1988). The Magothy aquifer is characterized by a sand and gravel lithology, largely devoid of carbonate minerals. Where the aquifer is unconfined and flow paths are short (Region I, Fig. 4–38), the dilute waters of the system evolve primarily by silicate mineral dissolution. Bicarbonate concentrations in this part of the flow system are relatively constant with increasing length of flow path (Fig. 4–39). In Region II (Fig. 4–38), the Magothy aquifer is confined beneath sediments of the Aquia Formation, which contains carbonate minerals. Recharge to the Magothy aquifer, which occurs by downward leakage through the Aquia Formation, is therefore a calcium–bicarbonate facies water with higher bicarbonate concentrations than Region I. Flow in Region II evolves to a higher bicarbonate concentration through subregions IIb and IIc (Fig. 4–39) accompanied by a sodium for calcium exchange reaction leading to a sodium–bicarbonate facies. The reaction differs from that of Eq. (4–48) in

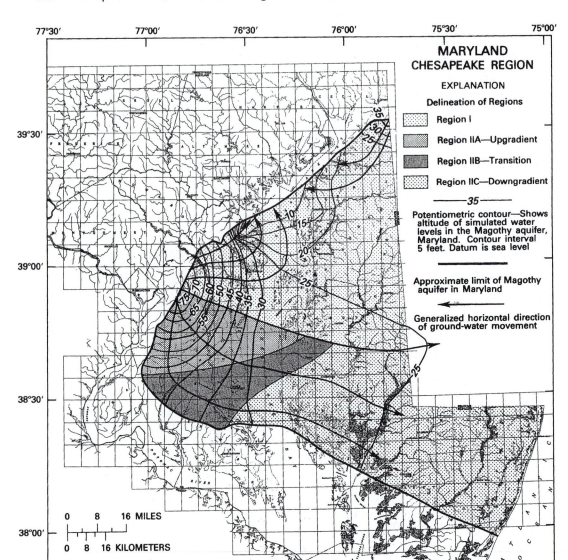

FIGURE 4–38 Simulated prepumping potentiometric surface and locations of Regions I, IIA, and IIC in the Magothy aquifer, Maryland (from Knobel and Phillips, 1988).

that incongruent dissolution of plagioclase yields calcium ions to exchange for sodium along the flow path to produce the increase in sodium in Region IIc (Fig. 4–40). The carbon dioxide driving plagioclase dissolution is traced to the oxidation of small amounts of lignite within the aquifer. The system is thus open with respect to carbon dioxide throughout the flow system in a confined aquifer.

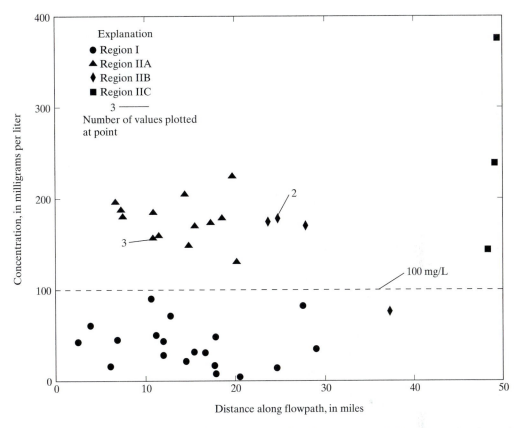

FIGURE 4–39   Concentrations of HCO_3^- with distance along the flow path in the Magothy aquifer. Ground water in Regions II and III evolves to higher bicarbonate concentration (from Knobel and Phillips, 1988).

Saltwater intrusion is often a problem along low-lying coastal areas. Pumpage of ground water for water supplies exacerbates the problem and causes upconing of saline water toward the pumping wells. One solution to deteriorating water quality is to increase recharge to the area, perhaps by infiltration of river water, to reverse the incursion of saltwater. Along the coast of the Netherlands, the application of this refreshening process produces a zonation in water chemistry facies due to ion exchange reactions occurring as the freshwater displaces the saltwater (Appelo and Postma, 1993). Near-surface ground water is calcium bicarbonate type, consistent with the infiltrating surface water (Fig. 4–41). Below the calcium-bicarbonate zone, the infiltrating water is converted to the sodium-bicarbonate facies by replacement of the calcium ions with previously adsorbed sodium ions. The sodium-bicarbonate zone, which can be fresh or brackish, is transitional to saline sodium-chloride water below that lies beneath the interface between saltwater and fresher ground water above.

Compared to natural ion-exchange processes, the introduction of contaminant solutions with cation concentrations much different than background waters is likely to

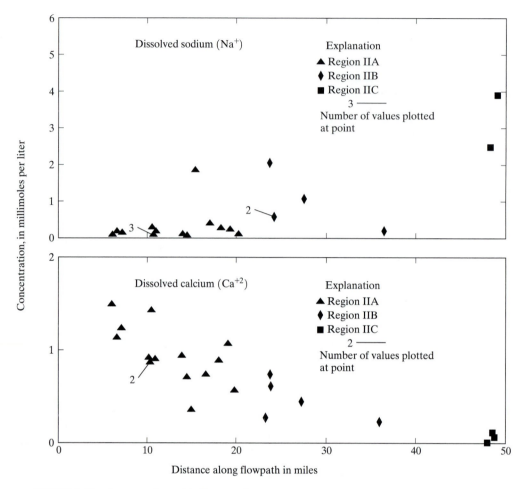

FIGURE 4–40   Concentrations of Ca^{2+} and Na^+ with distance along flow path in Magothy aquifer. Increasing trend of Na^+ and decreasing trend of Ca^{2+} explained by ion exchange (from Knobel and Phillips, 1989).

produce dramatic changes in ion concentrations over very short flow paths. A field experiment illustrating cation exchange reactions from a simulated contaminant solution was conducted by Dance and Reardon (1983). The experiment consisted of injection of a discrete volume of a solution (0.68 m^3) into a sand aquifer followed by very detailed spatial and temporal monitoring of downgradient ground water. The chemical characteristics of the injection solution and background ground water are shown in Table 4–8. In natural ground water at the site calcium was the dominant cation, exceeding other cation concentrations by at least an order of magnitude. In the injection solution, cation concentrations were much higher than background and varied by about 20% or less. The aquifer was composed of carbonate-rich sand with an extremely low cation exchange capacity of about 0.5 meq/100 g. Calcium was the dominant exchangeable ion present on aquifer solids.

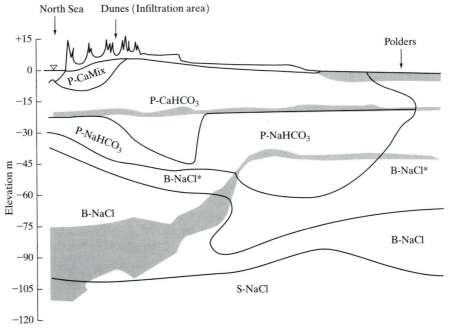

FIGURE 4–41   Cross section through coastal dunes area in Netherlands showing chemical facies of ground water. P = fresh, B = brackish, and S = saline. Low permeability layers shown by shading. Ion exchange during artificial recharge has refreshened ground water (from Appelo and Postma, 1993).

TABLE 4–8   Ionic concentrations (in mM) in solution injected in five wells and in background ground water in tracer injection study conducted by Dance and Reardon (1983).

	Injection Wells					Background Ground Water
Solute	IW1	IW2	IW3	IW4	IW5	
Ca^{2+}	5.09	4.54	4.32	4.99	5.06	1.01
Mg^{2+}	5.07	5.10	5.15	5.20	5.12	0.089
Na^+	3.93	4.15	4.00	4.06	3.97	0.081
K^+	3.84	3.64	3.39	3.58	3.59	0.011
HCO_3^-	3.76	3.35	1.81	4.02	3.96	1.93
SO_4^{2-}	4.23	3.61	3.45	4.01	3.88	0.21
Cl^-	16.6	16.1	16.2	16.3	16.3	0.051
pH				7.6		7.3
SI_{cal}				0.7		−0.6

Calcite saturation index [SI_{cal}] computed by WATEQF (Plummer et al., 1976).

Breakthrough curves for cation concentrations at a distance of 0.75 m from the injection point are shown in Fig. 4–42. For each ion, the relative concentration, C/C_0, is plotted as a function of time after injection. A C/C_0 value of 1.0 means that the concentration is exactly the same as the injected concentration of the ion. Chloride is included as an

FIGURE 4–42   Breakthrough curves for various ions at a distance of 0.75 m from injection wells in tracer experiment by Dance and Reardon (1983). Relative concentration C/C_0 shown on vertical axis. C/C_0 of 1 = concentration at the injection well (from Dance and Reardon, 1983). Permission not received yet.

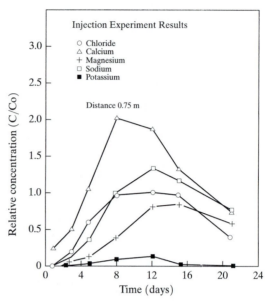

assumed conservative ion for comparison with cation behavior. The peak concentration of chloride passed the monitor point after about 12 days and had a relative concentration of 1.0. Among the cations, magnesium and potassium reached the point at relative concentrations of less than 1, much less than 1 in the case of potassium. Sodium and calcium, on the other hand, had peak relative concentrations greater than 1, with calcium attaining a peak C/C_0 of 2. These results indicate that magnesium and potassium were adsorbed by aquifer solids in exchange for sodium and calcium. Even though selectivity coefficients indicate a preference for calcium over magnesium, sodium, and potassium, calcium was released to solution because of the much higher-than-background concentrations of the other cations.

The breakthrough curves also indicate that the calcium peak reached the measurement point before chloride, the conservative parameter. The magnesium peak reached the sampling point slightly after chloride, which indicates its movement is retarded with respect to chloride, which moves at the average ground water flow velocity. The presence of elevated calcium concentrations at the leading edge of landfill leachate plumes that result from ion exchange (a "hardness halo") has been described by Griffin et al., 1976.

In the event that a contaminant source is continuous rather than a slug, which was simulated in the Dance and Reardon study, cation exchange processes produce a chromatographic pattern in which the cation of least adsorption affinity travels at a velocity approaching the average ground water flow velocity and the distribution of other cations is based on their selectivity by aquifer solids. Each separate cation "plume," particularly those that have a high sorption affinity for aquifer solids, has a sharp downgradient front that moves slower than the average ground water flow velocity. The front of the plume cannot move downgradient until all available exchange sites are filled with the cation. A sorption front was recognized by Kehew et al. (1984) downgradient from a municipal sewage lagoon (Fig. 4–43). The extent of dissolved ammonium was

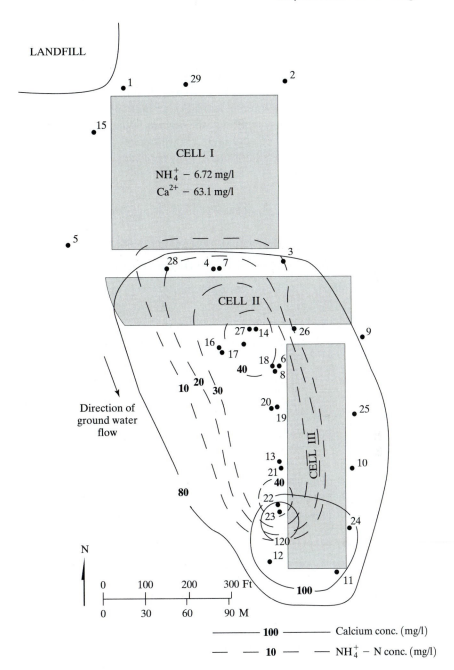

FIGURE 4–43 Contours of calcium and ammonium in ground water around waste stabilization cells at McVille, North Dakota. Black dots are locations of monitor wells. Concentrations shown in Cell I are from wastewater. Higher concentrations of ammonium in ground water are caused by mixing and reduction from nitrate derived from landfill in upper left (modified from Bulger et al., 1989. Reprinted by permission of the National Ground Water Association).

significantly retarded relative to conservative parameters such as chloride. Ammonium concentrations, which were surprisingly higher in ground water than in the wastewater in the lagoon, may have been augmented from the reduction of nitrate to ammonium in leachate from an adjacent landfill (Bulger et al., 1989). Dissolved ammonium dropped from approximately 40 mg/l in MW-22 to less than 0.1 mg/l in MW-12. In the vicinity of this sorption front, calcium concentrations made a sudden jump to approximately twice the wastewater concentration in the lagoon. The complete ion exchange reaction also included the sorption of potassium and the release of magnesium. The increase in calcium and magnesium ($\sim$2.5 meq/l) was equal to the decrease in ammonium and potassium concentrations.

PROBLEMS

1. The muscovite-gibbsite incongruent dissolution reaction is

$$KAl_3Si_3O_{10}(OH)_2 + H^+ + 9H_2O \rightleftharpoons 3Al(OH)_3 + K^+ + 3H_4SiO_4.$$

Which mineral is more stable in equilibrium with ground water with $a_{H_4SiO_4} = 10^{-4}$, $a_{K^+} = 10^{-3}$, pH $= 7$, T $= 25°C$?

2. The dissolution of serpentine can be written as

$$Mg_3Si_2O_5(OH)_4 + 6H^+ \rightleftharpoons 3Mg^{2+} + 2H_4SiO_4 + H_2O.$$

If a ground water at 25°C is in equilibrium with serpentine and has a Mg^{2+} molarity of $10^{-3.38}$ and a pH of 8.5, what is the H_4SiO_4 concentration in mg/l? Assume a $=$ m.

3. If the selectivity coefficient $K_{Na/K} = 0.2$ and $K_{Na/NH_4} = 25$, what is the value of K_{K/NH_4}? Write a reaction for this exchange and state which ion is preferentially adsorbed and why.

4. Discuss the dissolution mechanisms of the silicate and oxide minerals.

5. How does the solubility of quartz differ from that of the aluminosilicate minerals?

6. Describe the approach used by Garrels to reconstruct the weathering reactions occurring in flow systems of ephemeral springs in the Sierra Nevadas.

7. What are the two types of surface charge and how are they acquired?

8. Describe the individual components of net surface charge.

9. Summarize examples from the literature of ion exchange in uncontaminated and contaminated aquifers.

Redox Reactions and Processes

Acid–base reactions were defined in Chapter 3 in terms of the transfer of hydrogen ions from proton donors (acids) to proton acceptors (bases). We now turn to a class of reactions that can be explained by the transfer of electrons between reactants and products. In this branch of chemistry, *oxidation* is defined as a loss of electrons and *reduction* is a gain of electrons. Unlike the existence of hydrogen ions in a solution, free electrons do not occur; each electron that is lost from a species in solution must be gained by another. In these reactions, we will be dealing with electron donors and acceptors.

To grasp the significance of oxidation–reduction, or *redox*, reactions, we need only to look into the mirror. Our very life and survival is based on our ability to derive energy from the oxidation of organic compounds that we eat. The production of these compounds involves the thermodynamically unfavorable process of photosynthesis (Gibbs free energy increases!), in which plants convert carbon dioxide to organic carbon and give off oxygen. Humans belong to the class of organisms that turns the organic carbon back into carbon dioxide in order to harvest the energy stored within.

Redox reactions are enormously important in hydrogeology. The presence of iron in ground water, for example, implies a certain redox environment, one that is different from an aquifer that contains nitrate. Redox conditions directly or indirectly control the species and mobility of many elements. When an organic contaminant is released to a soil or an aquifer, a series of redox reactions is initiated. Many remedial methods rely on microorganisms indigenous to the aquifer to oxidize or reduce the contaminant, a group of processes known as biodegradation. These processes can be enhanced by changing the redox conditions in the subsurface. An understanding of contaminant hydrogeology is not possible without a thorough knowledge of redox processes.

ELECTRON TRANSFER REACTIONS

Oxidation Number

Oxidation, the loss of electrons, can also be defined as a positive change in oxidation number of an element. Under this convention, reduction is a negative change in oxidation number. It is therefore necessary to be able to determine the oxidation number of each element in a compound so that changes during reactions can be assessed. For free ionic species such as Ca^{2+}, the oxidation number is easily determined by the valence. Oxidation numbers are also equal to the charge on the ion in ionic compounds like NaCl.

In covalent compounds, including organic compounds, the electronegativity of the elements is used to ascertain the oxidation number, with the more electronegative element being assigned a negative oxidation number. For example, Table 1–2 shows that chloride is more electronegative than hydrogen and therefore would have an oxidation number of $-I$ in HCl. Hydrogen has to have an oxidation number of $+I$ because the sum of the oxidation numbers in this compound is zero. In methane, CH_4, carbon is the more electronegative compound and therefore has an oxidation number of $-IV$.

Many elements present in natural and contaminated ground waters can have different oxidation numbers in various compounds, depending on the redox conditions in the aquifer. Some common examples are illustrated in Table 5–1. Note that oxidation numbers are designated by Roman numerals.

Redox Reactions

By analogy with acid–base reactions, redox reactions can be generalized to show the oxidized and reduced species in the reaction:

$$Ox_1 + Red_2 \rightleftharpoons Ox_2 + Red_1. \tag{5-1}$$

In Eq. (5–1), Ox_1 gains electrons from Red_2 and becomes Red_1. Red_2, serving as the electron donor, loses electrons and becomes Ox_2. An actual reaction is shown in Eq. (5–2):

$$O_2 + 4Fe^{2+} + 4H^+ \rightleftharpoons 4Fe^{3+} + 2H_2O. \tag{5-2}$$

In this reaction, free oxygen, O_2 (Ox_1), is reduced to O^{-2} in water (Red_1) by accepting electrons from Fe^{2+} (Red_2), which in turn becomes Fe^{3+} (Ox_2). The reaction in Eq. (5–2) can be separated into its oxidation and reduction components by writing *half reactions*, which are then balanced to obtain the full redox reaction. The two half reactions for Eq. (5–2) are

TABLE 5–1   Oxidation numbers of elements in common compounds.

Element (Oxidation Number)	Example Compounds
C(+IV)	HCO_3^-, CO_3^{2-}
C(+II)	CCl_2CCl_2
C(0)	CH_2O
C(−IV)	CH_4
N(+V)	NO_3^-
N(+III)	NO_2^-
N(0)	N_2
N(−III)	NH_4^+, NH_3
Fe(+III)	Fe^{3+}, $Fe(OH)_3$
Fe(+II)	Fe^{2+}, $FeCO_3$
S(+VI)	SO_4^{2-}
S(−II)	H_2S, HS^-

$$O_2 + 4H^+ + 4e^- \rightleftharpoons 2H_2O \tag{5-3}$$

and

$$Fe^{2+} \rightleftharpoons Fe^{3+} + e^-. \tag{5-4}$$

Before we can add the two half reactions together to produce the full reaction, we must multiply Eq. (5–4) by 4 so that the same numbers of electrons are included:

$$4Fe^{2+} \rightleftharpoons 4Fe^{3+} + 4e^-. \tag{5-5}$$

At this point, the addition of Eqs. (5–5) and (5–3) will yield (5–2). Although half reactions imply that free electrons are present, this is not possible. Any electrons released by one species must be accepted by another.

Electron Activity

When redox reactions are expressed as half reactions, another analogy to acid–base reactions can be utilized. The pH of a solution, defined as pH $= -\log a_{H^+}$, is an indication of the tendency of the solution to donate hydrogen ions. In an acid solution (low pH), this tendency is strong. Keeping in mind that there are no free electrons in a solution, it is nevertheless useful to define a quantity known as the *electron activity*. This hypothetical quantity can be expressed in the same way as pH:

$$pe = -\log a_{e^-}. \tag{5-6}$$

The pe is an indication of the tendency of the solution to donate electrons. In an environment in which many species are in their reduced form—a solution with a high concentration of organic compounds, for example—there is a strong tendency to donate electrons, and the pe is low.

The pe of a half reaction can be determined in the following manner, using Eq. (5–4) as an example:

$$Fe^{3+} + e^- \rightleftharpoons Fe^{2+}.$$

The equilibrium constant for the reaction can be expressed in the usual way:

$$K = \frac{a_{Fe^{2+}}}{a_{Fe^{3+}} a_{e^-}}. \tag{5-7}$$

Solving for a_{e^-} yields

$$a_{e^-} = \frac{a_{Fe^{2+}}}{a_{Fe^{3+}} K}. \tag{5-8}$$

If we take the logarithm of both sides of Eq. (5–8) and multiply by -1, the result will be

$$-\log a_{e^-} = -\log\left(\frac{a_{Fe^{2+}}}{a_{Fe^{3+}}}\right) + \log K. \tag{5-9}$$

Because pe $= -\log a_{e^-}$,

$$pe = \log K - \log\left(\frac{a_{Fe^{2+}}}{a_{Fe^{3+}}}\right). \tag{5-10}$$

At 25°C, the value of K can be calculated in the usual way as

$$\log K = \frac{-\Delta G_R^0}{5.708} = \frac{-\left(\Delta G_{f-Fe^{2+}}^0 - \Delta G_{f-Fe^{3+}}^0\right)}{5.708}$$

$$= \frac{-(-90.0 - (-16.7))}{5.708} = 12.8. \tag{5–11}$$

Thus,

$$pe = 12.8 - \log\left(\frac{a_{Fe^{2+}}}{a_{Fe^{3+}}}\right). \tag{5–12}$$

Values of pe have little meaning unless they can be compared to a reference point. Such a point is provided by the reaction in which hydrogen gas is oxidized to the hydrogen ion:

$$1/2\,H_2\,(g) \rightleftharpoons H^+ + e_s^-. \tag{5–13}$$

By convention, the free energy of formation of an electron and the hydrogen ion are both zero, and the free energy of formation of H_2 is zero because it is in its standard state. Therefore, the free-energy change of this reaction is zero, and K is equal to 1 because $RT \ln K = 0$ and $\ln K = 0$. When we rearrange the equation to solve for pe, we obtain

$$K = \frac{a_{H^+} a_{e^-}}{p^{1/2}_{H_2}};$$

$$\log a_{e^-} = \log K + 1/2 \log p_{H_2} - \log a_{H^+};$$

and

$$pe = -\log K - 1/2 \log p_{H_2} + \log a_{H^+}. \tag{5–14}$$

If this half reaction were to occur in a solution in which the partial pressure of hydrogen gas was maintained at 1 atmosphere and the activity of the hydrogen ion was equal to 1 (pH = 0), pe would be equal to zero because we have already determined that K is 1. Eq. (5–13) is therefore defined as the reference pe for comparison with other reactions. This reference reaction can be combined with other reactions without altering the calculated value. For example, Eq. (5–4) is joined with Eq. (5–13) as follows:

$$Fe^{3+} + e^- + 1/2\,H_2 \rightleftharpoons H^+ + e^- + Fe^{2+}. \tag{5–15}$$

When the electrons are removed by subtraction and the standard free-energy change and equilibrium constant are calculated, the results are the same as obtained in Eq. (5–11) because of the free energies of formation of H^+ and H_2. Thus,

$$\log K = \frac{-\Delta G_R^0}{5.708} = \frac{-\left(\Delta G^0_{f-Fe^{2+}} + \Delta G^0_{f-H^+} - \Delta G^0_{f-Fe^{3+}} - 1/2\Delta G^0_{f-H_2}\right)}{5.708},$$

$$= \frac{-\left(\Delta G^0_{f-Fe^{2+}} - \Delta G^0_{f-Fe^{3+}}\right)}{5.708}. \tag{5–16}$$

The calculated pe of any half reaction, therefore, is equal to the pe calculated using the hydrogen reaction [Eq. (5–13)] to complete the redox reaction.

The Standard Hydrogen Electrode

An actual cell could be set up in the laboratory to produce the conditions of our reference reaction, $1/2 H_2(g) \rightleftharpoons H^+ + e^-$. As shown in Fig. 5–1, this cell would consist of a platinum electrode immersed in an acid solution in which $a_{H^+} = 1 (pH = 0)$. Hydrogen gas would be bubbled over the electrode at a pressure of 1 atm. As calculated earlier, the pe in the cell under these conditions would be zero, the reference point for all other redox reactions. It is therefore known as the *standard hydrogen electrode (SHE)*. If the reaction occurring in the cell is going to the right, the oxidation of the hydrogen gas will produce hydrogen ions and electrons. The electrons will accumulate on the platinum electrode because they cannot go into solution.

As long as conditions are constant in the standard hydrogen electrode, no reaction occurs. However, if we now connect the standard hydrogen electrode to a cell containing a different solution, as shown in Fig. 5–2, electrons may flow from one cell to another. The second cell contains a solution of Fe^{3+} and Fe^{2+}. It also has a platinum electrode which, along with the one in the SHE, is connected to a voltmeter. An electrical circuit is completed by adding a salt bridge, which contains a concentrated salt such as KCl that can conduct electricity. We have now constructed an electrochemical cell. If the pe values in the two cells differ, electrons will tend to flow from one cell to another, and the voltmeter will measure the difference in electrical potential between the two cells in volts.

The pe of the cell on the right is calculated with free-energy data using the procedure followed in Eqs. (5–7) through (5–12). The expression is

$$pe = 12.8 - \log\left(\frac{a_{Fe^{2+}}}{a_{Fe^{3+}}}\right).$$

FIGURE 5–1 The standard hydrogen electrode, in which $a_{H^+} = 1$ and $p_{H_2} = 1$ atm.

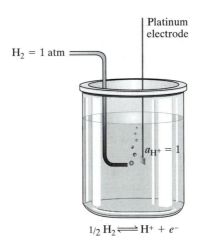

Platinum electrode

$H_2 = 1$ atm

$a_{H^+} = 1$

$1/2\ H_2 \rightleftharpoons H^+ + e^-$

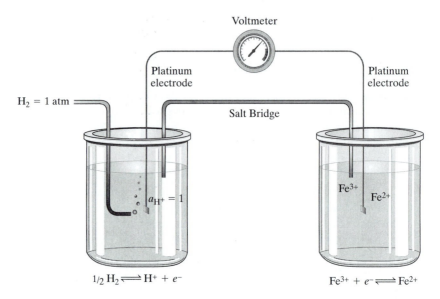

FIGURE 5–2 The standard hydrogen electrode connected by a voltmeter to a cell containing Fe^{2+} and Fe^{3+}. Electrons will flow from the cell with the higher electron activity to the cell with lower electron activity.

The pe is the same when the reactions in both cells are combined to form a complete redox reaction, as previously shown. If we assume that the activities of Fe^{2+} and Fe^{3+} in the right-hand cell are both equal to 1, the pe in the cell will be equal to 12.8. We now have a case in which the pe in the SHE is zero and the pe in the other cell is 12.8. From the definition of pe, the electron activity in the SHE is 1 and $10^{-12.8}$ in the other cell. Electrons will flow from a higher electron activity to a lower activity. Thus, in this case, H_2 is oxidized to H^+ and electrons in the SHE flow through the platinum electrode and the circuit leading to the platinum electrode in the second cell. There they are accepted by Fe^{3+}, which is reduced to Fe^{2+}. If the initial concentrations of Fe^{2+} and Fe^{3+} are not equal, as we assumed, the pe will also differ. For example, if $a_{Fe^{2+}}/a_{Fe^{3+}} = 10$, then, by Eq. (5–12),

$$\text{pe} = 12.8 - \log\left(\frac{a_{Fe^{2+}}}{a_{Fe^{3+}}}\right) = 12.8 - \log(10) = 12.8 - 1 = 11.8.$$

Under these conditions, electrons would still flow from the SHE, where the electron activity is 1, to the second cell, where the electron activity is $10^{-11.8}$.

The laboratory apparatus shown in Fig. 5–2 gives a method of actually measuring the redox potential of an unknown solution relative to the SHE. This value, measured in volts or millivolts, is an alternative to the use of pe for expressing a redox potential. It is known as *Eh*, which is defined as the potential relative to the SHE. Because both pe and Eh are really measuring the same thing, one must be able to be converted to the other. This conversion can be done using the Faraday constant, F, which has a value of 96.42 kJ per volt gram equivalent. The relationship is

$$pe = \frac{nF \times Eh}{2.303RT},\tag{5-17}$$

in which n is the number of electrons, R is the gas constant, and T is the absolute temperature. At 25°C, the conversion is simplified to

$$pe = 16.9Eh,$$

or

$$Eh = 0.059 \, pe.$$

A calculated pe or Eh value of zero indicates that the solution has the same electron activity as the SHE. Positive values of pe or Eh denote more oxidizing conditions (the solution accepts electrons from the SHE), and negative values of pe or Eh represent more reducing conditions than the SHE (the solution donates electrons to the SHE).

Determining Eh from Redox Couples

Although measurement of the Eh of a solution containing two members of a redox couple with an electrochemical cell is possible, the Eh can also be calculated using an equation known as the *Nernst equation*. For an example, consider the familiar iron reaction,

$$Fe^{3+} + 1/2 \, H_2 \rightleftharpoons Fe^{2+} + H^+.\tag{5-18}$$

To derive an expression for Eh for this reaction, we will make use of the relationship between Eh and the Gibbs free energy. First, the free-energy change of the reaction is determined as

$$\Delta G_R = \Delta G_R^0 + RT \ln\left(\frac{a_{Fe^{2+}} a_{H^+}}{a_{F^{3+}} P^{1/2}_{H_2}}\right),\tag{5-19}$$

in which $a_{H^+} = P^{1/2}_{H_2} = 1$ (the SHE). Therefore,

$$\Delta G_R = \Delta G_R^0 + RT \ln\left(\frac{a_{Fe^{2+}}}{a_{Fe^{3+}}}\right).\tag{5-20}$$

To convert the free energies to Eh, the relationship

$$\Delta G = -nFEh\tag{5-21}$$

is used, where n is the number of electrons and F is the Faraday constant. In the most commonly employed convention, electrons in the half reaction must appear on the left-hand side of the equation; that is, the oxidized species must be on the left, and the reaction is shown as a reduction. If we had not assumed that the SHE was involved, Eq. (5–18) would reduce to

$$Fe^{3+} + e^- \rightleftharpoons Fe^{2+},$$

which has the electrons on the left. Written this way, the relationship between Eh and ΔG is

$$\text{Eh} = \frac{-\Delta G}{nF}. \tag{5-22}$$

To convert Eq. (5–20) to Eh, we must divide both sides by $-nF$. Thus,

$$\frac{-\Delta G_R}{nF} = \frac{-\Delta G_R^0}{nF} - \frac{RT}{nF} \ln\left(\frac{a_{Fe^{2+}}}{a_{Fe^{3+}}}\right). \tag{5-23}$$

Substituting for Eh from Eq. (5–22), and defining $-\Delta G_R^0/nF$ as E^0, the *standard electrode potential*, we have

$$\text{Eh} = \frac{E^0 - RT}{nF} \ln\left(\frac{a_{Fe^{2+}}}{a_{Fe^{3+}}}\right), \tag{5-24}$$

or

$$\text{Eh} = \frac{E^0 + 2.303 \, RT}{F} \log\left(\frac{a_{Fe^{3+}}}{a_{Fe^{2+}}}\right). \tag{5-25}$$

E^0, the standard electrode potential, is the potential the Eh cell would have if all species were in their standard states. For reactions occurring at 25°C, Eq. (5–25) can be generalized for any reaction, as long as electrons appear on the left:

$$\text{Eh} = \frac{E^0 + 0.059}{n} \log\left(\frac{\text{activity product of oxidized species}}{\text{activity product of reduced species}}\right). \tag{5-26}$$

This equation is a statement of the Nernst equation. It gives us a way to calculate Eh in volts when the concentrations of two members of a redox couple are known. Because of the relationship between pe and Eh [Eq. (5–17)], either can be used to characterize the redox conditions and can easily be converted to the other system of measurement. The use of pe rather than Eh has gained popularity in recent years.

EXAMPLE 5–1

Samples are collected from a monitor well located just downgradient from an unlined municipal landfill. The ground water temperature is 25°C and the pH is 7.5. The concentration of H_2S is 2.4 mg/l, and the concentration of sulfate ion (SO_4^{2-}) is below the detection limit of 1 mg/l. Estimate the Eh and pe of the water.

Solution. In this problem, although two members of a redox couple were measured, one of them was nondetect. We cannot, therefore, use the Nernst equation to get a precise calculation of Eh. It is possible to obtain an estimate of Eh by choosing an arbitrarily small concentration of SO_4^{2-}. As it turns out, there are so many problems in interpreting redox measurements and calculations (as will be discussed later), that an estimate made in this way is probably about as useful as a more precise calculation. We will therefore assume a sulfate concentration of 0.1 mg/l.

The first step is to write a balanced half reaction with electrons on the left-hand side. We will start by putting the oxidized species, SO_4^{2-}, on the left with an unknown number of electrons and the reduced species, H_2S, on the right:

$$SO_4^{2-} + \_e^- \rightleftharpoons H_2S.$$

First we check to see if the sulfur molarities balance on both sides, and in this case they do. If they don't, a coefficient must be used to balance them. Next, using the rules for assigning oxidation numbers discussed earlier, we determine that sulfur in sulfate has an oxidation number of $+VI$ and that sulfur in hydrogen sulfide has an oxidation number of $-II$. This represents a difference of eight electrons, which must be shown as the coefficient of e^- in the reaction:

$$SO_4^{2-} + 8e^- \rightleftharpoons H_2S.$$

We then turn our attention to the hydrogen and oxygen in the reaction. These are balanced using H^+ and H_2O. This can be done by adding four waters to the right to balance the oxygen and 10 hydrogen ions to the left to balance the hydrogen. The final half reaction is

$$SO_4^{2-} + 8e^- + 10H^+ \rightleftharpoons H_2S + 4H_2O.$$

To determine the Eh, we start with the free-energy change for the reaction:

$$\Delta G_R = \Delta G_R^0 + RT \ln\left(\frac{a_{H_2S}}{a_{SO_4^{2-}} a_{H^+}^{10}}\right).$$

Next, we divide both sides by $-nF$ because $\Delta G = -nF Eh$, with electrons on the left, and $Eh = -\Delta G / nF$:

$$\frac{-\Delta G_R}{nF} = \frac{-\Delta G^0}{nF} - \frac{RT}{nF} \ln\left(\frac{a_{H_2S}}{a_{SO_4^{2-}} a_{H^+}^{10}}\right),$$

$$Eh = \frac{E^0 + 0.059}{8} \log\left(\frac{a_{SO_4^{2-}} a_{H^+}^{10}}{a_{H_2S}}\right).$$

From the free energies of formation,

$$E^0 = \frac{-\Delta G_R^0}{nF} = \frac{-(-232.23)}{(8)(96.48)} = +0.3v$$

and

$$Eh = 0.3 + 0.0074 \log\left(\frac{a_{SO_4^{2-}} a_{H^+}^{10}}{a_{H_2S}}\right).$$

By assuming that $a = m$, the activities of sulfate and hydrogen sulfide are

$$a_{SO_4^{2-}}(0.1 \text{ mg/l}) = 10^{-5}$$

and

$$a_{H_2S}(2.4 \text{ mg/l}) = 10^{-4.15}.$$

Substituting these numbers, we obtain

$$Eh = 0.3 + .0074 \log\left(\frac{10^{-5} 10^{-75}}{10^{-4.15}}\right)$$

$$= -0.26 \text{ volts and pe} = -4.39.$$

The estimated Eh and pe indicate that the redox potential is very low and the contaminant plume emanating from the landfill constitutes a highly reducing environment. This interpretation is probably more significant in developing a geochemical understanding of the contaminant plume than having a very exact Eh based on the sulfate–sulfide redox couple. Reducing conditions in landfill contaminant plumes are very common because of the high content of biodegradable organic compounds in municipal waste.

FIELD MEASUREMENT OF EH

Equipment and Procedures

Measurement of the redox potential of ground water can be a difficult and time-consuming task. In addition, the measured values are subject to questions about their meaning and usefulness. Values calculated with the Nernst equation are generally preferable to field measurements if enough redox couples are known. Despite these problems, field Eh measurements are relatively inexpensive and can give investigators a general sense of the redox conditions in an aquifer. This is particularly useful in contaminant plumes with strong redox gradients.

Redox measurements are highly unstable and must be made in a closed environment not in contact with the atmosphere. The most common procedure is to connect the discharge tubing of a submersible sampling pump to a *flow-through cell,* also known as a flow cell (Fig. 5–3). Flow cells are usually constructed of plexiglass or other transparent material. Ports on the top of the cell are available for the insertion of various probes

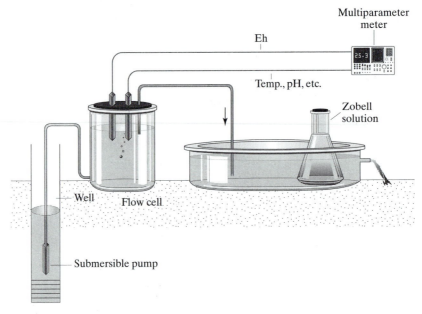

FIGURE 5–3   Field apparatus for measuring redox potential in a well.

and electrodes, including Eh probes. The principle of field Eh measurement is similar to the electrochemical cell (Fig. 5–2) containing the standard hydrogen electrode. In practice, however, combination electrodes with a different reference electrode are utilized. The combination electrode is composed of the platinum sensing electrode connected to the reference electrode, which is commonly made of calomel (mercury in a mercurous chloride solution) or silver enclosed in a silver-chloride solution.

Because the SHE is not being used in the field, a method of converting the readings of the reference electrode to the hydrogen electrode must be used. The normal procedure is to measure the potential of a solution with a known Eh, usually a solution known as Zobell's solution, which is a standard solution of potassium ferric–ferro cyanide (Wood, 1976). The theoretical Eh of Zobell's solution, which varies with temperature, is shown in Fig. 5–4. A bottle of this solution is carried to the field and brought to the sample temperature by immersing the bottle in circulating ground water (Fig. 5–3). The potential of the Zobell solution with respect to the calomel or silver–silver chloride reference electrodes is shown in Fig. 5–5.

With the submersible pump circulating ground water through the flow cell, the redox probe continuously measures the potential of the solution relative to the reference electrode. Readings are taken on a voltmeter connected to the probe (Fig. 5–3). Often a multiparameter meter is used that measures pH, temperature, and other parameters

FIGURE 5–4 The potential of Zobell solution relative to the standard hydrogen electrode at various temperatures (from Wood et al., 1976).

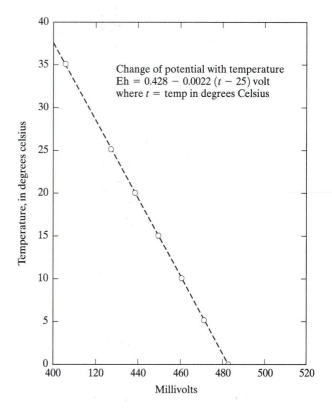

Change of potential with temperature
Eh = 0.428 − 0.0022 (t − 25) volt
where t = temp in degrees Celsius

FIGURE 5–5   The potential of Zobell solution relative to the silver–silver chloride and calomel electrode systems (from Wood et al., 1976).

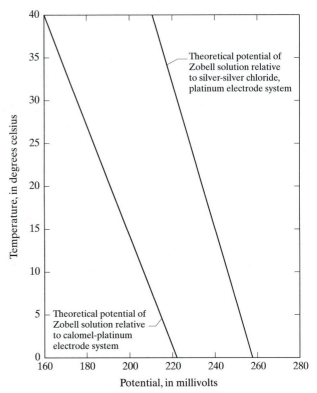

in addition to the redox potential. Because the readings will fluctuate significantly during purging of the well, the well must be thoroughly purged prior to recording the final reading, sometimes a time-consuming process.

Once the potential readings have stabilized, the final reading may be taken. This reading is the potential of the sample relative to the reference electrode in the probe. The following formula can be used to correct the reading to Eh:

$$Eh_{sys} = E_{obs} + Eh_{Zobell} - E_{Zobell\text{-}observed}. \tag{5–27}$$

Here,

Eh_{sys} = the Eh of the ground water sample,

E_{obs} = the measured potential of the sample with respect to the reference electrode,

Eh_{Zobell} = the theoretical Eh of the Zobell solution (Fig. 5–4),

and

$E_{Zobell\text{-}observed}$ = the measured potential of the Zobell solution relative to the reference electrode.

Note that Eh represents a measurement relative to the SHE, whereas E represents the potential relative to the actual reference electrode being used. For example, the term

$E_{\text{Zobell-observed}}$ is the potential of the Zobell solution as measured by the reference electrode when the Zobell solution is brought to sample temperature. This value is checked against the theoretical potential for that electrode (Fig. 5–5), and if the measured value is ±10 mv of the theoretical value, the *measured* value is used in Eq. (5–27). If the measured value is not within 10 mv of the theoretical value, the electrode may not be functioning correctly or the Zobell solution may be defective. The result of a calculation using Eq. (5–27) is a potential relative to the SHE measured for the ground water in the vicinity of the well screen.

EXAMPLE 5–2

A monitor well is purged until a stable reading of 34 mv is obtained in a flow cell with a silver–silver chloride reference electrode. The temperature is 10°C, and the potential of the Zobell solution is 243 mv. What is the Eh of the sample?

> **Solution.**   Before Eq. (5–27) is applied, the $E_{\text{Zobell-observed}}$ must be checked against the theoretical potential of the Zobell solution at 10°C (Fig. 5–5) for a silver–silver chloride electrode. The measured valued, 243 mv, is within 10 mv of the theoretical value of 250 mv. Therefore, we can proceed with the calculation, using a value of $\text{Eh}_{\text{Zobell}}$ from Fig. 5–3.
>
> $$\text{Eh}_{\text{sys}} = E_{\text{obs}} + \text{Eh}_{\text{Zobell}} - E_{\text{Zobell-observed}} = 34 + 460 - 243 = +253 \text{ mv}$$

Pitfalls of Using Field Eh Measurements

Although the measurement of Eh from wells using the procedure described in the preceding section is relatively straightforward, there are several inherent problems with the interpretation of these data. First, ground water solutions, like all natural waters, contain numerous redox couples. If sufficient dissolved species are analyzed, Eh or pe values can be calculated using the Nernst equation from more than one redox couple. Lindberg and Runnells (1984) examined a large data base containing many analyses with multiple redox couples and field-measured Eh values. When the calculated values were plotted against the measured values (Fig. 5–6), there was little correlation. Instead of plotting along a diagonal line across the diagram showing a correlated change in both parameters, the values from individual redox couples tended to plot in narrow vertical bands. These bands had a narrow range of calculated values, but a wide range in field measured values. The position of the vertical bands along the axis of computed redox potentials was controlled more by the E^0 values of the half reactions than by the variation in concentrations of the members of the redox couples. The lack of correlation between the field-measured and calculated potentials can be attributed to the fact that members of many common redox couples, including compounds of nitrogen, carbon, oxygen, hydrogen, and sulfur, are not *electroactive*. This means that the compound does not oxidize or reduce readily at the surface of the platinum electrode. Thus, those redox couples do not contribute to the measured values. For a single measured Eh value to characterize a solution, all redox couples in the solution would have to be in equilibrium. In many natural waters, disequilibrium is a common occurrence. Therefore, the concept of a single redox potential for a solution containing multiple redox couples is not

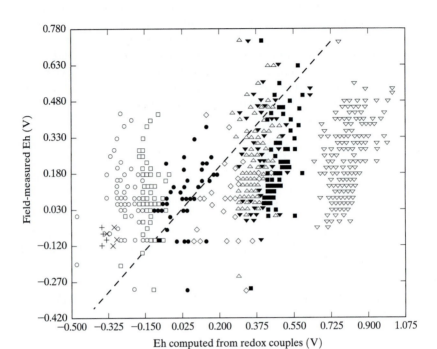

Symbol	Redox Couple
◇	Fe^{3+}/Fe^{2+}
▽	$O_{2\,aq}/H_2O$
○	HS^-/SO_4^{2-}
□	$HS^-/S_{rhombic}$
■	NO_2^-/NO_3^-
▼	NH_4^+/NO_3^-
△	NH_4^+/NO_2^-
+	$CH_{4\,aq}/HCO_3^-$
×	$NH_4^+/N_{2\,aq}$
●	$Fe^{2+}/Fe(OH)_{3\,(s)}$

FIGURE 5–6   Plot of Eh values computed from the Nernst equation for redox couples in solution vs. field-measured Eh values. Values for each couple tend to plot in a vertical band, despite wide variation in the field-measured Eh values. The dashed line represents the hypothetical trend of points if the two types of data are correlated (from Lindberg and Runnells, 1984. Reprinted with permission from Ground Water Redox Reactions: An Analysis of the Equilibrium State Applied to Eh Measurements and Geochemical modeling. *Science*, v. 225, pp. 925–927. Copyright 1984. American Association for the Advancement of Science).

valid. Other explanations for the discrepancies between measured and calculated values include measurement error and the tendency of the platinum electrode to become poisoned by hydrogen sulfide or other compounds.

Barcelona and Holm (1991) suggested a threefold redox classification of natural waters. Waters with measurable dissolved oxygen are *oxic*. Waters that do not contain oxygen or sulfide, but do have a significant amount of dissolved iron (more than about a tenth of a mg/l), are *suboxic*; and waters that contain both dissolved iron and sulfide are known as *reducing*. In a study of ground water in pristine and contaminated aquifers in Illinois, Barcelona et al. (1989) analyzed samples for a large number of redox couples for comparison with field-measured values. In oxic settings, there was poor agreement between the measured values and values calculated from the traditionally used O_2/H_2O couple based on measured dissolved oxygen readings (Fig. 5–7). There was, however, much better agreement between the measured values and calculated values using the O_2/H_2O_2 couple. Hydrogen peroxide, which is not normally analyzed in ground water studies, is an unstable intermediate in the reduction of dissolved oxygen to water. This provides evidence that the presence of dissolved oxygen and its reduction to hydrogen peroxide controls the redox potential in oxic aquifers similar to the one studied by Barcelona et al. (1989). In suboxic settings, Barcelona et al. (1989) determined that the Fe^{3+}/Fe^{2+} couple had a close correspondence to the measured values (Fig. 5–7), a result that is consistent with the electroactive behavior of the iron species. In reducing aquifers, however, Barcelona et al. (1989) were not able to find a redox couple yielding calculated values similar to the measured values (Fig. 5–7). They concluded that in reducing waters, the measured values are of little use in quantifying the redox potentials.

With all of the problems encountered in the interpretation of field-measured redox potentials, it is reasonable to ask if these measurements are worth the time and effort it takes to obtain them. The answer to this question may depend on the intended use of the data. If quantitatively accurate redox potentials are needed, the field-measured values may not be reliable, particularly in reducing conditions. If, however, the objective is to generally characterize a contaminant plume or changes along a ground water flow path, field-measured values may be useful as relative values. For example, the iso-redox potential map produced by Bulger et al. (1989) at a municipal sewage lagoon site (Fig. 5–8) correlates well with plume maps made using specific chemical parameters from monitor well samples. At this site, seepage to ground water of organic-rich wastewater was occurring from cells I and II. The map of field-measured pe values shows a distinct low beneath these cells with an increasing pe downgradient to the south caused by loss of organics due to biodegradation and increases in redox potential along the flow path. The field-measured pe values, which correlate well with the dissolved organic carbon (DOC) distribution, adequately define the redox conditions of the plume at a lower cost than analyzing water samples for a large number of parameters.

PE–PH DIAGRAMS

Relationships between species containing oxidized and reduced members of redox couples are commonly displayed on diagrams known as pe–pH diagrams. These diagrams show stability fields for solid and/or aqueous species as functions of pe and pH. As a

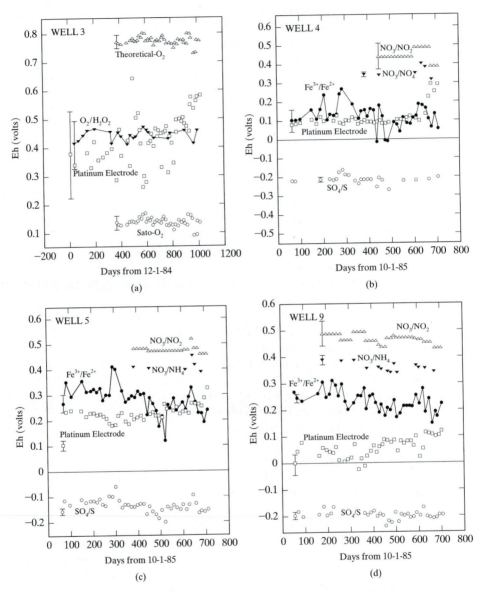

FIGURE 5–7 Data from four wells in Illinois, showing platinum electrode field measurements of Eh and various calculated values from redox couples. In oxic settings (Well 3), the platinum electrode measurements fell closest to the O_2/H_2O_2 couple. In suboxic waters (Wells 4 and 5), the platinum electrode measurements correlated well with the Fe^{3+}/Fe^{2+} couple. The platinum electrode measurements did not correlate with any measured redox couples in reducing waters (Well 9) (from Barcelona et al., *Water Resources Research*, v. 25, pp. 991–1003, 1989, Copyright by the American Geophysical Union).

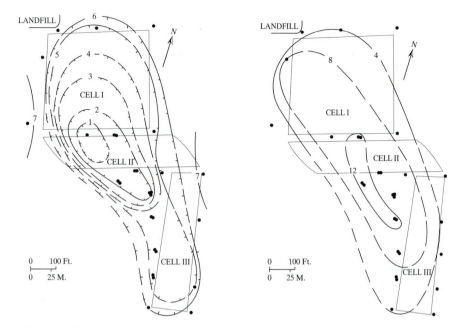

FIGURE 5–8   Contours of mean field-measured pe from wells in ground water around waste stabilization lagoons in North Dakota (left). Contours of dissolved organic carbon in mg/l(DOC) (right) show a similar pattern. Black dots show locations of monitor wells. (From Bulger et al., Reprinted with the permission of the National Ground Water Association. Copyright 1989).

general tool for predicting whether oxidized or reduced species will occur in a particular environment, pe–pH diagrams are very useful. However, the difficulties of using field-measured pe values to characterize a body of water, as already discussed, should not be forgotten.

pe–pH diagrams are constructed by writing half reactions and making assumptions to eliminate unwanted variables so that boundaries can be drawn on the diagram. The boundaries delineate stability fields for the species used in constructing the diagram. Selection of the species is critical in the interpretation of the final diagram, because species that are not used may be more stable than those used in the diagram.

Stability Limits of Water

On all pH diagrams, the stability field of water is first determined. This can be done by drawing boundaries between O_2 and H_2O and between H_2O and H_2. For the first of these two, the correct half reaction is written

$$1/2O_2 + 2e^- + 2H^+ \rightleftharpoons H_2O. \qquad (5\text{--}28)$$

In the normal fashion, the equilibrium constant K_{eq} is expressed as

$$K_{eq} = \frac{1}{p^{1/2}_{O_2} a^2_{e^-} a^2_{H^+}}. \qquad (5\text{--}29)$$

Taking the logarithm of both sides, we obtain

$$\log K_{eq} = -1/2 \log p_{O_2} - 2 \log a_{e^-} - 2 \log a_{H^+},$$

or

$$\log K_{eq} = -1/2 \log p_{O_2} + 2pe + 2pH.$$

Solving for pe, we get

$$pe = 1/2 \log K_{eq} + 1/4 \log p_{O_2} - pH.$$

The resulting equation has three variables, which cannot be plotted on a two-dimensional pe–pH diagram. To eliminate one variable, we can choose a value of p_{O_2} that serves as an upper limit for the partial pressure of oxygen. Because the near-surface environment of the earth has a pressure of one atmosphere, the partial pressure of oxygen could never exceed that value. By choosing a value of 1 atmosphere for p_{O_2}, the second term will drop out of the equation and the equation becomes

$$pe = 1/2 \log K_{eq} - pH.$$

From the usual method for determining K_{eq}, we have

$$\log K_{eq} = \frac{-\Delta G_R^0}{2.303 \, RT} = \frac{237.1}{5.708} = 41.54,$$

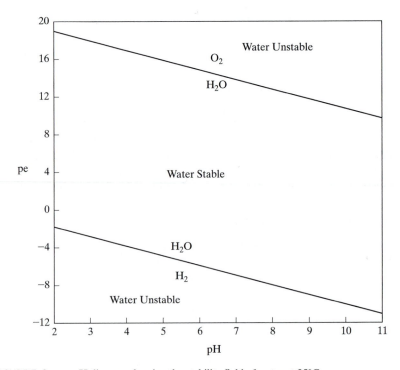

FIGURE 5–9   pe–pH diagram showing the stability field of water at 25°C.

and

$$pe = 20.77 - pH. \tag{5-30}$$

A line on the pe–pH diagram (Fig. 5–9) can now be constructed by choosing values of pH and solving for pe. The significance of the boundary is that it represents an upper limit for near-surface reactions. Above this line, water would be unstable and would oxidize to free oxygen.

Similarly, a boundary representing the lower stability limit of water can be constructed by determining the point at which water would be reduced to hydrogen gas by the reaction

$$H^+ + e^- \rightleftharpoons 1/2\, H_2. \tag{5-31}$$

Following the same steps as before yields

$$K_{eq} = \frac{p^{1/2}_{H_2}}{a_{H^+}\, a_{e^-}},$$

or

$$\log K_{eq} = 1/2 \log p_{H_2} + pe + pH.$$

The standard free energy of reaction for Eq. (5–31) is zero, according to conventions that have been previously discussed, and as a result, $\log K_{eq} = 0$. Using the same reasoning as for the upper stability limit of water, we see that p_{H_2} could not exceed 1 atmosphere. With this value, $\log p_{H_2} = 0$ and

$$pe = -pH. \tag{5-32}$$

This boundary, as plotted in Fig. 5–9, provides a lower limit for all redox reactions and conditions that will occur in shallow ground water.

System Fe—O₂—H₂O

The system $Fe-O_2-H_2O$ provides an excellent example of the construction of a pe–pH diagram. The species that will be included are $O_{2(g)}$, $H_{2(g)}$, H_2O, Fe^{2+}, Fe^{3+}, $Fe(OH)_3$, and $Fe(OH)_2$. Other species could be used in the diagram, but iron hydroxides and free ions are most common in shallow aquifers. Amorphous iron hydroxide, $Fe(OH)_3$, is the most typical iron species present in near-surface aquifers containing measurable dissolved oxygen. When dissolved iron is present in ground water within the moderate (6–8) pH range, it is commonly assumed to be the ferrous (Fe^{2+}) ion.

The first boundary to be drawn, other than the stability boundaries for water that are common to all diagrams, is the $Fe(OH)_3/Fe(OH)_2$ boundary. The balanced reaction is written, and then a mass law expression is set up and solved for pe. We have

$$Fe(OH)_3 + e^- + H^+ \rightleftharpoons Fe(OH)_2 + H_2O; \tag{5-33}$$

$$K_{eq} = \frac{1}{a_{e^-}\, a_{H^+}};$$

$$\log K_{eq} = -\log a_{e^-} - \log a_{H^+} = pe + pH;$$

$$pe = \log K_{eq} - pH. \tag{5-34}$$

The equilibrium constant must be obtained from the standard free energy of reaction. A temperature of 25°C is commonly used for ease of calculation. For temperatures near 25°C, not much error in the position of the boundary is introduced by using 25°. The calculation is as follows:

$$\Delta G_R^0 = \Delta G_{f-Fe(OH)_2}^0 + \Delta G_{f-H_2O}^0 - \Delta G_{f-Fe(OH)_3}^0$$

$$= -486.5 - 237.1 - (-696.5)$$

$$= -27.1;$$

$$\log K_{eq} = \frac{+27.1}{5.708} = 4.75,$$

$$pe = 4.75 - pH. \tag{5-35}$$

Equation (5–35) can be plotted as a line, as shown in Fig. 5–10 (boundary 1). The dashed portion of the line will be removed later, after all boundaries have been drawn. The boundary indicates that $Fe(OH)_3$, the hydroxide containing ferric iron, is stable in waters that plot above the line and $Fe(OH)_2$ would be stable in waters that plot below the line.

Next, we will construct a boundary between Fe^{3+} and $Fe(OH)_3$. The same procedure is followed. We have

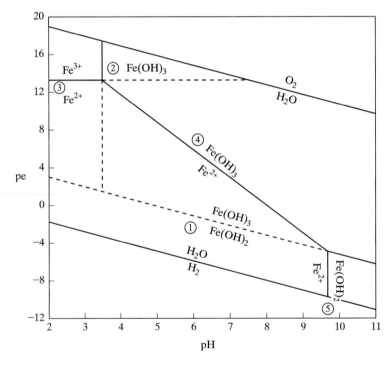

FIGURE 5–10   Construction of boundaries for selected species in the system $Fe-O_2-H_2O$ at 25°C. $a_{Fe^{2+}}$ and $a_{Fe^{3+}}$ values assumed to equal 10^{-6} for boundary construction.

$$Fe(OH)_3 + 3H^+ \rightleftharpoons Fe^{3+} + 3H_2O \tag{5–36}$$

and

$$K_{eq} = \frac{a_{Fe^{3+}}}{a^3_{H^+}},$$

So

$$\log K_{eq} = \log a_{Fe^{3+}} + 3pH.$$

In this case, both species are present in the Fe^{3+} oxidation state. Because the only variable is pH, the boundary will be vertical (boundary 2). Thus,

$$pH = 1/3 \log K_{eq} - 1/3 \log a_{Fe^{3+}}$$

and

$$\begin{aligned}
\Delta G^0_R &= \Delta G^0_{f-Fe^{3+}} + 3\Delta G^0_{f-H_2O} - \Delta G^0_{f-Fe(OH)_3} \\
&= -16.7 + 3(-237.1) - (-696.5) \\
&= -31.5.
\end{aligned}$$

So

$$\log K_{eq} = +31.5/5.708 = 5.5$$

and

$$pH = 1/3(5.5) - 1/3\big(\log a_{Fe^{3+}}\big). \tag{5–37}$$

In order to plot a line on our diagram, we must assign a value to the activity of ferric iron to represent a concentration so low that it is assumed to be insoluble. Although this is somewhat arbitrary, an activity of 10^{-6} is commonly chosen. Using this value, we get

$$pH = 1.8 - 1/3\,(-6) = 3.8. \tag{5–38}$$

A point that plots to the right of this boundary in Fig. 5–10 means that Fe^{3+} is insoluble and ferric hydroxide is the stable form of Fe^{3+}. A point that plots to the left implies that Fe^{3+} is soluble. Because of the activity assumption, the boundary technically means that the activity of Fe^{3+} is below 10^{-6} to the right and above 10^{-6} to the left. The activity value of 10^{-6} is low enough that ferric iron reasonably can be considered to be insoluble at activities lower than this.

As yet, we have not defined the region of the diagram in which Fe^{2+} will be stable. To do this, we must construct boundaries between Fe^{2+} and aqueous Fe^{3+}, and between Fe^{2+} and the solid hydroxides $Fe(OH)_3$ and $Fe(OH)_2$. The first of these, boundary 3 in Fig. 5–10, is defined by the equation

$$Fe^{3+} + e^- \rightleftharpoons Fe^{2+}. \tag{5–39}$$

From Eq. (5–12),

$$pe = 12.8 - \log\left(\frac{a_{Fe^{2+}}}{a_{Fe^{3+}}}\right).$$

Because pH is not in the equation, the relation will plot as a horizontal line on the diagram. A convenient assumption that can be made to simplify the expression is that $\left(\dfrac{a_{Fe^{2+}}}{a_{Fe^{3+}}}\right) = 1$. The horizontal line is then fixed at a pe of 12.8.

An important boundary on the diagram separates Fe^{2+} and $Fe(OH)_3$ (boundary 4, Fig. 5–10). The reaction is

$$Fe(OH)_3 + e^- + 3H^+ \rightleftharpoons Fe^{2+} + 3H_2O, \tag{5–40}$$

with

$$K_{eq} = \frac{a_{Fe^{2+}}}{a_{e^-} a_{H^+}^3}.$$

Thus,

$$\log K_{eq} = \log a_{Fe^{2+}} + pe + 3pH,$$

or

$$pe = \log K_{eq} - \log a_{Fe^{2+}} - 3pH.$$

Also,

$$\begin{aligned}
\Delta G_R^0 &= \Delta G^0{}_{f-Fe^{2+}} + 3\Delta G^0{}_{f-H_2O} - \Delta G^0{}_{f-Fe(OH)_3} \\
&= -90.0 + 3(-237.1) - (-696.5) \\
&= -104.8
\end{aligned}$$

and

$$\log K_{eq} = \frac{+104.8}{5.708} = 18.4,$$

or

$$pe = 18.4 - (-6) - 3pH = 24.4 - 3pH. \tag{5–41}$$

The activity value of Fe^{2+} chosen to represent aqueous solubility is the same as the value for $a_{Fe^{3+}}$ selected earlier, namely, 10^{-6}. With this assumption, the $Fe(OH)_3/Fe^{2+}$ boundary must intersect the junction of the Fe^{2+}/Fe^{3+} and the $Fe^{3+}/Fe(OH)_3$ boundaries. It is sometimes difficult to decide which side of a boundary to place a certain species. For this boundary, for example, $Fe(OH)_3$ contains ferric iron—the more oxidized form of iron—and therefore belongs above the boundary in the portion of the diagram where pe is higher.

The last boundary, between $Fe(OH)_2$ and Fe^{2+} (boundary 5, Fig. 5–10), is constructed as follows:

$$Fe(OH)_2 + 2H^+ \rightleftharpoons Fe^{2+} + 2H_2O; \tag{5–42}$$

$$K_{eq} = \frac{a_{Fe^{2+}}}{a_{H^+}^2};$$

$$\log K_{eq} = \log a_{Fe^{2+}} + 2pH;$$

$$pH = 1/2 \log K_{eq} - 1/2 \log a_{Fe^{2+}};$$

$$\Delta G_R^0 = \Delta G^0_{f-Fe^{2+}} + 2\Delta G^0_{f-H_2O} - \Delta^0 G_{f-Fe(OH)_2}$$

$$= -90.0 + 2(-237.1) - (-486.5)$$

$$= -77.7;$$

$$\log K_{eq} = +77.7/5.708 = 13.61;$$

$$pH = 1/2(13.61) - 1/2(-6) = 6.81 + 3 = 9.81. \tag{5–43}$$

Equation (5–43) plots as a vertical line at pH = 9.81.

When the unneeded boundary segments are removed, the final diagram is shown in Fig. 5–11. This diagram illustrates that iron is insoluble as a hydroxide solid under moderate to high pe levels over a fairly high range of pH values. If the pe drops below the $Fe(OH)_3/Fe^{2+}$ boundary, iron becomes soluble and is commonly detected in water analyses. The iron that occurs in solution dissolves from aquifer solids similar to $Fe(OH)_3$. If a water sample from a domestic well or monitor well contains dissolved iron above about a tenth of a milligram per liter, it can be concluded that the redox potential falls

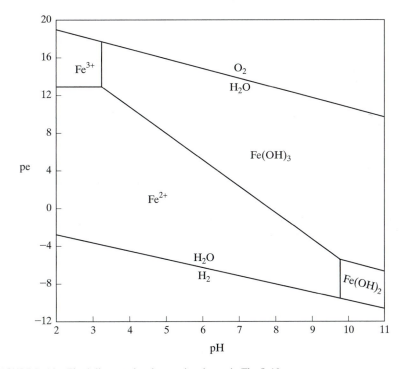

FIGURE 5–11 Final diagram for the species shown in Fig. 5–10.

below the $Fe(OH)_3/Fe^{2+}$ boundary. This is important geochemical information about the aquifer. It may mean that the aquifer is not in contact with oxidizing recharge water from land surface or that the presence of some strong electron donor—an organic contaminant, for example—has lowered the redox potential in the aquifer to the point where iron will dissolve into solution in the more soluble ferrous form from aquifer solids containing iron in the ferric form.

System $N-O_2-H_2O$

The widespread occurrence of aquifers contaminated by nitrate, particularly in agricultural areas, requires a close look at redox transformations of nitrogen in ground water. Nitrogen, an essential nutrient for crop growth, also occurs in human and animal waste and is the most abundant gas in the atmosphere. Nitrogen occurs in various oxidation states from nitrate (+V) to ammonia (−III). Transformations between species are almost exclusively facilitated by microorganisms. Thus, the equilibrium relationships implied by a pe–pH diagram must be understood to be controlled by microbial kinetic factors.

In the soil, nitrogen gas (N_2) is *fixed* by bacteria and, therefore, available for the formation of proteins through the activities of legumes and other plants. Proteins are incorporated into living organisms during growth. When the organism dies, the nitrogen is converted to ammonium through the process of *ammonification.* In an oxidizing environment, the ammonium will be oxidized to nitrate, a process known as *nitrification.* This occurs in the soil zone above the water table to decaying organic matter, as well as to excess ammonia applied to the soil as fertilizer. Once converted to the anion nitrate, infiltration to the water table is rapid. Nitrate will travel in ground water conservatively unless the redox potential drops along with the removal of dissolved oxygen. Nitrate then will function as an electron acceptor and be reduced to N_2 in the process of *denitrification.*

The stability fields of the major species are shown in Fig. 5–12. The stability of nitrate is defined by the upper limit of water stability and by the boundary between nitrate and nitrogen gas. The equation for denitrification is

$$NO_3^- + 6H^+ + 5e^- \longrightarrow 1/2\,N_2 + 3H_2O. \tag{5-44}$$

Note that this equation is not reversible: Nitrogen gas cannot be oxidized to nitrate, which is fortunate for us because, if the nitrogen gas in the atmosphere were to oxidize, it would consume the oxygen present there and lower the pH of the oceans to less than 2! The equation for the boundary is developed as follows:

$$K_{eq} = \frac{p^{1/2}_{N_2}}{a_{NO_3^-}\, a^6_{H^+}\, a^5_{e^-}};$$

$$\log K_{eq} = 1/2 \log p_{N_2} - \log a_{NO_3^-} + 6pH + 5pe,$$

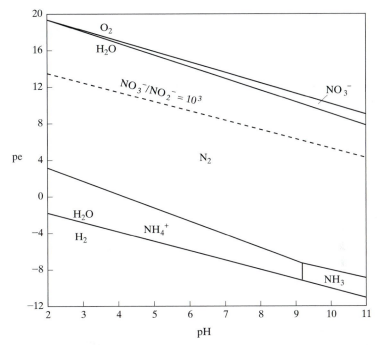

FIGURE 5–12   pe–pH diagram for selected species in the system $N-O_2-H_2O$ at 25°C. The NO_3^-/NO_2^- boundary is dashed because both species are unstable relative to N_2 in that part of the diagram.

or

$$\text{pe} = 1/5 \log K_{eq} - 1/10 \log p_{N_2} + 1/5 \log a_{NO_3^-} - 6/5 \text{ pH};$$

$$\Delta G_R^0 = 3\Delta G^0{}_{f-H_2O} - \Delta G^0{}_{f-NO_3^-}$$
$$= 3(-237.1) - (-110.8) = -600.5;$$

$$\log K_{eq} = \frac{+600.5}{5.708} = 105.2.$$

Values for p_{N_2} and $a_{NO_3^-}$ must be chosen to plot the last equation on the pe–pH diagram. The atmospheric value of 0.77 for p_{N_2} is reasonable, and an $a_{NO_3^-}$ value of 10^{-3}, which is near the drinking water standard for NO_3^--N of 10 mg/l, is selected. Thus,

$$\text{pe} = 1/5 \,(105.2) - 1/10 \,(-.11) + 1/5 \,(-3) - 6/5 \text{ pH},$$

or

$$\text{pe} = 20.45 - 6/5 \text{ pH}. \qquad (5\text{–}45)$$

This line lies very close to the stability boundary for water, implying that nitrate is stable only under very oxidizing conditions in which free oxygen is present in the aquifer.

We would not expect to find nitrate in waters of moderate-to-low redox potential. With reference to Fig. 5–11, we would also not expect to find dissolved iron and nitrate co-existing in the aquifer. If the redox potential were low enough so that Fe^{+2} was the stable form of iron, nitrate should be denitrified to N_2.

Ammonium and ammonia are the stable forms of nitrogen near the lower stability boundary of water. The relationship between these two species can be examined through the following equations:

$$NH_3 + H^+ \rightleftharpoons NH_4^+; \qquad (5\text{--}46)$$

$$K_{eq} = \frac{a_{NH_4^+}}{a_{NH_3} a_{H^+}};$$

$$\log K_{eq} = \log a_{NH_4^+} - \log a_{NH_3} + pH,$$

so

$$pH = \log K_{eq} - \log a_{NH_4^+} + \log a_{NH_3};$$

$$\Delta G_R^0 = \Delta G^0{}_{f-NH_4^+} - \Delta G^0{}_{f-NH_3}.$$

$$= -79.4 - (-26.50) = -52.9;$$

$$\log K_{eq} = +52.9/5.708 = 9.27.$$

With the assumption that both $a_{NH_4^+}$ and a_{NH_3} are equal to 10^{-3}, we obtain

$$pH = 9.27. \qquad (5\text{--}47)$$

This boundary shows that NH_4^+ is the stable form of reduced nitrogen in an aqueous environment below a pH of 9.27, a result that has important implications for the transport of the cationic ammonium in ground water. Attenuation by sorption and ion exchange will play an important role whenever ammonium is present in a ground water flow system.

The boundaries between N_2 and NH_4^+ and N_2 and NH_3 are established by following the same procedures. The reactions are

$$1/2\, N_2 + 4H^+ + 3e^- \rightleftharpoons NH_4^+ \qquad (5\text{--}48)$$

and

$$1/2\, N_2 + 3H^+ + 3e^- \rightleftharpoons NH_3. \qquad (5\text{--}49)$$

The same assumptions as made before are utilized to derive the boundaries.

Other boundaries could also be shown on the nitrogen diagram. For example, the denitrification of NO_3^- to N_2 entails several intermediate steps. Nitrite (NO_2^-) is the first intermediate in this reduction process. A boundary between NO_3^- and NO_2^- is shown as a dashed line in Fig. 5–12. The calculations are as follows:

$$NO_3^- + 2H^+ + 2e^- \longrightarrow NO_2^- + H_2O; \qquad (5\text{--}50)$$

$$\log K_{eq} = \log a_{NO_2^-} - \log a_{NO_3^-} + 2pH + 2pe;$$

$$pe = 1/2 \log K_{eq} - 1/2 \log a_{NO_2^-} + 1/2 \log a_{NO_3^-} - pH;$$

$$\Delta G_R^0 = \Delta G_{f-NO_2^-}^0 + \Delta G_{f-H_2O}^0 - \Delta G_{f-NO_3^-}^0$$

$$-37.2 + (-237.1) - (-110.8) = -163.5;$$

$$\log K_{eq} = \frac{+163.5}{5.708} = 28.64.$$

The assumption used for the activities of nitrate and nitrite is that $NO_3^-/NO_2^- = 10^3$. Other values of this ratio could be chosen, but the concentrations of nitrite are commonly much less than the concentrations of nitrate. We get

$$pe = 14.32 - 1/2\,(-6) + 1/2\,(-3) - pH$$

$$= 15.82 - pH. \tag{5–51}$$

The boundary is plotted as a dashed line because it plots well below the NO_3^-/N_2 boundary, indicating that NO_2^- is unstable with respect to N_2. We can conclude from this that NO_2^-, along with several other intermediates of denitrification, such as $NO_{(g)}$ and $N_2O_{(g)}$, are unstable species whose concentrations are controlled by kinetics rather than by equilibrium stability.

System S—O$_2$—H$_2$O

Our final example of pe–pH diagram construction involves sulfur, another one of the most important redox elements. In ground water, sulfur most commonly occurs with the $+VI$ oxidation number as sulfate, SO_4^{2-}. When sulfate concentrations are moderate to high, particularly approaching an activity of 10^{-2}, it is usually an indication that the aquifer contains gypsum or anhydrite. Sulfate can also occur by the oxidation of sulfide minerals, such as pyrite. Sulfate serves a very important function in waters contaminated with organic compounds. An electron acceptor in biodegradation reactions after free oxygen, nitrate, and ferric iron have been consumed, sulfate is reduced to sulfide ($-II$ oxidation number) in these reactions. A rotten egg smell can often be observed in a well that contains sulfide. Sulfide can be removed from the aqueous phase by degassing of hydrogen sulfide (H_2S) if it occurs near the water table. Another removal mechanism is the precipitation of metal sulfide solids, which have very low solubilities. For example, when ferrous iron is available, it will precipitate with sulfide at low temperatures as amorphous FeS (Morse et al., 1987). Over time, this solid can be converted to crystalline pyrite. Several trace metals also form sulfide minerals of low solubility.

The relationship between the oxidized and reduced species of sulfur is shown in Fig. 5–13. The boundary between H_2S and SO_4^{2-} is calculated as follows:

$$SO_4^{2-} + 8e^- + 10H^+ \rightleftharpoons H_2S + 4H_2O; \tag{5–52}$$

$$\log K_{eq} = \log\left(\frac{a_{H_2S}}{a_{SO_4^{2-}}}\right) + 8pe + 10pH.$$

Assuming that the activities of hydrogen sulfide and sulfate are equal at the boundary, we obtain the following equations:

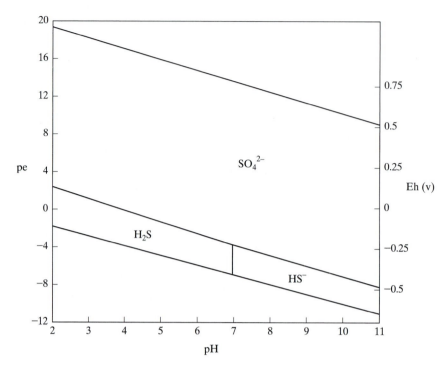

FIGURE 5–13   pe–pH diagram for several species in the system S—O_2—H_2O at 25°C.

$$pe = 1/8 \log K_{eq} - 5/4 \, pH;$$

$$\Delta G_R^0 = \Delta G^0{}_{f-H_2S} + 4\Delta G^0{}_{f-H_2O} - \Delta G^0{}_{f-SO_4^{2-}}$$

$$= -27.83 + 4(-237.1) - (-744.0) = -232.2;$$

$$\log K_{eq} = \frac{+232.2}{5.708} = 40.69;$$

$$pe = 5.09 - 1.25pH. \tag{5–53}$$

The calculations for the boundary between bisulfide (HS^-) and sulfate are similar. The boundary between H_2S and HS^- is a vertical line with a pH of 6.99, derived in the usual way from the equation

$$H_2S \rightleftharpoons H^+ + HS^-. \tag{5–54}$$

When total sulfur activities exceed 10^{-3}, a wedge-shaped field separating H_2S and SO_4^{2-} is necessary for the presence of solid elemental sulfur (S^0). This field, which is widest at low pH values, narrows as the pH increases. The field becomes progressively larger as the total sulfur concentrations increase. Drever (1997) describes the inclusion of this field in the diagram. The field is very narrow in the mid-to-high pH range when total sulfur concentrations in the system are low.

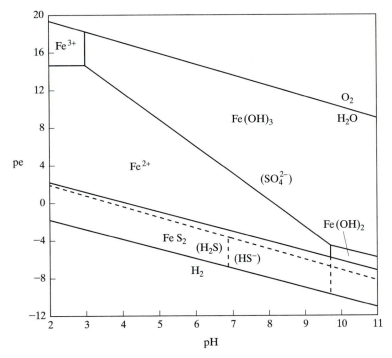

FIGURE 5–14 Combination of Figs. 5–13 and 5–11. Dashed boundaries are SO_4^{2-}/H_2S and SO_4^{2-}/HS^- from Fig. 5–13 and $Fe(OH)_3/Fe(OH)_2$. New boundaries for Fe^{2+}/FeS_2 and $Fe(OH)_2/FeS_2$ are added.

In reducing systems containing both iron and sulfur, the precipitation of sulfide minerals is likely. This can be demonstrated by adding the boundaries shown in Fig. 5–13 to Fig. 5–11 and constructing new boundaries between Fe^{2+} and pyrite (FeS_2) and between $Fe(OH)_2$ and pyrite (Fig. 5–14). The Fe^{2+}/FeS_2 boundary is developed using the following reaction, with the assumptions that $a_{SO_4^{2-}} = 10^{-3}$ and $a_{Fe^{2+}} = 10^{-6}$:

$$2SO_4^{2-} + Fe^{2+} + 16H^+ + 14e^- \rightleftharpoons FeS_2 + 8H_2O; \tag{5–55}$$

$$\log K_{eq} = -2\log a_{SO_4^{2-}} - \log a_{Fe^{2+}} + 16pH + 14pe;$$

$$\Delta G_R^0 = \Delta G^0_{f-FeS_2} + 8\Delta G^0_{f-H_2O} - 2\Delta G^0_{f-SO_4^{2-}} - \Delta G^0_{f-Fe^{2+}}$$

$$= -160.1 + 8(-237.1) - 2(-744.0) - (-90.0)$$

$$= -478.9;$$

$$\log K_{eq} = \frac{+478.9}{5.708} = 83.9;$$

$$pe = 1/14\,(83.9) + 2/14\,(-3) + 1/14\,(-6) - 16/14\,pH$$

$$= 5.13 - 1.14pH. \tag{5–56}$$

As shown in Fig. 5–14, this boundary plots slightly above the sulfate–sulfide boundary, reflecting the oversaturation and precipitation of pyrite even when only small amounts of sulfide are present. The figure explains the presence of pyrite and other metal sulfide minerals in coals and other reducing environments such as landfill leachate plumes. The low solubility of sulfide minerals commonly keeps metal concentrations low if sulfide is available.

NATURAL REDOX CONDITIONS IN AQUIFERS

Redox conditions in aquifers vary over a wide range. The factors that determine these conditions at a particular point include the distance and travel time from the recharge area; but more importantly, redox conditions are controlled by the relative abundance of electron donors and acceptors. When these compounds are present in the solid or aqueous phases of an aquifer, redox reactions will progress until either the electron donors or electron acceptors have been consumed. Therefore, the factors that control the abundance of the electron donors and acceptors, along with the kinetics of the re-actions, determine the Eh that is measured or calculated. One final ingredient is neces-sary to establish aquifer Eh values—microorganisms. The transfer of electrons from donor to acceptor is much more rapid when the reaction is utilized by bacteria for growth energy. Over the past few years, it has been shown conclusively that bacteria which can facilitate most redox reactions are widely distributed in the subsurface. In this section, we will assume that redox conditions are relatively constant at the measurement point. Later, we will consider the effects of kinetics in redox reactions in the context of biodegradation of contaminants.

Given the problems with direct measurement of the redox potential, it is fortu-nate that the chemical composition of ground water can be used to qualitatively infer redox conditions or quantitatively calculate Eh or pe via the Nernst equation. The ap-proach used is to measure the concentrations of electron acceptors and reaction prod-ucts. Because of the sequence of electron acceptors that are consumed by microorgan-isms, this is a very useful strategy.

The most common electron donor in nature is organic carbon. Whenever ground water or aquifer solids contain labile (reactive) organic carbon, reactions with electron acceptors will begin immediately. In humid, temperate regions with shallow water tables, water that reaches the water table during recharge events is commonly at or near satu-ration with dissolved oxygen. In agricultural or suburban areas, nitrate derived from fertilization is also abundant near the water table. Other electron acceptors originate in the aquifer solids. Sulfate may be present because of gypsum or anhydrite dissolution or from the oxidation of pyrite in aquifer solids. Ferric iron and manganese are present in oxide or silicate minerals that may be contained in the solid phases of the aquifer. Car-bon dioxide, which also may serve as an electron acceptor, may be derived from car-bonate equilibria or as a reaction product of the oxidation of organic carbon. The relative abundance of organic carbon and these electron acceptors governs the progression of redox reactions in the system.

The sequence of changes in concentration, either with time or distance, in a flow system is shown schematically in Fig. 5–15 for a hypothetical aquifer that contains organic

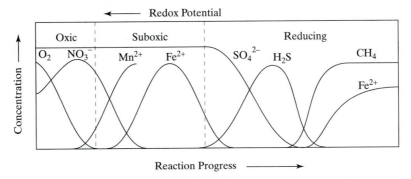

FIGURE 5–15   Conceptual changes in concentrations of redox parameters with time or distance along a flow path when organic carbon and appropriate electron acceptors are present (reprinted from: *Geochemistry, Ground Water, and Pollution*, revised edition, Appelo, C. A. J. and D. Postma, 1996, A. A. Balkema, Rotterdam).

carbon and the major electron acceptors. As long as free oxygen is present, it will be the favored electron acceptor by aerobic bacteria. Since the saturation levels of oxygen are fairly low, the oxygen will be rapidly depleted and, as the level falls to and then below 1 mg/l, nitrate can be used as an electron acceptor if it is present. Nitrate is consumed by microorganisms that are *facultative*, which is the ability to function both aerobically and anaerobically. With reference to the discussion associated with Fig. 5–12, nitrate is reduced through the unstable intermediate product, nitrite, to N_2 in the process known as denitrification. When all the oxygen is consumed, any further reactions are considered to be anaerobic. As the nitrate nears depletion, the system enters the suboxic range, which contains measurable iron, because ferric iron is the next major electron acceptor following denitrification. Figure 5–11 shows that the redox boundary between $Fe(OH)_3$ and Fe^{2+} occurs at a pe value of about 2 for a pH of 7. The reduction of ferric iron oxide coatings on aquifer grains yields the more soluble ferrous iron, which is an excellent redox indicator. The irritating iron stains and deposits in household plumbing fixtures indicate that the ground water intercepted by the well screen is in the suboxic range.

With further reaction, the available supplies of reactive iron are consumed, and the microbial community begins to utilize sulfate as the dominant electron acceptor. It is generally assumed that the anaerobic bacteria for sulfate reduction are present in the aquifer, perhaps at very low concentrations in more reducing microenvironments, until the pe drops low enough for populations to expand to dominance. Concurrent with the decline in sulfate concentrations associated with its reduction to hydrogen sulfide is a decrease in iron concentrations caused by the precipitation of sulfides. These conditions characterize reducing ground waters. Hydrogen sulfide rises in concentration and then falls because of sulfide precipitation.

If sufficient organic carbon remains in the system after sulfate has been consumed, the lowest redox level can be achieved, which is characterized by the process of *methanogenesis*. In this fermentative process, carbon dioxide becomes the electron acceptor and is reduced to methane. The process is marked by increasing methane concentrations, and iron as well, if hydrogen sulfide has been depleted by sulfide precipitation.

Identification of TEAPs

The sequence of redox reactions described in the previous section suggests that it would be possible to characterize the *terminal electron accepting process* (TEAP) in an aquifer by measuring the concentrations of appropriate electron acceptors and metabolic by-products (Chappelle et al., 1995). For example, if sulfate reduction is the TEAP in an aquifer, one would expect to observe sulfate concentrations decreasing along a flow path along with increases in hydrogen sulfide. These changes would provide a fair degree of confidence in the diagnosis of sulfate reduction as the TEAP. However, suppose that the aquifer contained a source of sulfate, so that sulfate concentrations could be replenished as reduction occurred. Or, suppose that dissolved sulfide concentrations failed to increase because of sulfide mineral precipitation. Either of these circumstances would reduce the level of confidence in the TEAP diagnosis.

An analytical parameter that is useful in the characterization of redox conditions is H_2, an unstable intermediate product that is both produced and consumed in anaerobic metabolism of organic carbon (Chappelle et al., 1995). H_2 is an effective indicator of the electron-accepting process because each TEAP has a characteristic range of H_2 concentration (Fig. 5–16). H_2 measurements can thus be used to supplement concentrations of electron acceptor and metabolic by-product compounds to assess the TEAP at a particular point in an aquifer. The drawback with H_2 measurements is that they must be made in the field with a specially equipped gas chromatograph.

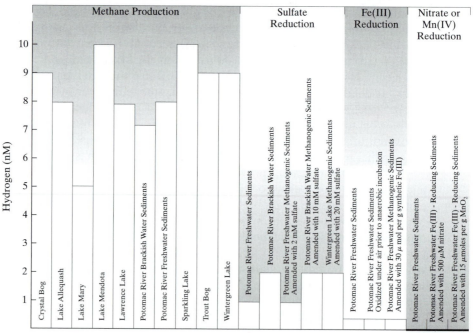

FIGURE 5–16 H_2 concentrations in redox environments with different TEAPs. (reprinted from Geochim. Cosmochim. Acta, v. 52, Lovely, D. R. and Goodwin, S., Hydrogen concentrations as an indicator of the predominant terminal electron—accepting reactions in aquatic sediments, pp. 2993–3003, Copyright, 1988, with permission from Elsevier Science).

Redox Buffering

The sequential utilization of electron acceptors maintains the redox potential at specific levels, a phenomenon known as *redox buffering* (Drever, 1997). As shown in Fig. 5–17, as long as an electron acceptor is available, the redox potential is fixed at a certain value. When the electron acceptor has been consumed, the redox potential drops immediately to the next electron acceptor available in the system. Eh is said to be buffered at these specific levels while the electron acceptor remains. It is important to remember that the electron acceptor species may be either dissolved or contained within the aquifer solids.

Examples of Aquifer Redox Conditions and Processes

A wide variety of redox conditions occurs in aquifers, depending upon the relationship between electron donors and acceptors in the aqueous and solid phases. Gradual, systematic changes in redox potential and dissolved species can be traced through long regional flow paths in deep, confined aquifers. Recharge water to the aquifers is commonly oxic and near saturation in dissolved oxygen. For example, in the Madison aquifer in Montana, Wyoming, and South Dakota (Plummer et al., 1990), oxic recharge water enters the aquifer through surface outcrops, and the redox potential gradually declines to the level of sulfate reduction along the flow path.

The Fox Hills–Basal Hell Creek aquifer in southwestern North Dakota (Thorstenson et al., 1979) spans a wider range of redox levels from the recharge to the discharge area (Fig. 5–18). Redox conditions decline in this aquifer to the level of methanogenesis in the transition zone and discharge zones. Selected redox parameters are given in Table 5–2. Lignitic carbon in the aquifer serves as the electron donor. Iron and sulfate

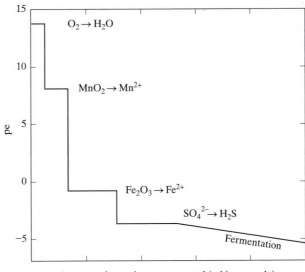

FIGURE 5–17 Redox buffering by prominent electron acceptors. The redox potential remains at each plateau until the electron acceptor is consumed (from *The Geochemistry of Natural Waters* 3/e by Drever © 1997).

Amount of organic matter reacted (arbitrary scale)

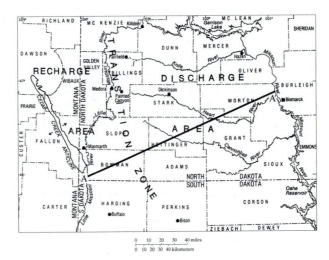

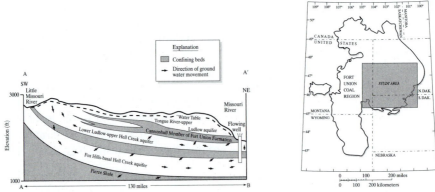

FIGURE 5–18 Location and schematic flow system along cross section line A–A' in the Fox Hills–Basal Hell Creek aquifer system in North Dakota (From Thorstenson, D. C., Fisher, D. W. and Croft, M. G., *Water Resources Research*, v. 15, pp. 1479–1498, 1979, Copyright by the American Geophysical Union).

TABLE 5–2 Selected redox parameters from the Fox Hills–Basal Hell Creek Aquifer (From Thorstenson, D. C., Fisher, D. W. and Croft, M. G., *Water Resources Research*, v. 15, pp. 1479–1498, 1979, Copyright by the American Geophysical Union).

	Recharge Area	Transition Area	Discharge Area
Depth, m	239	552	274
Temp, °C	13.5	20.5	13.2
pH	8.6	8.80	8.60
Titration alkalinity, meq/l	11.8	15.7	19.1
SO_4^{2-}, mmol/l	6.14	0.30	0.031
Fe^{2+}, mmol/l	<0.02	<0.02	<0.023
CH_4, mmol/l	0.044	NA	2.75
CO_2 (measured), mmol/l	0.10	NA	0.16
HS^-	ND	ND	<0.001

ND: Non detect

NA: value not determined.

are depleted in the transition and discharge zones due to the presumed precipitation of pyrite.

Shallow, near-surface aquifers may have short flow paths, but may also show significant redox changes. Starr (1988) and Starr and Gilham (1989) demonstrated that a shallow aquifer could be depleted in dissolved oxygen by the downward transport of soil organic carbon when the water table is within a few feet of the ground surface. Barcelona et al.(1989) pointed out that the vertical redox gradients in shallow aquifers are often much greater than the horizontal gradients. They illustrated this with several piezometer nests in an uncontaminated aquifer in Illinois (Fig. 5–19). In a vertical distance of 30 m, oxygen levels declined from saturation at the water table to nearly nondetect levels, along with simultaneous increases in dissolved iron.

Kehew et al. (1996), in a study of an unconfined glacial drift aquifer in Michigan (Fig. 5–20), recognized shallow flow cells in contact with lakes and wetlands overlying an intermediate flow system isolated from surface water interactions. They plotted well depth vs. iron concentration to determine whether iron concentrations would increase

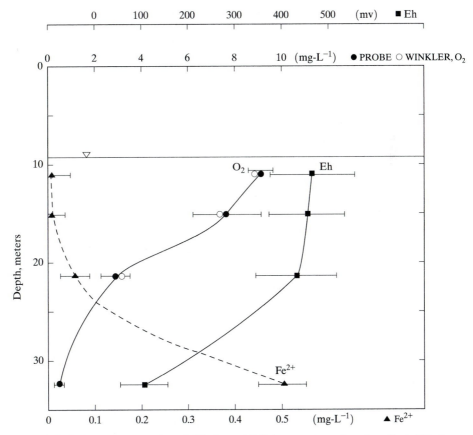

FIGURE 5–19   Vertical gradients in Eh, O_2, and Fe^{2+} in an uncontaminated aquifer in Illinois (from Barcelona et al., *Water Resources Research*, v. 25, pp. 991–1003, 1989. Copyright by the American Geophysical Union).

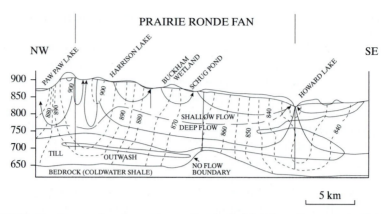

FIGURE 5–20   Cross section of ground water flow in glacial-drift aquifer system in Michigan. Shallow flow systems consist of cells that interact with lakes and wetlands (from Kehew et al., 1996. Reprinted with the permission of the National Ground Water Association).

in deeper wells that presumably sampled older ground water with longer flow paths. To the contrary, some of the highest iron concentrations occurred in very shallow wells (Fig. 5–21). These wells were located downgradient of wetlands and ponds in the shallow flow systems where the surface water bodies contributed recharge to the aquifer. Flow through the reducing muck at the base of the wetlands lowered the redox potential of the water to at least the level of iron reduction.

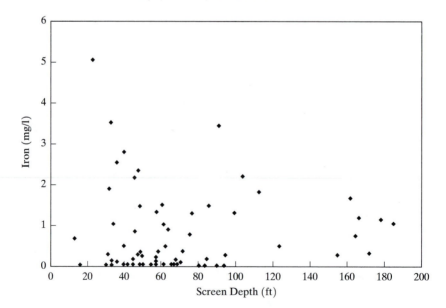

FIGURE 5–21   Plot of iron concentration vs. well depth in area of Fig. 5–20. Iron did not increase with well depth; some of the highest iron concentrations occur in shallow flow systems downgradient from wetlands and lakes because of ground water recharge through reducing sediments (from Kehew et al., 1996. Reprinted with the permission of the National Ground Water Association).

PROBLEMS

1. Consider the following half redox reaction involving NO_3^- at a concentration of 10^{-5} M and ammonium at a concentration of 10^{-3} M in ground water with a pH of 8. Assume that a = m.

$$1/8\ NO_3^- + 5/4\ H^+ + e^- = 1/8\ NH_4^+ + 3/8\ H_2O$$

What are the pe and Eh of the system at 25°C?

2. The couple O_2/H_2O is found to be the dominant couple producing a pe of 12.0. By assuming redox equilibrium, determine how $(Fe)_T = 10^{-5}$ M would be distributed as Fe^{2+} and Fe^{3+}. Assume that a = m.

3. In a field redox measurement in which the temperature of the ground water sample is 10°C, the observed potential of the sample is $E_{obs} = 52$ mv, and the observed potential of the Zobell solution is +245 mv. The reference electrode system is silver–silver chloride. What is the E_h of the sample?

4. Construct a pe–pH diagram for the system $Mn—O_2—H_2O$. Include the following species in the diagram: Mn^{2+}, MnO_2, Mn_2O_3, Mn_3O_4, and $Mn(OH)_2$.

5. A ground water sample from a shallow unconfined aquifer overlain by a coarse, sandy vadose zone has a dissolved iron concentration of 1.5 mg/l. The sample is taken from a well installed in a heavily fertilized agricultural field. Where would you expect this water to plot on Fig. 5–11, assuming a near-neutral pH value? Would you expect this water to exceed the drinking water standard for $NO_3—N$? Why or why not?

6. Give several possible explanations for elevated iron concentrations in shallow unconfined aquifers with high recharge rates through sandy vadose zones.

C H A P T E R   6

Structures, Properties, and Occurrence of Organic Compounds in Ground Water

Probably the most important advance in ground water chemistry in the past several decades was the discovery that organic compounds derived primarily from inadvertent releases of fuels and industrial compounds were pervasive in many aquifers, particularly in urbanized areas. This discovery, which was really a gradual process throughout the 1960s, '70s, and '80s, was accompanied by sensationalized media reports and raised great concern among the general population about the safety of drinking water. The steadily increasing concern about the presence of these compounds in ground water spawned a set of stringent regulations at all levels of government, and a massive buildup in commercial analytical laboratories and hydrogeological consulting firms formed to investigate and remediate this contamination. Increasing sophistication in analytical instrumentation, including a progressive lowering of analytical detection limits, led to a growing list of organic compounds required for testing by site owners. In the 1990s, after several decades of delineating and tracking organic contaminant plumes, the environmental industry began to concentrate on remedial actions. Changes in regulations, including the adoption of risk-based corrective action (RBCA), raised the contaminant concentrations necessary to meet cleanup standards at many sites.

The importance of organic compounds as contaminants in ground water necessitates that hydrogeologists become familiar with these chemicals and their properties. In this book, the properties and structures of organic compounds most likely to be found in ground water are stressed. The physical properties of organic compounds—their vapor pressure, Henry's constant, solubility, and octanol–water partitioning coefficient—are critical in predicting the migration of compounds in the subsurface. These properties control the partitioning of a compound between solid, aqueous, and gaseous phases that may exist in the subsurface environment. The selection of a remedial technology is also governed by the physical properties of the compound. Air stripping, for example, is an excellent choice for the more volatile compounds, but is ineffective for compounds of low volatility.

PROPERTIES THAT INFLUENCE PHASE PARTITIONING OF ORGANIC COMPOUNDS

Vapor Pressure

In a closed system, molecules of a liquid will evaporate into a contiguous gas phase until equilibrium between the liquid and gas is achieved. The equilibrium pressure in the gas phase is the *vapor pressure*, P^0. The vapor pressure, which is not a constant, increases with a rise in temperature. For comparison purposes, a temperature of 25°C is generally used. Values are reported in atmospheres or mm of mercury (1 atm = 760 mm of mercury). Vapor pressures at 25°C for most organic compounds range from about 1 atm to 10^{-12} atm. If the vapor pressure of a compound at a certain temperature is equal to 1 atm, vapor bubbles will form spontaneously in the liquid and rise to the liquid surface. This particular temperature is defined as the *boiling point* for that compound, because the compound exists in the gas phase above the boiling point and in the liquid phase below the boiling point.

The vapor pressure is an indication of the *volatility* of a compound—that is, the tendency of the compound to evaporate from a pure liquid of the compound (Table 6–1). Keep in mind, however, that this refers to evaporation from the pure liquid. In the case of NAPLs (dense nonaqueous phase liquids), which include many solvent compounds frequently encountered in ground water, separate-phase solvent may actually be present in the subsurface above or below the water table. Above the water table, the solvent compound could evaporate directly into the vadose atmosphere. It is also common, however, for the contaminant to be dissolved in an aqueous phase in contact with the vadose atmosphere either at the water table or in soil water above the water table. When the compound is dissolved in water, evaporation is described by Henry's constant, which is described subsequently.

Solubility

The solubility of organic compounds in water is defined exactly as is the solubility of inorganic compounds: the mass or number of moles of the compound dissolved in a unit volume or mass of water. Like vapor pressure, the solubilities of organic compounds vary over many orders of magnitude, from *miscible*, or completely soluble compounds, to compounds that can barely be measured in water when they dissolve to equilibrium. Many organic compounds of importance in hydrogeology are *immiscible* with water; that is, they separate to form pure or nearly pure organic phases when mixed with water. When an organic compound of this type is mixed with water, however, there is always

TABLE 6–1   Relative volatilities of organic compounds.

Volatility	Vapor Pressure (atm)	Compounds
Volatile	$>10^{-4}$	Light hydrocarbons; halogenated hydrocarbons
Semivolatile	10^{-4}–10^{-11}	Larger hydrocarbons (C12–C28; aromatics (3–4 rings); most compounds with O, N, S, P functional groups.
Nonvolatile	$<10^{-11}$	Other organics

some dissolution of the compound in the aqueous phase and some dissolution of water in the organic phase.

The factors that govern the solubility of many organics depend on the structure and nature of the bonds within the compound. Organics tend to be less polar than many natural inorganic compounds. As described in Chapter 1, this means that electronegativities of the molecules making up the compound are fairly close in magnitude. In general, the less polar a compound, the lower is its aqueous solubility. The reason for this has to do with the phenomenon of hydrogen bonding in water. As discussed earlier in the text, these bonds give water a weak structure, and nonpolar molecules are difficult to force into this framework. Many organic compounds do have polar components, particularly those containing oxygen and nitrogen, and these compounds have significantly higher solubility. Nonpolar compounds of low water solubility are known as *hydrophobic*, and polar compounds that readily dissolve in water are known as *hydrophilic*.

Another factor that influences solubility is the molecular weight of compounds with similar structures. Table 6–2 shows that three similar aromatic hydrocarbons have solubilities that are inversely proportional to their molecular weight.

Density

The primary importance of the density of organic compounds in hydrogeology relates to the behavior of free-phase compounds in the subsurface. Compounds with a specific gravity less than 1.0 ("floaters") tend to pool on the capillary fringe if sufficient quantities are present. Compounds that are denser than water ("sinkers") may penetrate the water table as a free-phase liquid and descend within the aquifer until a low permeability barrier is reached. Both floaters and sinkers dissolve in water encountered above and below the water table, depending upon their solubilities. In the aqueous phase, their densities are less important, and they move with ground water flow. Texts such as Fetter (1993) describe the physical aspects of hydrophobic compounds in the subsurface in greater detail.

Henry's Constant

Although the vapor pressure of a compound is an indication of its volatility, the term refers only to evaporation from the pure compound. In hydrogeology, most organic compounds are dissolved in an aqueous phase in the subsurface. The tendency of a compound to be present in either the aqueous phase or the gas phase is dependent upon the value of a *partitioning coefficient*, which is simply the ratio of the abundance of the compound in air to the abundance of the compound in the aqueous phase at equilibrium. The

TABLE 6–2   Solubility of three compounds of similar structure.

Compound	Molecular Weight	Solubility (g/m^3)
Benzene	78.0	1,780
Toluene	92.0	515
O-xylene	106.0	175

gas–water partitioning coefficient used is known as *Henry's constant*, and we have already used it in our discussion of the carbonate system for CO_2 gas, where it was called K_{CO_2}.

Henry's constant is defined in general for any compound as

$$K_H = \frac{p_i}{C_w} \ (\text{atm l mol}^{-1}), \qquad (6\text{–}1)$$

where p_i is the partial pressure of the compound, in atmospheres, and C_w is the molarity of the compound in aqueous solution. The constant is also commonly reported in dimensionless form when the equilibrium gas concentration is measured in moles per liter of air:

$$K_H = \frac{C_A}{C_W} \ (\text{mol l}_A^{-1} \ \text{mol}^{-1} \ \text{l}_W). \qquad (6\text{–}2)$$

Although some values of Henry's constant are experimentally determined, estimation using vapor pressure and solubility is a common practice (Hounslow, 1995; Schwarzenbach et al., 1993). The two factors involved in Henry's constant—the air and water concentrations—indicate that the vapor pressure is not the only factor involved in air–water partitioning. Compounds with both a high vapor pressure and high solubility will have only moderate Henry's constants. The compounds that are most likely to partition from aqueous solutions into gas phases (those with the highest Henry's constants) have high vapor pressures and low solubilities. Figure 6–1 shows the relationships among solubility, vapor pressure, and Henry's constant.

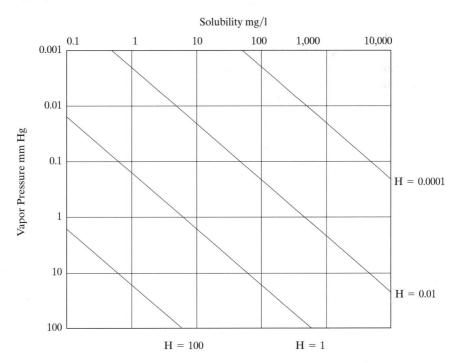

FIGURE 6–1 Graphical relationship between solubility, vapor pressure, and Henry's constant (sloping lines). Henry's constant is highest when vapor pressure is high and solubility is low (from Hounslow, 1995. Reprinted by permission of CRC Press).

Octanol–Water Partitioning Coefficient

Other partitioning coefficients are necessary to describe the distribution of organic compounds among solid, aqueous, and gaseous phases in the subsurface. A commonly used coefficient is the *octanol–water partitioning coefficient*, K_{ow}. This coefficient, which describes the partitioning of an organic compound between two immiscible liquids—octanol, an organic solvent, and water—is defined as

$$K_{ow} = \frac{C_s}{C_w} \; (\mathrm{mol}\,1_s^{-1}\,\mathrm{mol}^{-1}\,1_w), \tag{6–3}$$

in which C_s is the concentration of the compound in octanol and C_w is the concentration in water. The coefficient, which is dimensionless, is often reported as the base 10 logarithmic value. Nonpolar organic compounds partition preferentially into octanol and therefore have high K_{ow} values. Polar organics prefer water, yielding low K_{ow} values.

The octanol–water partitioning coefficient is useful in estimating other properties. Water solubility is one parameter that is related to K_{ow}. Empirical relationships between K_{ow} and solubility have been developed for a wide variety of compounds. Chiou et al. (1982) derived the general relationship

$$\log K_{ow} = 7.30 - 0.747 \log S_b \tag{6–4}$$

for several different types of compounds. Similar equations were developed for individual classes of organic compounds.

The K_{ow} value is also useful in predictions of adsorption of dissolved organics. For nonpolar organics, adsorption differs from the processes described in Chapter 4. Rather than undergoing an electrical interaction with the surfaces of charged particles, these hydrophobic compounds are most strongly adsorbed to neutral solid organic particles in the subsurface. Since what we are really talking about here is partitioning between solid organic carbon and water, a partitioning coefficient is useful. K_{oc}, which describes the partitioning between organic carbon and water, is defined as

$$K_{oc} = \frac{C_{ad}(\text{g solute adsorbed/g soil organic carbon})}{C_w(\text{g solute/m}^3 \text{ solution})}. \tag{6–5}$$

K_{oc} can be estimated using K_{ow} by means of equations in the form of Eq. (6–4). Two examples are

$$\log K_{oc} = -0.21 + \log K_{ow} \;\text{(Karickhoff et al., 1979)} \tag{6–6}$$

and

$$\log K_{oc} = 0.49 + 0.72 \log K_{ow} \;\text{(Schwarzenbach and Westall, 1981).} \tag{6–7}$$

K_{oc} values can be converted to K_d values similar to those discussed in Chapter 4 for inorganic compounds by taking into account the amount of organic carbon in an aquifer. In most aquifers, this is quite a small amount. If some reasonable value can be chosen for the weight fraction of organic carbon, f_{oc}, K_{oc} can be converted to K_d using the equation

$$K_d = K_{oc} f_{oc}. \tag{6–8}$$

Organic Compound Partitioning

Using the properties and coefficients just described and others similar to them, we can estimate the partitioning of organic compounds in the subsurface. The relative distribution of compounds between the solid, liquid, and gas phases is extremely important in predicting the migration of organic compounds, and even more important in choosing and comparing remedial technologies.

An example of the partitioning of several organic compounds among solid aqueous, and gas phases from Davis (1997), based on the equations of Feenstra et al. (1991), is shown in Table 6–3. The percentages of the compound in the solid, aqueous, and gas phases are determined by calculating the mass of the compound in a cubic cm of soil that contains both water and air in its void volume. For the solids,

$$C_s = K_d C_w, \tag{6-9}$$

where C_s is the concentration adsorbed to the soil, K_d is the distribution coefficient (cm^3/g), and C_w is concentration in the aqueous phase. For the dissolved concentration, C_w, any arbitrary value at or less than the solubility can be used, and K_d can be estimated as

$$K_d = 0.6 f_{oc} K_{ow} \text{ (Karickhoff et al., 1979),} \tag{6-10}$$

in which f_{oc} is the fraction of organic carbon in the soil. For the examples of high and low organic matter in Table 6–3, 2% and 1%, respectively, are used. Once the concentration in the solids has been determined by Eq. (6–9), the mass of the compound, M_s, in a 1-cm^3 volume of soil is obtained from Eq. (6–11) as

$$M_s = C_s \rho_b, \tag{6-11}$$

where ρ_b is the bulk density of the soil. For the high organic soil in Table 6–3, 1.28 g/cm^3 was chosen, and for the low organic soil, 1.86 g/cm^3 was used. Calculation of the mass of organic compounds in water requires a porosity value and a value for volumetric water saturation. The soils in Table 6–3 were assigned a porosity of 30% and a water saturation of 50%. The mass of water is given as

TABLE 6–3　Partitioning of three solvents into air, aqueous, and solid phases of media with high (2%) organic matter and low (1%) organic matter. Two values of K_{ow} used for 1,1,1-TCA estimates and two temperatures used for TCE (from Davis, 1997).

	High Organic Matter			Low Organic Matter		
	Air	Water	Solids	Air	Water	Solids
1,1,1-Trichloroethane $K_{ow} = 300$	5%	5%	90%	26%	23%	51%
1,1,1-Trichloroethane $K_{ow} = 147.9$	6%	6%	88%	35%	31%	34%
Trichloroethene 20°C	1.6%	4.6%	93.8%	14%	35%	52%
Trichloroethene 90°C	30%	7%	63%	73%	16%	11%
Acetone	0.07%	85%	15%	0.08%	99%	1%

$$M_w = C_w n_w, \tag{6-12}$$

where n_w is the water-filled porosity, which is 50% of 0.30, or 0.15. The final component, the mass of the organic compound in the gas phase, can be determined as follows: The concentration in air, C_a, is

$$C_a = C_w K_H. \tag{6-13}$$

The mass of the compound in air is then found to be

$$M_a = C_a n_a. \tag{6-14}$$

The air-filled porosity, n_a, is also equal to half the total porosity, or 0.15. When the masses of the compound in the solid, water, and gas phases have been calculated by the preceding procedure, their percentages can be determined.

The examples in Table 6–3 illustrate some of the possibilites of partitioning. In a medium with high organic content, the mass of 1,1,1-trichloroethane (1,1,1-TCA), an industrial solvent, is partitioned almost entirely into the aquifer solids. This has important implications for remediation, because, excluding excavation and removal of the contaminated soil, it is much more difficult to remove the contaminant from the solid phase than from the aqueous and gas phases. By contrast, the example shown for low organic matter assumes an organic carbon content of 1% and a bulk density of 1.86 g/cm^3. In this case, significantly more of the mass is contained in the aqueous and gas phases, making it easier to remove from the soil.

Trichloroethene (TCE) is a solvent that has a K_{ow} value similar to 1,1,1-TCA, but a Henry's constant less than 1,1,1-TCA. Therefore, although the compound is strongly partitioned to the solid phase, the fraction contained in the aqueous phase is greater than that of 1,1,1-TCA.

Acetone, a polar compound with low octanol–water and Henry's constants, displays a greater contrast in partitioning. As a result, most of the acetone is concentrated in the aqueous phase. The bad news suggested by these calculations is that acetone is very mobile and will potentially migrate a long distance from the source. The good news, however, is that if the plume can be captured by one or more purge wells, acetone can be effectively removed from the aquifer.

The examples given in Table 6–3 illustrate the importance of the partitioning coefficients and other physical properties of organic compounds in subsurface migration. The properties of some of the most common organic compounds that have been detected in ground water are shown in Table 6–4. It is now necessary to examine the structures of the organic compounds and consider their fate and transport in ground water in more detail.

CLASSES OF ORGANIC COMPOUNDS AND THEIR OCCURRENCE IN GROUND WATER

Structure of Organic Compounds

Organic compounds, by definition, must contain carbon. The number of other elements contained in common organic compounds is surprisingly small. For the most part, the compounds of interest to us are covalently bonded structures of carbon, oxygen,

TABLE 6–4   Properties of common organic ground water contaminants (from Davis, 1997).

Contaminant	BP °C	Density CP 25°C	Water Solubility mg/l 25°C	Vapor Pressure mm Hg 10°C	Henry's Constant (dimensionless) 25°C	K_{ow}
Methylene Chloride	40	1.3182	20,000	260.9	0.105 ± 0.008	17.78
Acetone	56.3	0.7899	∞	121.7	0.000842	1.74
1,1,1-Trichloroethane	74.1	1.3303	4400	67.4	1.13 ± 0.016	309
Carbon Tetrachloride	76.8	1.5833	800	58.3	0.807 ± 0.161	676.1
Benzene	80.1	0.88	1770	47.8	0.22 ± 0.01	134.90
Trichloroethene	87.3	1.4578	1100	37.6	0.397	195
Tetrachloroethene	121.3	1.613	150	400	0.928 ± 0.161	400
2-Butanone	79.6	0.7994	26,800	314.3	0.001	1.820

Source: Davis, 1997.

hydrogen, and nitrogen, with the occasional addition of other elements. The number of covalent bonds between these elements is determined by the number of electrons in the outer shell of the atom, which are called *valence* electrons. Hydrogen, with one valence electron, can fill its outer shell with one additional electron and therefore can form one covalent bond. Oxygen has six valence electrons in its outer shell. Because the outer shell can contain a total of eight electrons, oxygen can form two covalent bonds. Nitrogen has five valence electrons and can therefore form three covalent bonds. To complete our basic list of elements, carbon, with four valence electrons, can form four covalent bonds with carbon or other elements.

The composition of organic compounds can be illustrated in several ways. The most informative is the *Lewis structure*, which shows all valence electrons for the elements in the compound (Fig. 6–2). When two valence electrons form a covalent bond, they are shown in the Lewis structure as a straight line, and when they are not involved in a covalent bond, they are shown as two adjacent dots. Two parallel lines are used to show bonds formed by sharing two pairs of valence electrons (a *double bond*), and three lines indicates the occurrence of a *triple bond*. The problem with Lewis structures is that they are somewhat cumbersome and time consuming to write for every compound. As an alternative, *condensed structural formulas* and *molecular formulas* are used. In condensed structural formulas, some of the bonds are omitted. For example, the bonds between the hydrogens attached to a carbon are left out, as are the dots for unshared valence electrons. All lines representing bonds are omitted in molecular formulas. Examples of the three types of nomenclature are shown in Fig. 6–2.

The condensed structural formula will be the most frequently used type of formula throughout this text. Molecular formulas are ambiguous because compounds with different structures can have the same molecular formulas. For example, Fig. 6–3 shows two compounds with the molecular formula C_2H_6O. If the oxygen atom is bonded to one carbon and one hydrogen, the compound is ethanol, an alcohol, whereas if the oxygen atom is bonded to two carbons, the compound is dimethyl ether. If two compounds have the same molecular formulas, but different structures, they are known as *structural isomers.*

Compound	Lewis Structure	Condensed Structural Formula	Molecular Formula
ammonia	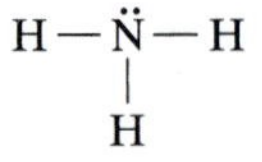 $H-\ddot{N}-H$ with H below	NH_3	NH_3
methane	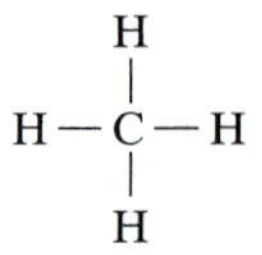 $H-C-H$ with H above and below	CH_4	CH_4
ethylene (ethene)	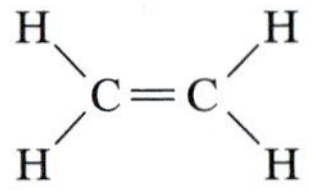 $C=C$ with H's	$CH_2=CH_2$	C_2H_4
methanol	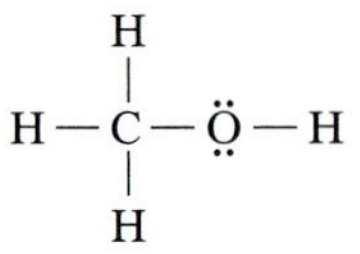 $H-C-\ddot{O}-H$ with H above and below	CH_3-OH	CH_4O
acetic acid	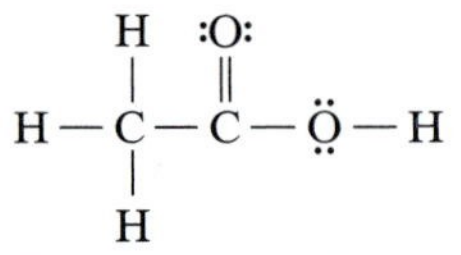 $H-C-C-\ddot{O}-H$ with H above, H below, and $:\!O\!:$ double bonded	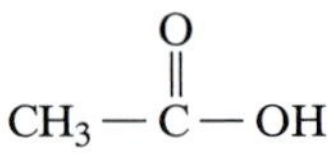 CH_3-C-OH with O double bonded	$C_2H_4O_2$

FIGURE 6–2   Structures of some common compounds.

FIGURE 6–3   Structural isomers with molecular formula C_2H_6O.

$$CH_3-CH_2-OH \qquad CH_3-O-CH_3$$

ethanol   dimethyl ether

The compounds shown in Fig. 6–3 can be used to illustrate the basis for the classification of organic compounds. The characteristic feature of ethanol that makes it an alcohol is that the hydroxyl ion, OH, is bonded to a carbon atom. This part of the structure of ethanol is known as a *functional group*, because its presence identifies the compound as a member of a class of compounds—the alcohols, in this case. The alcohol functional group is represented as —OH, where the line designates a covalent bond with a carbon atom. The remainder of the structure can vary appreciably, but the compound will still have the basic properties of an alcohol. A general formula for the alcohol group can be written as *R*—OH, where *R* represents a carbon atom or a carbon chain.

Ethers, another basic group of organic compounds, are defined by the bonding of an oxygen to two carbon atoms. The functional group in this case is shown as —O—, and a general formula for ethers can be written as *R*—O—*R′*. A group of the most important functional groups is shown in Fig. 6–4.

Functional Group Name	Functional Group (attached to carbon)	Example	
acid anhydride	$\overset{\displaystyle :O:}{\overset{\displaystyle \|}{-C}}-\ddot{O}-\overset{\displaystyle :O:}{\overset{\displaystyle \|}{C}}-$	$CH_3-\overset{:O:}{\overset{\|}{C}}-\ddot{O}-\overset{:O:}{\overset{\|}{C}}-CH_4$	acetic anhydride
alcohol and phenol	$-\ddot{O}H$	$CH_3CH_2-\ddot{O}H$	ethyl alcohol
aldehyde	$-\overset{:O:}{\overset{\|}{C}}-H$	$CH_3CH_2-\overset{:O:}{\overset{\|}{C}}-H$	propanal
alkene or olefin	$\overset{\diagdown}{\diagup}C=C\overset{\diagup}{\diagdown}$	$\overset{H}{\overset{\diagdown}{\diagup}}C=C\overset{H}{\overset{\diagup}{\diagdown}}$ (with H's)	ethylene
alkyne	$-C\equiv C-$	$H-C\equiv C-H$	acetylene
amide	$-\overset{:O:}{\overset{\|}{C}}-\ddot{N}-$	$CH_3-\overset{:O:}{\overset{\|}{C}}-\ddot{N}H_2$	acetamide
amine, primary	$-\ddot{N}H_2$	$CH_3-\ddot{N}H_2$	methylamine
amine, secondary	$-\ddot{N}H-$	$CH_3-\ddot{N}H-CH_3$	dimethylamine
amine, tertiary	$-\ddot{N}-$	$CH_3-\overset{}{\underset{CH_3}{\overset{\|}{N}}}-CH_3$	trimethylamine
carboxylic acid	$-\overset{:O:}{\overset{\|}{C}}-\ddot{O}-H$	$CH_3CH_2-\overset{:O:}{\overset{\|}{C}}-\ddot{O}-H$	propanoic acid
ester	$-\overset{:O:}{\overset{\|}{C}}-\ddot{O}-$	$CH_3-\overset{:O:}{\overset{\|}{C}}-\ddot{O}-CH_3$	methyl acetate

FIGURE 6–4 Important organic functional groups.

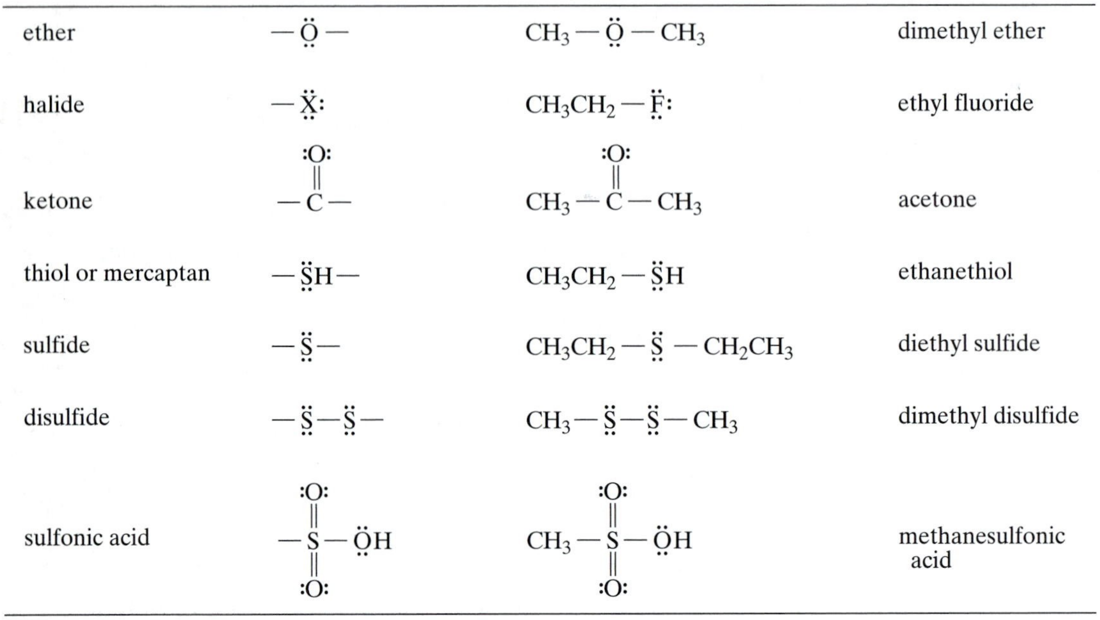

ether	$-\ddot{O}-$	$CH_3-\ddot{O}-CH_3$	dimethyl ether
halide	$-\ddot{X}:$	$CH_3CH_2-\ddot{F}:$	ethyl fluoride
ketone	$-\underset{\overset{\displaystyle :O:}{\|}}{C}-$	$CH_3-\underset{\overset{\displaystyle :O:}{\|}}{C}-CH_3$	acetone
thiol or mercaptan	$-\ddot{S}H-$	$CH_3CH_2-\ddot{S}H$	ethanethiol
sulfide	$-\ddot{S}-$	$CH_3CH_2-\ddot{S}-CH_2CH_3$	diethyl sulfide
disulfide	$-\ddot{S}-\ddot{S}-$	$CH_3-\ddot{S}-\ddot{S}-CH_3$	dimethyl disulfide
sulfonic acid	$-\underset{\underset{\displaystyle :O:}{\|}}{\overset{\overset{\displaystyle :O:}{\|}}{S}}-\ddot{O}H$	$CH_3-\underset{\underset{\displaystyle :O:}{\|}}{\overset{\overset{\displaystyle :O:}{\|}}{S}}-\ddot{O}H$	methanesulfonic acid

FIGURE 6–4   (*continued*)

With the preceding abbreviated introduction to organic chemistry, we now focus on individual types of compounds, beginning with compounds composed of carbon and hydrogen—the *hydrocarbons*.

Aliphatic Hydrocarbons

Alkanes.   Because of the simplicity of their structure, the *aliphatic hydrocarbons* are a convenient place to start our discussion of organic compounds. *Aliphatic* refers to compounds not containing the benzene ring—compounds that are composed of straight, branched, or cyclical chains of carbon atoms. There are several subdivisions of the aliphatic hydrocarbons. Compounds with only single covalent bonds between carbon atoms are known as *alkanes, saturated hydrocarbons*, or *paraffins*. The general formula for alkanes is C_nH_{2n+2}. In straight-chain alkanes, the carbon atoms form the links of the chain, with hydrogens occupying the other available bond locations. Alkane compounds form a progressive series in which one carbon is added for each compound, beginning with methane, the most basic alkane. The first 10 members of the series are named in Table 6–5, and the structures of the first 5 are shown in Fig. 6–5. Notice that each carbon is bonded to either two, three, or four hydrogens by single covalent bonds. Hydrocarbons that contain single bonds only are referred to as *saturated hydrocarbons*.

TABLE 6–5   Physical properties of C_1–C_{10} alkanes.

Compound	Condensed Structural Formula	MP (°C)	BP (°C)	Density, g/ml (20°C)	Water Solubility, mg/l (25°C)	Log K_{ow}	VP mm/Hg	K_H atm-m³/mol
Methane	CH_4	−182.6	−161.4	0.424	2.44E+001	1.09	4.66E+005	6.58E−001
Ethane	CH_3CH_3	−172	−88	0.546	6.02E+001	1.81	3.15E+004	5.00E−001
Propane	$CH_3CH_2CH_3$	−187.6	−42.1	0.501	6.24E+001	2.36	7.15E+003	7.07E−001
Butane	$CH_3(CH_2)_2CH_3$	−138.2	−0.5	0.579	6.12E+001	2.89	1.82E+003	9.50E−001
Pentane	$CH_3(CH_2)_3CH_3$	−129.7	36.1	0.626	3.80E+001	3.39	5.14E+002	1.25E+000
Hexane	$CH_3(CH_2)_4CH_3$	−95.3	69	0.659	9.50E+000	3.90	1.51E+002	1.80E+000
Heptane	$CH_3(CH_2)_5CH_3$	−90.6	98.4	0.684	3.40E+000	4.66	4.60E+001	2.00E+000
Octane	$CH_3(CH_2)_6CH_3$	−56.8	125.6	0.703	6.60E−001	5.18	1.41E+001	3.21E+000
Nonane	$CH_3(CH_2)_7CH_3$	−53.5	150.7	0.718	2.20E−001	4.76	4.45E+000	3.40E+000
Decane	$CH_3(CH_2)_8CH_3$	−29.7	174.1	0.730	5.20E−002	5.01	1.43E+000	5.15E+000

FIGURE 6–5   Alkane structures.

methane

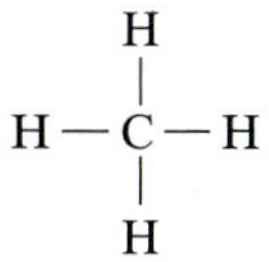

ethane

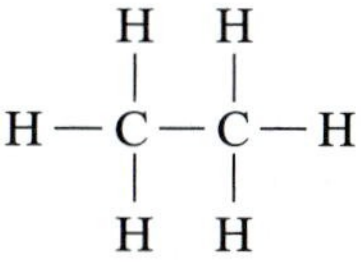

propane

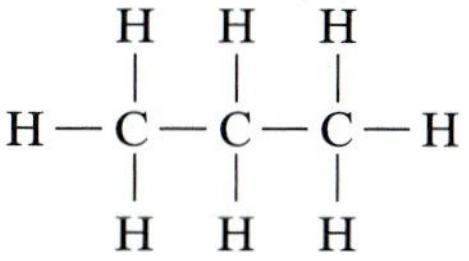

butane

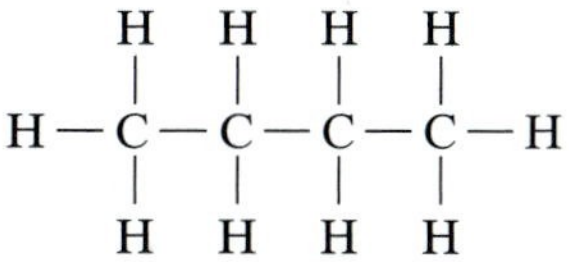

pentane

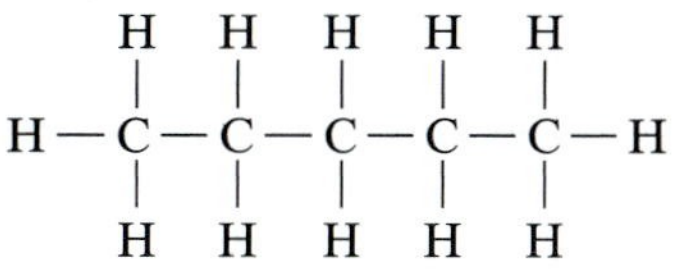

FIGURE 6–6   Structures of branched-chain alkanes.

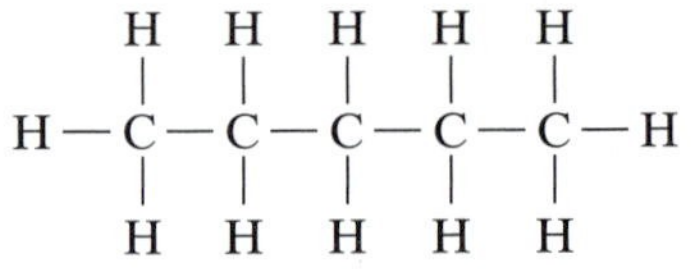

pentane, C_5H_{12}

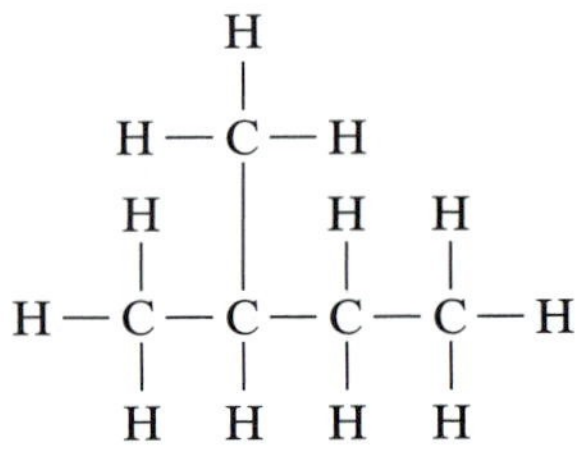

2-methylbutane, C_5H_{12}

In addition to the straight-chain alkanes, branched-chain structures can be formed. With the exception of methane, ethane, and propane, which can be formed in only one way, alkanes with higher carbon numbers can form structural isomers. For example, in Fig. 6–6, the structures of pentane and 2-methylbutane are shown. These compounds have the same molecular structure, but because of their different structures, they have slightly different physical properties. In 2-methylbutane, the carbon atom that is attached to the parent chain is known as a *substituent*. The substituent is named *methyl* because it has one carbon atom, and the 2 in the name signifies that the substituent is attached to the second carbon in the parent chain. If the substituent is derived from an alkane compound, it is also known as an *alkyl group*. The most common alkyl groups are shown in Fig. 6–7.

Naming of Alkanes.   Organic compounds are named using rules adopted by the International Union of Pure and Applied Chemistry (IUPAC). IUPAC names describe the structure of the compound in a manner so that structural formulas can be written. Unfortunately, long-standing names predating the IUPAC system still persist for some compounds. This can make recognition of the type of compound difficult. The rules for naming alkanes are as follows (Brown, 1982):

1. The general name for the saturated hydrocarbon is *alkane*.
2. The longest chain of carbon atoms in a branched-chain hydrocarbon is designated the parent chain, and the root name of the compound is that of the parent chain.
3. Each substituent group attached to the parent chain is given a name and number. The number indicates the carbon atom of the parent chain to which the substituent is attached.
4. If the same substituent occurs more than once, the number of each carbon of the parent chain on which the substituent occurs is given. In addition, the number of

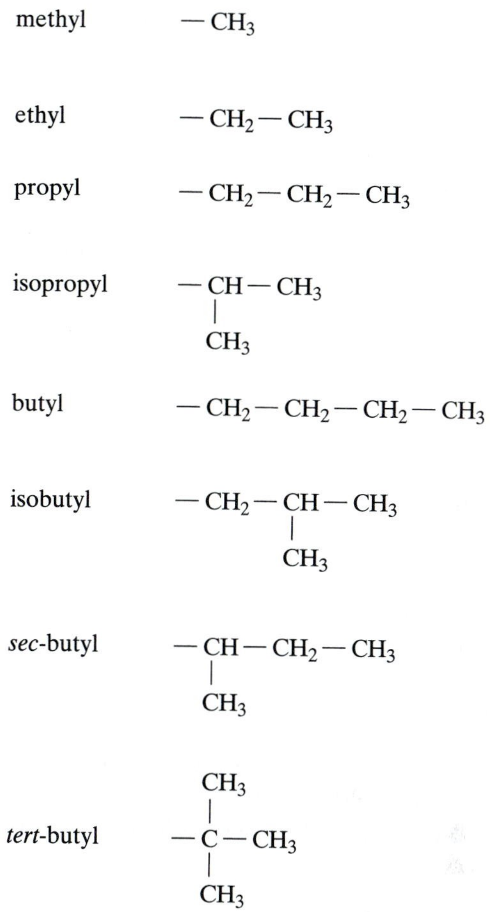

FIGURE 6–7   Alkyl substituent groups.

| methyl | $-CH_3$ |

| ethyl | $-CH_2-CH_3$ |

| propyl | $-CH_2-CH_2-CH_3$ |

isopropyl

$$-CH-CH_3$$
$$\quad |$$
$$\quad CH_3$$

| butyl | $-CH_2-CH_2-CH_2-CH_3$ |

isobutyl

$$-CH_2-CH-CH_3$$
$$\qquad\quad |$$
$$\qquad\quad CH_3$$

sec-butyl

$$-CH-CH_2-CH_3$$
$$\quad |$$
$$\quad CH_3$$

tert-butyl

$$\qquad\quad CH_3$$
$$\qquad\quad |$$
$$-C-CH_3$$
$$\quad |$$
$$\quad CH_3$$

times the substituent group occurs is indicated by a prefix—for example, di-, tri-, penta-, or hexa-.

5. If there is one substituent, the parent chain is numbered from the end that gives the substituent the lowest number. If there are two or more substituents, the parent is numbered from the end that gives the lowest number to the substituent encountered first.

6. If there are two or more different alkyl substituents, they are listed in alphabetical order.

Example 6–1.

Name the following alkanes.

a.

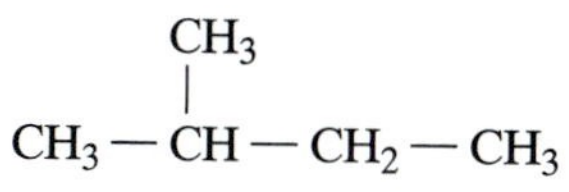

$$\qquad\qquad CH_3$$
$$\qquad\qquad |$$
$$CH_3-CH-CH_2-CH_3$$

The parent chain has four carbons, and a methyl substituent group is attached to the second carbon. The alkane is 2-methylbutane.

b.

$$CH_3 - \underset{\underset{CH_3}{|}}{\overset{\overset{CH_3}{|}}{C}} - CH_2 - CH_2 - CH_3$$

The parent chain has five carbons, and two methyl groups are attached to the second carbon. The alkane is 2,2-dimethylpentane.

c.

$$CH_3 - CH_2 - CH_2 - \underset{\underset{CH_2 - CH_2 - CH_3}{|}}{\overset{\overset{CH_3}{\overset{|}{CH - CH_3}}}{\underset{|}{C}}} - CH_2 - CH_2 - CH_3$$

The parent chain has seven carbons. An isopropyl and a propyl group are attached to the fourth carbon. The alkane is 4-isopropyl-4-propylheptane.

Cycloalkanes.   Alkanes can also be joined together in rings of various sizes. These structures are known as *cycloalkanes*. The root name is the name of the straight-chain alkane with the prefix *cyclo*. If there is one substituent group on the ring, it is unnumbered; but if there is more than one, they all must be numbered. Polygons with the same number of sides as carbon atoms can be used as a shorthand way of writing the compounds without showing the hydrogen atoms. Representative cycloalkanes are shown in Fig. 6–8.

Structure of Alkenes and Alkynes.   The hydrocarbons known as *alkenes* fall into one of the three categories of unsaturated hydrocarbons, the other two being the *alkynes* and the *aromatic hydrocarbons*. Alkenes are hydrocarbon chains containing one or more double bonds, and alkynes are chains with one or more triple bonds. The structures of the first and second alkene compounds are shown in Fig. 6–9. Both ethene and ethylene are acceptable names for the two-carbon alkene. It is apparent from these two structures that the general formula for the alkenes is C_nH_{2n}. Longer alkenes form isomers similar to those of the alkanes, although the position of the double bond can now vary. Both branched and straight chains are present.

The naming convention for alkenes is to determine the longest carbon chain containing the double bond and change the corresponding suffix from -ane to -ene. Thus, pentane becomes pentene. The position of the double bond is specified by numbering the chain from the direction that gives the double bond its lowest number. Thus, $CH_3—CH_2—CH=CH—CH_3$ is 2-pentene because the double bond occurs after the second carbon. Branched chains are named according to the same rules as alkanes. For example, the compound with the structure

$$CH_3 - \underset{\underset{CH_3}{|}}{CH} - CH = CH - CH_2 - CH_3$$

Compound	Structure	Symbol
cyclopropane	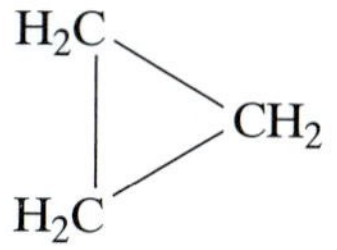 H_2C $\quad$ CH_2 H_2C	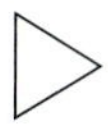 △
cyclobutane	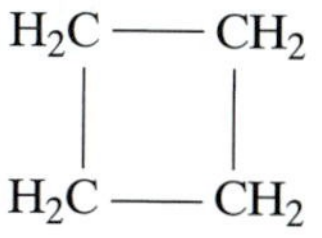 H_2C —— CH_2 H_2C —— CH_2	 □
cyclopentane	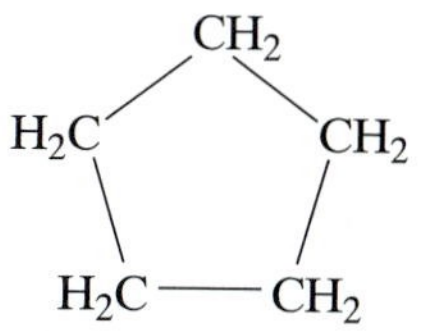 CH_2 $H_2C \qquad CH_2$ H_2C —— CH_2	 ⬠
1,2-dimethylcyclohexane	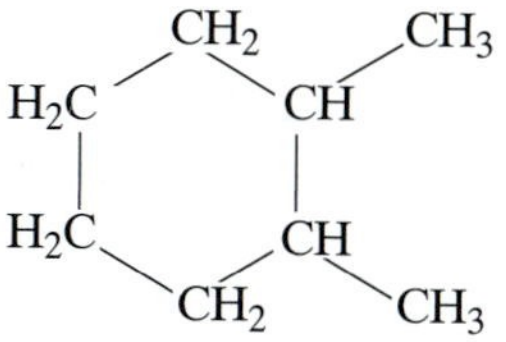 $CH_2 \quad CH_3$ $H_2C \qquad CH$ $H_2C \qquad CH$ $CH_2 \quad CH_3$	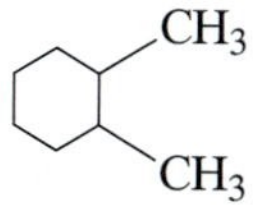 CH_3 CH_3

FIGURE 6–8   Structures of cycloalkanes.

FIGURE 6–9   Structures of C_2 and C_3 alkenes.

Compound Name	Structure
ethene (ethylene)	$\begin{array}{cc} H & H \\ \diagdown C=C \diagup \\ H & H \end{array}$ $CH_2 = CH_2$
propene (propylene)	$\begin{array}{c} H \\ \mid \\ H-C-C=C \\ \mid \ \mid \quad H \\ H \ H \end{array}$ $CH_3 - CH = CH_2$

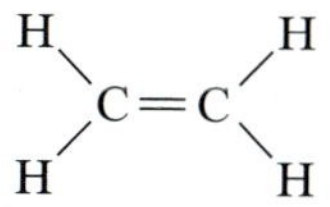

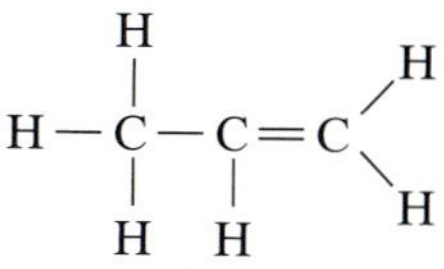

is named 2-methyl-3-hexene. Alkenes containing more than one double bond are called *dienes* if there are two double bonds, *trienes* if there are three double bonds, and so forth. An example is 1,3-butadiene, which has the structure: $CH_2=CH-CH=CH_2$. The common name for this compound, butadiene, is used extensively, as are the common names for propene (propylene) and 2-methylpropene (isobutylene). Formal IUPAC names are used for other alkenes.

One or more carbon–carbon triple bonds is the distinguishing characteristic of the *alkynes*. The suffix -yne replaces -ane in the alkane name, to form the parent chain of alkynes. Except for the first alkyne, which goes by its common name, *acetylene*, the nomenclature follows that of the alkenes. The structure of acetylene is $H-C\equiv C-H$. The second member of the series is propyne, $CH_3-C\equiv C-H$, and the general formula for the alkynes is C_nH_{2n-2}.

Aromatic Hydrocarbons

Although the aliphatic hydrocarbons are important components of petroleum-based ground water contaminants, many of them are not significant contaminants because of their lower water solubilities and toxicities (Table 6–5). The *aromatic hydrocarbons*, on the other hand, include the most common contaminant compounds in ground water. They are ubiquitous around subsurface releases of petroleum fuels, which constitute the majority of contaminant sites in the United States.

The structure of the aromatic hydrocarbons, which, like the alkenes and alkynes, are unsaturated, is based upon the presence of the *benzene ring*, a six-sided ring composed of six carbons and six hydrogens (Fig. 6–10). Lewis structures of the benzene ring can be drawn to show alternating double bonds between carbon atoms to satisfy the valence requirements of the carbon and hydrogen atoms. This model, however, does not correctly describe the structure of benzene. Chemists have long recognized that the electrons which supposedly form the double bonds are not fixed in those locations, but instead are shared symmetrically around the ring. These electrons are referred to as *delocalized*. As a result, the benzene structure is alternatively drawn as a hexagon with an inscribed circle. The hydrogen atoms are omitted from the structure for convenience.

Benzene is rarely mentioned as a contaminant without including the related compounds *toluene, ethylbenzene,* and *xylene*. These compounds are so commonly associated with each other, that they are referred to by their initials as *BTEX*. The structures of toluene and ethylbenzene are formed by adding one methyl and one ethyl substituent,

FIGURE 6–10   Two representations of the benzene structure.

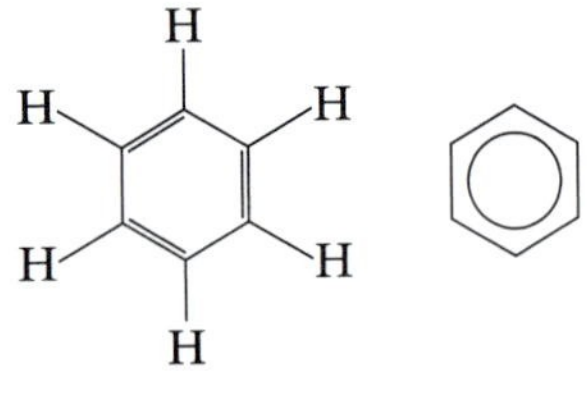

C_6H_6

respectively, to the benzene ring (Fig. 6–11). Xylene, which is a dimethylbenzene, has two substituent methyl groups. The positioning of the two methyls determines whether the name is prefixed by *ortho-*, *meta-*, or *para-*. Many other derivative compounds of benzene can be formed by adding substituent groups. The substituents' positions are numbered clockwise around the ring, and the prefixes *tri-*, *tetra-*, etc., are included in the name. An example is 2,4,6-trimethylbenzene.

Polycyclic Aromatic Compounds.   The compounds described in the previous section contain one benzene ring. It is also possible for benzene rings to be linked together to form an important class of compounds known as *polycyclic (polynuclear) aromatic hydrocarbons* (PAHs). The PAHs occur in several types of contaminated sites, including manufactured-gas sites (coal gasification), wood-preservative sites, and oil refineries, and in the routine underground tank and pipeline releases. Because they have lower solubilities and higher octanol–water partitioning coefficients, they are less mobile in the subsurface. Nevertheless, their concentrations commonly exceed regulatory limits at contaminant sites. The structures and physical properties of some of the PAHs are shown in Fig. 6–12.

Occurrence of Petroleum Hydrocarbons in Ground Water

There is no doubt that petroleum hydrocarbons have been the most commonly detected contaminants in ground water. Under Subtitle I of the Resource Conservation and Recovery Act (RCRA), all underground storage tank owners in the United States were mandated to upgrade, close, or replace all tanks. As of March, 1996, 300,000 confirmed releases from underground storage tanks (USTs), most of which involved petroleum fuels, had been reported (US EPA, UST home page). In addition to these sources, thousands of additional sites nationwide include petroleum refineries, pipeline spills and leaks, manufactured-gas plants, and wood-preservation plants. Contaminants released at these sites range from crude oil to various petroleum distillates such as gasoline. In the petroleum distillation process, crude oil is heated to high temperatures so that petroleum fractions with similar boiling points can be separated and collected. Because the boiling point of a compound generally corresponds to the molecular weight of the compound, each distillate has a characteristic range of molecular weights. Crude oil itself is a complex mixture of hundreds of compounds. At one of the best documented petroleum contaminant sites, the Bemidji, Minnesota, pipeline rupture, the crude oil was composed of 58–61% saturated hydrocarbons, 33–36% aromatics, 4–6% resins (including compounds of nitrogen, sulfur, and oxgyen), and 1–2% asphaltenes (Eganhouse et al., 1993). Common types of distillates are shown in Table 6–6.

Beneath the ground surface at petroleum release sites, a free-phase pool of petroleum liquid may accumulate near the top of the capillary fringe. This liquid, referred to as free product or light nonaqueous-phase liquid (LNAPL) because it is generally less dense than water, may depress the water table and also migrate downgradient. Organic compounds from the NAPL dissolve into the ground water and form dissolved plumes extending farther downgradient than the NAPL. Most studies of petroleum releases focus on the small number of these compounds that are of regulatory concern,

Compound Name	Structure	mp (°C)	bp (°C)	Water Solubility mg/l (25°C)	$\log K_{ow}$	P° (mm Hg)	K_H $\left(\dfrac{atm-}{m^3/mol}\right)$
benzene		5.53	80.09	1.79E+003	2.13	9.52E+001	5.55E−003
toluene	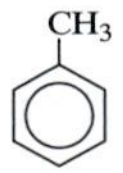	−94.991	110.63	5.26E+002	2.73	2.84E+001	6.64E−003
ethylbenzene	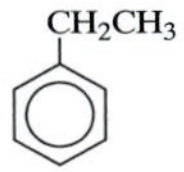	−94.975	136.193	1.69E+002	3.15	9.60E+000	7.88E−003
o-xylene	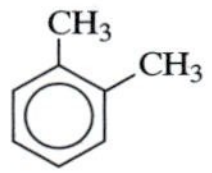	−25.182	144.429	1.78E+002	3.12	6.61E+000	5.18E−003
m-xylene	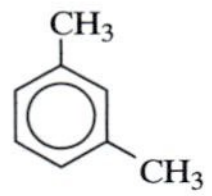	−47.872	139.12	1.61E+002	3.20	8.29E+000	7.18E−003
p-xylene	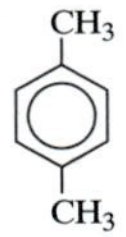	13.263	138.359	1.62E+002	3.15	8.84E+000	7.53E−003
1,2,4 trimethylbenzene	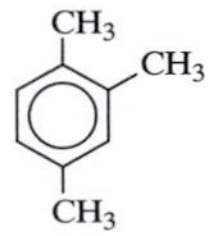	−43.80	168.89	5.70E+001	3.78	2.10E+000	6.16E−003
1,2,3,5 tetramethylbenzene	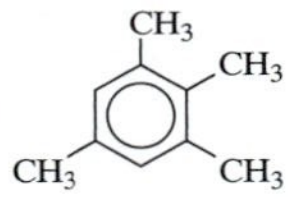	−24	197.9	2.79E+001	4.10	4.98E+001	7.99E−003

FIGURE 6–11 Structures and properties of benzene and some alkyl benzenes (values from Howard and Meylan, 1997). mp: melting point. bp: boiling

Compound Name	Structure	mp (°C)	bp (°C)	Water Solubility mg/l (25°C)	log K_{ow}	P° (mm Hg)	K_H $\left(\dfrac{\text{atm} - }{\text{m}^3/\text{mol}}\right)$
naphthalene	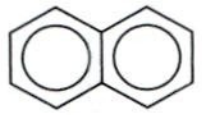	80.29	217.942	3.10E+001	3.30	8.50E−002	4.83E−004
acenaphthene	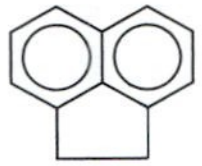	95	279	3.90E+000	3.92	2.50E−003	1.55E−004
acenaphthylene		89-91	280	1.61E+001	3.94	9.12E−004	1.14E−004
anthracene	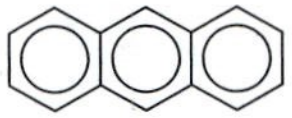	215	340	4.34E−002	4.45	2.67E−006	7.20E−004
fluorene	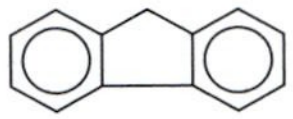	–	–	1.34E+000	4.02	1.26E−003	1.67E−004
chrysene	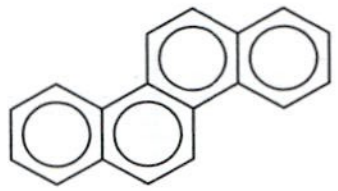	254-255	448	2.00E−003	5.50	6.23E−009	9.46E−005
benzo[a]pyrene		179.2	311	1.62E−003	5.97	5.49E−009	2.45E−006

FIGURE 6–12 Structure of polycyclic aromatic hydrocarbons (values from Howard and Meylan, 1997). mp: melting point. bp: boiling point. P°: vapor pressure. K_H: Henry's constant.

TABLE 6–6   Crude oil distillates.

Fraction	Number of Carbon Atoms/Molecule	Boiling-point Range (°C)	Use
Gases	C_1–C_4	<20	Cooking, home heating
Naphthas	C_4–C_{10}	20–200	Gasoline, raw materials for chemical industry
Kerosene	C_9–C_{15}	175–275	Fuel, heat, jet fuel
Gas oil	C_{15}–C_{25}	200–400	Diesel fuel, fuel oil
Lubricating oil, heavy fuel oil	C_{19}–C_{40}	>550	Grease, lubricating oil
Asphalt	>40	>560	Road surfaces

including BTEX, PAHs, and fuel additives. The distribution of individual compounds in the dissolved plume depends on a variety of attenuation processes, including adsorption and biotransformation, which leads to other types of organic compounds and, ultimately, to carbon dioxide or methane. The behavior of individual compounds can be explained by the values of their octanol–water partitioning coefficient, solubility, molecular weight, Henry's constant, and susceptibility to biodegradation. Of the BTEX compounds, benzene commonly occurs at higher relative concentrations near the leading edge of the plume resulting from its relatively high solubility and lower K_{ow}. Toluene is more susceptible to biodegradation and may be depleted rapidly in the plume, depending upon the availability of electron acceptors in the aquifer. These processes will be examined in greater detail in Chapter 7.

The investigation of the Bemidji pipeline rupture site is unique because of the effort devoted to characterizing petroleum compounds and their transformation products in ground water. Plan view and cross sections of the plume are shown in Fig. 6–13. The large LNAPL pool of crude oil is evident at the center of the cross section. The site is somewhat unusual in that oil was sprayed in the upgradient direction when the pipeline ruptured, creating a region of ground water degradation upgradient from the LNAPL pool. Otherwise, the LNAPL pool lying at the top of the water table and the dissolved plume extending downgradient are typical of petroleum releases. The distribution of 1,2,3,4-tetramethylbenzene is shown in plan and cross-sectional views of the site because of that compound's stability in ground water and, therefore, its usefulness as an indicator of the dissolution of petroleum hydrocarbons.

The composition of the dissolved organic carbon derived from the oil body is a complex mixture of many compounds that change in relative proportions with distance along the flow path. Figure 6–14 shows the bulk composition of the dissolved organic carbon in Zone II, the oil spray area upgradient of the oil pool, and Zone III, the anoxic zone closest to the oil body on the downgradient side. The dissolved organic carbon is divided into volatile dissolved organic carbon (VDOC) and nonvolatile dissolved organic carbon (NVDOC). Volatile hydrocarbons (VHCs) are identified as a component of VDOC, and low-molecular-weight organic acids are identified as a component of the NVDOC. In Zone II, the dissolved organic carbon is composed mostly of NVDOC,

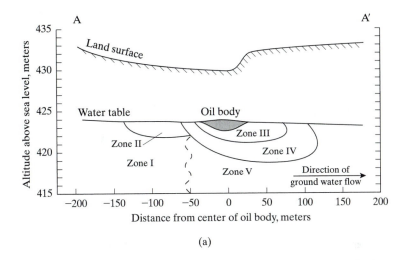

(a)

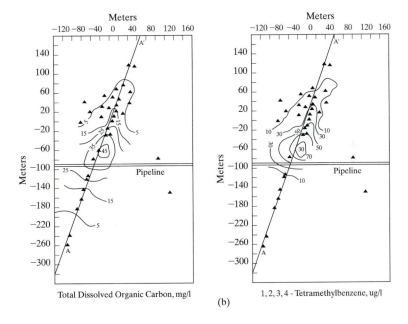

(b)

FIGURE 6–13 (a) Cross section of Bemidji oil-spill site, showing zonation of oil body and dissolved-phase plume. Zone I: oxic, native ground water. Zone II: oxic, oil spray area. Zone III: anoxic. Zone IV: suboxic transition. Zone V: oxic, downgradient. (b) Spatial distribution of contaminants at the water table. Total organic carbon, mg/l (left), and 1,2,3,4-tetramethylbenzene, μg/l (right). (c) Vertical distribution of total dissolved organic carbon (top) and 1,2,3,4-tetramethylbenzene (bottom) during 1987 (reprinted from *App. Geochem.*, v. 8, Eganhouse, R. P., Baedecker, M. J., Cozzarelli, I. M., Aiken, G. R., Thorn, K. A., and Dorsey, T. F., Crude oil in a shallow sand and gravel aquifer–II. Organic Geochemistry, pp. 551–567, Copyright 1993, with permission from Elsevier Science).

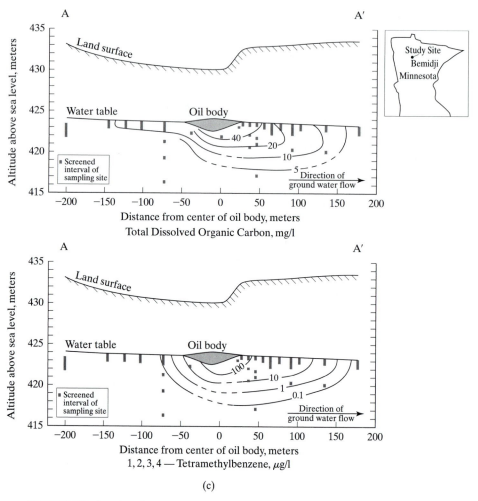

FIGURE 6–13c (*continued*)

because, when the oil was sprayed onto the ground surface, the volatile components were lost by volatilization to the atmosphere during migration downward from the land surface. The VDOC increases drastically in Zone III. Most of this fraction is composed of VHC. As the ground water moves downgradient from the location of the well represented by Fig. 6–14b to the well represented by Fig. 6–14c, VDOC, VHCs, and acids all decrease, most likely by biodegradation. Some of the VHCs are converted to NVDOC, which remains at a nearly constant concentration between the two wells.

Which compounds make up the volatile hydrocarbons in the downgradient anoxic zone? Figure 6–15 shows that in the area nearest the oil body (Fig. 6–15a), the volatile hydrocarbons are dominated by aromatics, including benzene, and related alkylbenzenes such as toluene, ethylbenzene and xylene (C_2—Bz), trimethylbenzene (C_3—Bz),

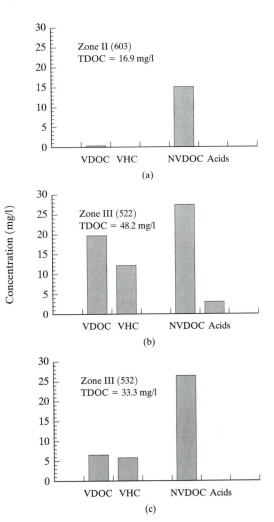

FIGURE 6–14 Concentration of dissolved organic carbon in ground water at the water table during 1987: (a) within Zone II (603), (b) within Zone III at the leading edge of the oil body (552), and (c) within Zone III, 21 m downgradient of the leading edge of the oil body (532). TDOC: Total dissolved organic carbon; VDOC: volatile dissolved organic carbon; VHC: volatile hydrocarbons; NVDOC: nonvolatile dissolved organic carbon; acids: low molecular weight organic acids (reprinted from *App. Geochem.*, v. 8, Eganhouse, R. P., Baedecker, M. J., Cozzarelli, I. M., Aiken, G. R., Thorn, K. A., and Dorsey, T. F., Crude oil in a shallow sand and gravel aquifer–II. Organic Geochemistry, pp. 551–567, Copyright (1993), with permission from Elsevier Science).

and tetramethylbenzene (C_4—Bz). Although alkanes and other compounds occur, the aromatics dominate the volatile hydrocarbons because of their relatively high solubility. Over a distance of 31 m along the flow path (Fig. 6–15b), dramatic changes in the composition of the volatile hydrocarbons occur. Although all compounds show decreases in concentration resulting from biodegradation of the volatile organics, benzene is much less affected, because it is more resistant to microbial degradation than are the other compounds. Biodegradation will be addressed in the next chapter.

Polycyclic aromatic compounds are widely detected in contaminant plumes at most petroleum release sites. Releases composed of the heavier petroleum distillates, such as diesel fuel, jet fuel, and heating oil, contain higher concentrations of PAHs than gasoline releases contain. In addition, PAHs in disposal sites from several types of chemical manufacturing and treatment plants are particularly abundant. Wood-preservative compounds

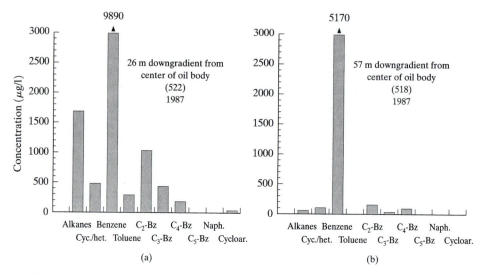

FIGURE 6–15 Concentrations of specific organic compounds in 1987 in a well 26 m downgradient from center of oil body (522) and 57 m downgradient from oil body (518). Benzene is more resistant to biodegradation than are other compounds (reprinted from *App. Geochem.*, v. 8, Eganhouse, R. P., Baedecker, M. J., Cozzarelli, I. M., Aiken, G. R., Thorn, K. A., and Dorsey, T. F., Crude oil in a shallow sand and gravel aquifer–II. Organic Geochemistry, pp. 551–567), Copyright 1993, with permission of Elsevier Science).

such as creosote contain as much as 85% PAHs (Godsy et al., 1992). Creosote production and application facilities produce waste that in the past was discharged to unlined pits or impoundments. Contaminant plumes from these sites are enriched in PAHs and many other compounds. Contaminant plumes at the site of former manufactured-gas plants are also known for high concentrations of PAHs. These plants, which produced gas from coal, were largely replaced by the use of natural gas by the 1950s. Waste from the plants, consisting of heavy petroleum fractions, was buried in trenches or pits (Jones, 1996). Low-molecular-weight PAHs are abundant in the dissolved ground water plumes emanating from these disposal pits.

Halogenated Hydrocarbons

Although perhaps not as abundant in ground water as the hydrocarbons, the *halogenated hydrocarbons* are of great concern to hydrogeologists as a result of their persistence and toxicity. Some members of this large group of compounds are less biodegradable than hydrocarbons in some geochemical environments, have significant solubilities, and are very mobile in aquifers. These characteristics, along with regulatory drinking water limits as low as one part per billion for some halogenated hydrocarbons, have cost government and industry billions of dollars in assessments and cleanup costs in the past several decades during which their presence in ground water has been recognized. The compounds are produced by the replacement of hydrogen atoms with halogen ions, the most common being chloride, bromide, and fluoride. Their usage encompasses raw materials for plastics and other products, as well as solvents and degreasers for many types of manufacturing processes. It is perhaps this function that has generated most of the ensuing environmental problems. In the past, waste solvents were commonly discharged to pits, dumped on the ground surface for disposal, or leaked from tanks or through cracks in floors. Improper handling of solvents during tank filling, transport, or usage led to countless spills on the ground or floors.

The most important difference between the halogenated and nonhalogenated hydrocarbons is the greater density imparted to the compounds by the halogen atoms. Because the specific gravity of most of these compounds is greater than water, free-phase liquids, known as dense, nonaqueous-phase liquids (DNAPLs), have the potential to sink through the water table and descend vertically into the aquifer. DNAPLs will continue to sink until a low permeability horizon is encountered (Fig. 6–16), whereupon the liquid may form a pool by displacement of the water from the aquifer pores. The DNAPL pool can migrate under its own weight along the contact of the low-permeability layer in whichever direction the contact slopes. Non-mobile (residual) DNAPLs will occur above the DNAPL pool, along the downward path of the sinking liquid. Residual DNALPs will dissolve into ground water passing through the aquifer, creating a dissolved plume controlled by the physical and geochemical properties of the aquifer as well as the partitioning properties of the DNAPLs. From the more mobile compounds, to be described in the next section, long, persistent plumes can be generated. Biodegradation of these compounds occurs, but under more restricted geochemical conditions than those of the nonhalogenated hydrocarbons. These processes will be addressed in the next chapter.

Halogenated Aliphatics.   The nomenclature of halogenated hydrocarbons is very similar to that of the compounds from which they are derived. One problem in recognition of the compounds, however, is the continued use of common names instead of IUPAC terms. The C_1 alkanes are a good example. Figure 6–17 illustrates the structure and nomenclature of the chlorinated methane compounds, which have a wide variety of

FIGURE 6–16   Migration of DNAPL contaminants in the subsurface (diagram courtesy of Stan Feenstra).

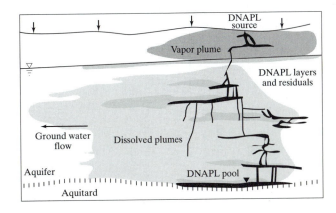

Compound (IUPAC Name)	Structure	Spec.[1] Gravity (temp.)	mp (°C)	bp (°C)	Water Solubility (mg/l 25°C)	log K_{ow}	P° (mm Hg)	K_H $\left(\text{atm} - \text{m}^3/\text{mol}\right)$
methyl chloride (chloromethane)	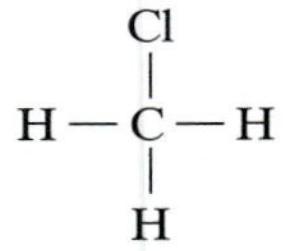 Cl—C—H with H below, H to left	0.916 (20°C)	−97.70	−24.20	5.32E+003	0.91	4.30E+003	8.82E−003
methylene chloride (dichloromethane)	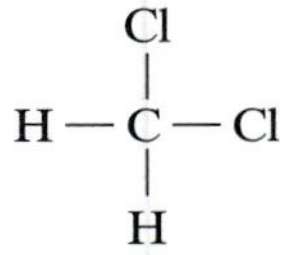 Cl—C—Cl with H below, H to left	0.327 (20°C)	−94.92	39.64	1.30E+004	1.25	4.35E+002	3.25E−003
chloroform (trichloromethane)	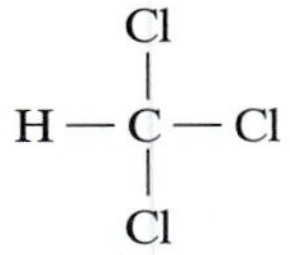 Cl—C—Cl with Cl below, H to left	1.489 (20°C)	−63.52	61.18	7.95E+003	1.97	1.97E+002	3.67E−003
carbon tetrachloride (tetrachloromethane)	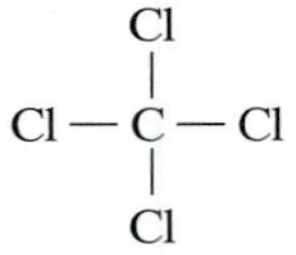 Cl—C—Cl with Cl below, Cl to left	1.594 (20°C)	285–287	76.7	7.93E+002	2.83	1.15E+002	2.76E−002

FIGURE 6–17 Structures and properties of chlorinated C_1 aliphatics (values from Howard and Meylan, 1997). mp: melting point. bp: boiling point. P°: vapor pressure. K_H: Henry's constant. [1]Montgomery and Welkom, 1989.

uses and occurrences, including solvents, refrigerants, and insecticides. Compounds known as *chlorofluorocarbons* (CFCs) are used as refrigerants and aerosol propellents and are of environmental concern because of their deleterious effects on the ozone layer of the atmosphere. One example of these compounds, Freon 12, has the following structure:

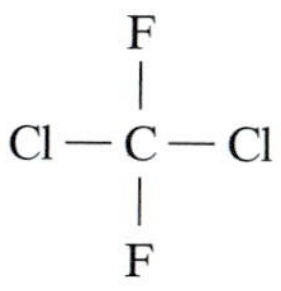

$$Cl - \underset{\underset{F}{|}}{\overset{\overset{F}{|}}{C}} - Cl$$

Halogenated methanes are also encountered in chlorinated drinking water, in which chlorine or other halogens added as disinfectants react with natural organics in the water to form compounds such as chloroform and bromoform. This is an example of a beneficial treatment producing a health risk by undesired reactions in the treated water.

Chlorinated ethanes are shown in Fig. 6–18. When more than one halogen ion is present, the distribution of all of them on the two carbons must be specified by numbers at the beginning of the name and the prefixes *di-*, *tri-*, etc.

Chlorinated ethenes (Fig. 6–19) include examples of some of the most common ground water contaminants resulting from their widespread use as solvents. Both *tetrachloroethene* (perchloroethene) and *trichloroethene* are common industrial solvents. The names of these compounds may end with the suffix *-ethylene* rather than *-ethene*. Tetrachloroethene is well known for its use as a dry-cleaning compound. Trichloroethene is perhaps the most common organic solvent found in ground water worldwide. The various forms of dichloroethene often occur as biodegradation products of tetrachloroethene and trichloroethene. The *cis* and *trans* forms of 1,2-dichloroethene constitute a special type of isomerism. The *cis-* isomer has both chloride ions on the same side of the carbon–carbon double bond, whereas they occur on opposite sides in the *trans-* isomer. Despite their similarity, like other isomers they are different compounds with different properties. Chloroethene, which is usually called by its common name, *vinyl chloride*, is one of the most condemned members of the chlorinated ethenes because of its carcinogenic effects, low regulatory concentration standards, and mobility and persistence in aquifers. It occurs as a biodegradation product in the chlorinated ethene series, but is also used as a raw material for polyvinyl chloride (PVC) and in other manufactured compounds.

Halogenated Aromatics.   Simple halogenated aromatics are formed by the substitution of chlorine for hydrogen on benzene or alkylated benzenes (Fig. 6–20). These organics are widely used in solvent and pesticide compounds. Their solubilities are low, but significant, and their log K_{ow} values are higher than those of the corresponding non-halogenated compounds.

The benzene ring can also serve as a functional group, called the -phenyl group, which can then combine with other functional groups. One important group of

Compound Name	Structure	Spec.[1] Gravity (temp.)	mp (°C)	bp (°C)	Water Solubility (mg/l at 25°C)	$\log K_{ow}$	P° (mm Hg)	K_H $\left(\frac{atm-}{m^3/mol}\right)$
chloroethane	$CH_3—CH_2Cl$	.92 (20°C)	–136.4	12.27	5.68E+003	1.43	1.01E+003	1.11E–002
1,1-dichloroethane	$CH_3—CHCl_2$	1.174 (20°C)	–96.96	57.30	5.06E+003	1.79	2.27E+002	5.62E–003
1,2-dichloroethane	$CH_2Cl—CH_2Cl$	–	–35.66	83.483	8.52E+003	1.48	7.89E+001	1.18E–003
1,1,1-trichloroethane	$CH_3—CCl_3$	1.44 (20°C)	–30.4	74.08	1.50E+003	2.49	1.24E+002	1.72E–002
hexachloroethane	$CCl_3—CCl_3$	2.09 (20°C)	185	186.8	7.70E+000	3.91	2.10E–001	3.89E–003

FIGURE 6–18   Structures and properties of C_2 chlorinated ethanes (values from Howard and Meylan, 1997). mp: melting point. bp: boiling point. P°: vapor pressure. K_H: Henry's constant. [1]Montgomery and Welkom, 1989.

Compound Name	Structure	Spec.[1] Gravity (temp.)	mp (°C)	bp (°C)	Water Solubility (mg/l at 25°C)	log K_{ow}	P° (mm Hg)	K_H (atm-m³/mol)
vinyl chloride (chloroethene)	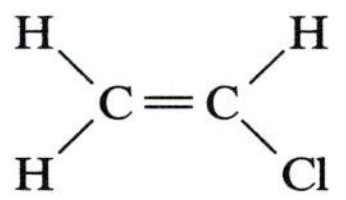 H₂C=CHCl (H, H / H, Cl)	0.91 (20°C)	−153.8	−13.37	8.80E+003	1.62	2.98E+003	2.78E−002
1,1-dichloroethene	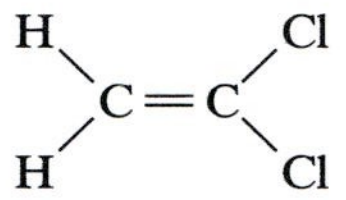 Cl₂C=CH₂ (H, H / Cl, Cl)	1.22 (20°C)	−122.56	31.56	2.25E+003	2.13	6.00E+002	2.61E−002
trans-1,2-dichloroethene	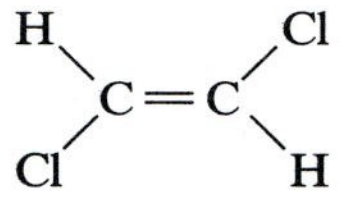 ClHC=CHCl (H, Cl / Cl, H)	1.26 (20°C)	−49.8	47.7	6.30E+003	2.09	3.31E+002	9.38E−003
cis-1,2-dichloroethene	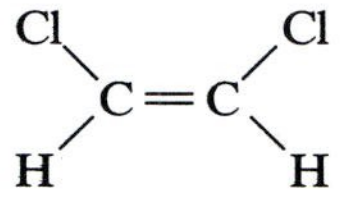 ClHC=CHCl (Cl, H / Cl, H)	1.28 (20°C)	−80.1	60.2	3.50E+003	1.86	2.01E+002	4.08E−003
trichloroethene	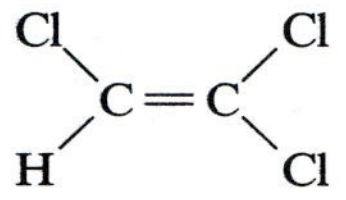 Cl₂C=CHCl (Cl, H / Cl, Cl)	1.46 (20°C)	−84.8	86.7	1.10E+003	2.42	6.90E+001	9.85E−003
tetrachloroethene (perchloroethene)	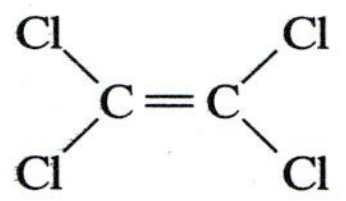 Cl₂C=CCl₂ (Cl, Cl / Cl, Cl)	1.62 (20°C)	−22.35	121.07	2.00E+002	3.40	1.86E+001	1.77E−002

FIGURE 6–19 Structures and properties of chlorinated ethenes (values from Howard and Meylan, 1997). mp: melting point. bp: boiling point. P°: vapor pressure. K_H: Henry's constant. [1]Montgomery and Welkom, 1989.

Compound Name	Structure	Spec. Gravity (temp.)	mp (°C)	bp (°C)	Water Solubility (mg/l at 25°C)	log K_{ow}	P° (mm Hg)	K_H $\left(\dfrac{atm-}{m^3/mol}\right)$
chlorobenzene	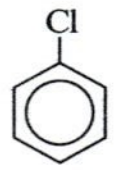	1.11 (20°C)	−45.58	131.687	4.98E+002	2.84	1.20E+001	3.77E−003
1,2-dichlorobenzene	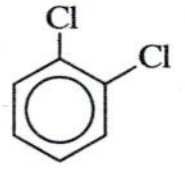	1.30 (20°C)	−17.01	180.48	1.56E+002	3.43	1.36E+000	1.90E−003
1,4-dichlorobenzene		1.25 (20°C)	53.13	174.12	7.60E+001	3.44	1.00E+000	2.40E−003
1,2,3-trichlorobenzene	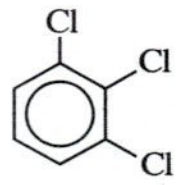	–	52.6	221	1.80E+001	4.05	2.10E−001	1.25E−003
hexachlorobenzene	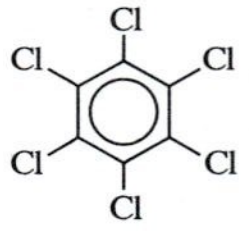	1.57 (20°C)	228.7	319.3	6.20E−003	5.73	1.80E−005	1.70E−003

FIGURE 6–20 Structures and properties of chlorinated benzenes (values from Howard and Meylan, 1997). mp: melting point. bp: boiling point. P° : vapor pressure. K_H: Henry's constant. [1]Montgomery and Welkom, 1989.

compounds of this type includes the chlorinated pesticide, 4,4′-DDT (dichlorodiphenyl-trichloroethane):

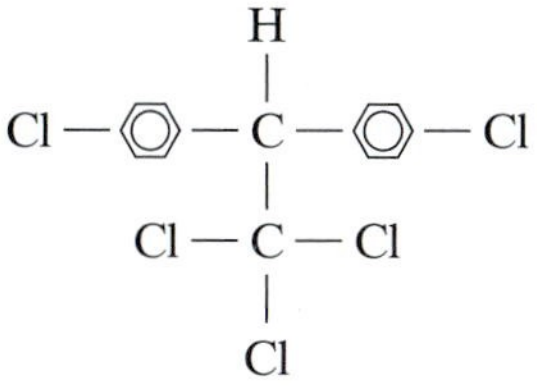

$$
\begin{array}{c}
\text{H} \\
| \\
\text{Cl} - \bigcirc - \text{C} - \bigcirc - \text{Cl} \\
| \\
\text{Cl} - \text{C} - \text{Cl} \\
| \\
\text{Cl}
\end{array}
$$

The phenyl groups are attached to one of the carbon atoms in trichloroethane, and each phenyl group contains a chlorine substituting for hydrogen. Several variants of the DDT structure exist. This pesticide was banned in the United States in 1972 because of its tendency to bioaccumulate and cause birth defects in wildlife.

Biphenyls consist of two phenyl groups joined by a single bond. Chlorinated biphenyls include 209 individual *congeners* known collectively as PCBs. Up to 10 chlorine atoms, designated by the numbering system shown in the following diagram, can replace hydrogens on the phenyl groups.

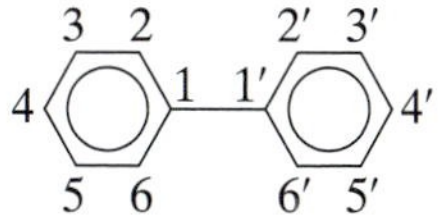

$$
\begin{array}{c}
3 \quad 2 \qquad 2' \quad 3' \\
4 \left\langle \bigcirc \right. ^{1} \,\, ^{1'} \left. \bigcirc \right\rangle 4' \\
5 \quad 6 \qquad 6' \quad 5'
\end{array}
$$

In the United States, PCBs are designated by the trade name Arochlor®, followed by a four-digit number. The first two digits are either 12 for biphenyls or 54 for *terphenyls* (three phenyl groups), and the second two digits represent the weight percent chlorine in the compound. PCBs are characterized by low solubility, low volatility, a high dielectric constant, and high thermal stability. Prior to being banned in the United States in 1977, they were used extensively as insulators in transformers and capacitors and in paint, plasticizers, and sealants. In the paper industry, they were used in de-inking during recycling of paper and for other uses. When released to waste impoundments or streams, they remain tightly bound to sediment particles and can be consumed by fish or organisms higher on the food chain, causing reproductive problems and other health effects. PCBs are resistant to chemical and biological breakdown, making them a serious long-term environmental problem.

Halogenated Hydrocarbons in Ground Water

Halogenated hydrocarbons, particularly the low-molecular-weight chlorinated ethanes and ethenes, are the most frequently detected organic compounds at hazardous waste disposal sites (Plumb, 1992). Coupled with their relatively high toxicity, mobility, and resistance to biodegradation, these compounds pose an enormous challenge to hydrogeologists. As with any contaminant, the site hydrogeology is critical in determining the scope of the problem and the difficulty of remediation. In homogeneous sand and gravel aquifers, dissolved-phase plumes of halogenated hydrocarbons are relatively easy to delineate and control. Plumes tend to be thin and elongated in shape, with well-defined

FIGURE 6–21 Cross sections showing distribution of TCE and PCE (a) and total dichlorobenzene isomers (b) at the Otis Air Force Base site on Cape Cod, MA (Reprinted from Barber, 1992, p. 103, by courtesy of Marcel Dekker, Inc.).

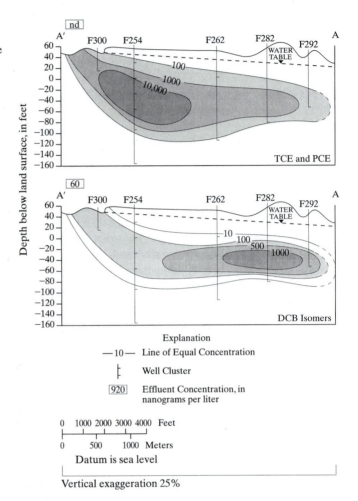

centers of mass (Fig. 6–21) that travel through the aquifer in the same fashion as conservative ions such as chloride. At low concentrations in a dissolved-phase plume, the density is not important, and plumes follow ground water flow lines. Under oxic conditions, biodegradation is very slow, and attenuation results, primarily from advection, dispersion, and sorption.

The total mass of the contaminant compounds is relatively small in many plumes. Mackay and Cherry (1989) compared the dimensions, water volumes, and contaminant masses of several well-documented plumes (Fig. 6–22). Calculations of contaminant mass indicate that a release from a small number of 55-gal drums of pure compound can produce a huge volume of contaminated ground water. The problem with calculated masses is that they represent the mass present in the dissolved phase only. As demonstrated earlier in the chapter, partitioning to the solid phase can represent a much greater mass than that in solution. The contaminated mass contained on the solid phase becomes

Site Location and Plume Map 0 ⊢———— 5 km ⊣ Flow ⟶	Presumed Sources	Predominant Contaminants	Plume Volume (Liters)	Contaminant Mass Dissolved in Plume (as Equivalent NAPL Volume in Liters or 55-gal Drums)
Ocean City, NJ 	chemical plant	TCE TCA PER	5,700,000,000	15,000 (72 drums)
Mountain View, CA 	electronics plants	TCE TCA	6,000,000,000	9800 (47 drums)
Cape Cod, MA 	sewage infiltration beds	TCE PER Detergents	40,000,000,000	1500 (7 drums)
Traverse City, MI 	aviation fuel storage	Toluene Xylene Benzene	400,000,000	1000 (5 drums)
Gloucester, ON Canada	special waste landfill	1,4 Dioxane Freon 113 DEE, THF	102,000,000	190 (0.9 drums)
San Jose, CA 	electronics plant	TCA Freon 113 1,1 DCE	5,000,000,000	130 (0.6 drums)
Denver, CO 	trainyard, airport	TCE TCA DBCP	4,500,000,000	80 (0.4 drums)

FIGURE 6–22 Size and characteristics of organic contaminant plumes in sand and gravel aquifers. TCE = trichloroethylene; TCA = 1,1,1-trichloroethane; PER = tetrachloroethylene; 1,1-DCE = dichloroethylene; DEE = diethyl ether; THF = tetrahydrofuran; DBCP = dibromochloropropane. Contaminant mass accounts only for dissolved phase, not contaminant sorbed on aquifer surfaces (from Mackay and Cherry, 1989. Reprinted with permission from *Environ. Sci. Technol*, v. 23, pp. 630–636. Copyright 1989 by American Chemical Society).

critical in aquifer remediation. Pumping from purge wells in the plume can deplete the higher concentrations contained in high-permeability beds and lenses quite rapidly.

Concentrations in the purge well often drop dramatically in a short period of time as pumping occurs. If, however, a substantial portion of the contaminant mass is contained on the solids, further cleanup is controlled by the slow partitioning of the contaminant back into the dissolved phase. Estimated durations of remediation range into the decades for sites of this type.

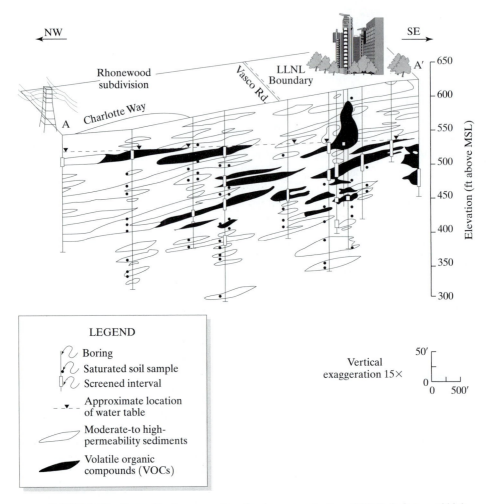

FIGURE 6–23   Generalized cross section showing concentration of VOCs in lenses of high hydraulic conductivity in alluvial sediments beneath LLNL (from Thorpe et al., 1990).

    In heterogeneous aquifers, the problems of delineation and remediation are multiplied. At the Lawrence Livermore National Laboratory in California, solvent releases over several decades into a heterogeneous stratigraphic sequence created a highly complex distribution of contaminants (Fig. 6–23), which accumulate in high hydraulic conductivity lenses and from there move into fine-grained sediment in the aquifer primarily by diffusion. Even after the more permeable parts of the aquifer are purged of contaminants, their removal by pumping alone from the sorptive, poorly permeable, fine-grained lenses in sites of this type may be a nearly endless task.

    Further complications arise when free-phase DNAPL accumulations are present in the aquifer (Fig. 6–16). DNAPL pools at the base or within an aquifer may slowly release the compounds to the aqueous phase by dissolution for very long periods of time.

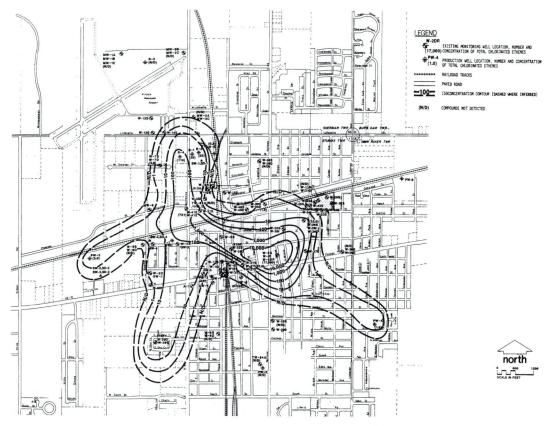

FIGURE 6–24   Map of Sturgis, Michigan, showing distribution of contaminants (μg/l) in lower glacial aquifer in 1989. Contours represent total chlorinated ethenes. Cones of depression from municipal and industrial well fields draw contamination toward them (from Warzyn, 1991).

Figure 6–24 illustrates a contaminant plume in a glacial-drift aquifer in Sturgis, Michigan. The regional stratigraphy consists of two thick sand-and-gravel units separated by one or more silty, clayey tills. Unfortunately, at the location of a major release at this site, the till units were thin or absent, and it is suspected that free-phase DNAPL sank into, and perhaps to the base of the lower aquifer. Through an additional unfortunate circumstance, the cones of depression of municipal and industrial well fields located around the edges of the city draw the contaminants toward the pumping wells in several different directions. Sites of this type indicate that halogenated hydrocarbons will remain problematic for the foreseeable future.

Heterocyclics

Heterocyclic compounds are composed of rings with one or more atoms of an element other than carbon in the ring. The most common heterocyclics contain oxygen, nitrogen, sulfur, or phosphorus in rings and can be saturated or unsaturated.

Oxygen heterocyclics include the *epoxides*, which are three-membered rings constructed of two carbons and one oxygen atom. The compound 1,2-epoxyethane (ethylene oxide) CH_2—CH_2 is a toxic epoxide used in pesticides and other chemicals.

The compound 1,4-dioxane is a six-membered ring with two oxygens that is used as a solvent. The epoxides and dioxane, along with dioxins and furan, are heterocyclics that also fall into the ether group of compounds and will be discussed in more detail later.

Examples of heterocyclics containing nitrogen are shown in Fig. 6–25. Although the hydrogen atoms bonded to carbons are omitted from the diagrams, hydrogen is shown if it is bonded to nitrogen. Because nitrogen forms three covalent bonds, hydrogens occur with nitrogen on the saturated compounds, but not on aromatic compounds, because of the delocalized double bonds in the ring. The pyrrole ring occurs in many natural compounds, including hemoglobin, chlorophyll, and nicotine. Pyrridine is a foul-smelling compound used in chemical synthesis. The triazine ring forms the basis for several widely used herbicides, such as atrazine. Large amounts of atrazine are used as a preemergent herbicide in the midwestern corn belt of the United States.

Alcohols

Compounds whose structure is similar to water, but in which one hydrogen is replaced by a carbon atom or chain, are known as *alcohols*.

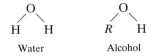

Water   Alcohol

In the general formula for the alcohols, *R*—OH, *R* represents a carbon chain. IUPAC names for alcohols are formed by selecting the longest carbon chain containing the OH group. The *-e* suffix of the *alkane* name is replaced by *-ol* to designate the compound as an alcohol, and the location of the OH group is denoted by a number. Common names are still retained for many of the simple compounds. Alcohols are further classified as primary, secondary, or tertiary, depending upon whether the carbon containing the OH group is bonded to one, two, or three other carbons (Fig. 6–26). Structures of some common alcohols, along with their IUPAC and common names, are shown in Fig. 6–27.

The polyhydroxyl alcohols contain more than one hydroxyl group. Examples of these compounds include ethylene glycol,

$$CH_2 - CH_2$$
$$|\qquad\ |$$
$$OH\quad OH$$

Compound Name	Structure	mp (°C)	bp (°C)	Water Solubility (mg/l at 25°C)	log K_{ow}	P° (mm Hg)	K_H $\left(\dfrac{\text{atm} - \text{m}^3/\text{mol}}{}\right)$
pyrrole	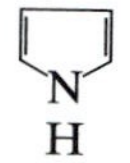	−23.4	129.8	4.70E+004	0.75	8.36E+000	1.80E−005
pyrrolidine	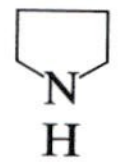	−57.8	86.5	1.00E+006	0.46	6.27E+001	2.39E−006
pyridine		−41.6	115-116	1.00E+006	0.65	2.08E+001	1.10E−005
triazine ring (sym–triazine)		86	114	5.47E+005	−0.73	7.98E+000	1.21E−006
atrazine	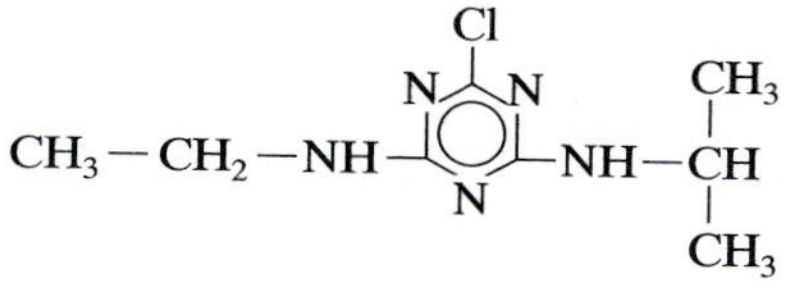	171-174	—	2.80E+001	2.61	3.00E−007	4.47E−009

FIGURE 6–25 Structures and properties of heterocyclics containing nitrogen (values from Howard and Meylan, 1997). mp: melting point. bp: boiling point. P°: vapor pressure. K_H: Henry's constant.

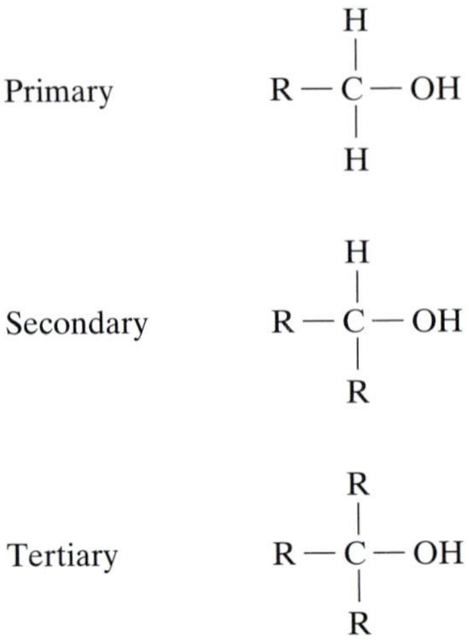

FIGURE 6–26   Structures of
primary, secondary, and tertiary
alcohols.

$$
\text{Primary} \qquad R-\underset{\underset{H}{|}}{\overset{\overset{H}{|}}{C}}-OH
$$

$$
\text{Secondary} \qquad R-\underset{\underset{R}{|}}{\overset{\overset{H}{|}}{C}}-OH
$$

$$
\text{Tertiary} \qquad R-\underset{\underset{R}{|}}{\overset{\overset{R}{|}}{C}}-OH
$$

and glycerol,

$$
\underset{\underset{OH}{|}}{CH_2}-\underset{\underset{OH}{|}}{CH}-\underset{\underset{OH}{|}}{CH_2}.
$$

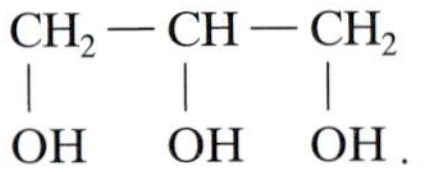

Ethylene glycol is used in antifreeze and to deice airplanes. The large amount of deicing compounds sprayed on airplanes in winter creates an obvious source of contamination involving infiltration through cracks in runways.

Aromatic alcohols include some important contaminant compounds. Examples are given in Fig. 6–28. Phenol is a constituent of creosote and petroleum compounds and is a common contaminant in industrial wastes. It differs from the aliphatic alcohols in that it is a weak acid $(K_A = 1.2 \times 10^{-10})$. The cresols include o-cresol as well as m- and p- isomers. These compounds are also found in coal tar and creosote. Chlorinated phenols, such as pentachlorophenol, are used as wood preservatives and therefore occur along with PAHs at creosote sites.

Simple alcohols are polar compounds because of the electronegativity difference between oxygen and carbon. As a result, they range from being soluble to infinitely soluble in water, which allows them to be highly mobile in ground water. The alcohols are intermediate oxidation products of aliphatic and aromatic hydrocarbons. They are produced by the biodegradation of hydrocarbons in contaminant plumes and then are themselves biodegraded to aldehydes and ketones.

Phenols are anaerobic biodegradation products of BTEX compounds. Phenol has a low vapor pressure $(2.6 \times 10^{-4}\ \text{atm})$, a low Henry's constant $(3.33 \times 10^{-7}\ \text{atm-}m^3/\text{mol})$, and a low log K_{ow} (1.46). These properties favor partitioning into the aqueous phase and mobility in ground water flow systems. At the KL Avenue Landfill Superfund site in Kalamazoo, Michigan, the distribution of phenol was used to define the extent of the plume (Fig. 6–29), much as chloride and other conservative ions are used for the same purpose at other landfill plumes.

Compound (IUPAC Name)	Structure	mp (°C)	bp (°C)	Water Solubility (mg/l at 25°C)	$\log K_{ow}$	P° (mm Hg)	$K_H \left(\dfrac{atm-}{m^3/mol} \right)$
methyl alcohol (methanol)	H—C—OH (with H above and H below the C)	−97.68	64.55	1.00E+006	−0.77	1.27E+002	4.55E−006
ethyl alcohol (ethanol)	H—C—C—OH (with H, H above and H, H below)	−114.49	78.293	1.00E+006	−0.31	5.93E+001	5.00E−006
n-propyl alcohol (1-propanol)	$CH_3 - CH_2 - CH_2 - OH$	−127	97.2	1.00E+006	0.25	2.10E+001	7.41E−006
isopropyl alcohol (2-propanol)	$CH_3 - CH - CH_3$ with OH below middle C	−88.0	82.24	1.00E+006	0.05	4.54E+001	8.10E−006
n-butyl alcohol (1-butanol)	$CH_3 - CH_2 - CH_2 - CH_2 - OH$	−88.62	117.73	6.32E+004	0.88	6.70E+000	8.81E−006
sec-butyl alcohol (2-butanol)	$CH_3 - CH_2 - CH - CH_3$ with OH below	−114.7	99.512	1.81E+005	0.61	1.83E+001	9.06E−006
isobutyl alcohol (2-methyl-1-propanol)	$CH_3 - CH - CH_2 - OH$ with CH_3 below	−108	107.886	8.50E+004	0.76	1.05E+001	9.78E−006

FIGURE 6–27 Structures and properties of common alcohols (values from Howard and Meylan, 1997). mp: melting point. bp: boiling point. P°: vapor pressure. K_H: Henry's constant.

Compound (IUPAC Name)	Structure	mp (°C)	bp (°C)	Water Solubility (mg/l at 25°C)	log K_{ow}	P° (mm Hg)	K_H $\left(\text{atm-} \atop \text{m}^3/\text{mol}\right)$
phenol	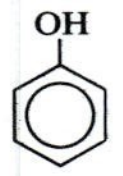	40.90	181.839	8.28E+004	1.46	3.50E-001	3.33E-007
catechol (1,2-benzenediol)	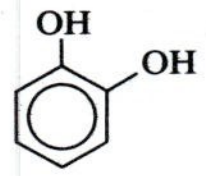	105	245.5	4.61E+005	0.88	1.00E-002	3.14E-009
hydroquinone (1,4-benzenediol)		172	287	7.33E+004	0.59	6.70E-004	1.32E-009
o-cresol (2-methylphenol)	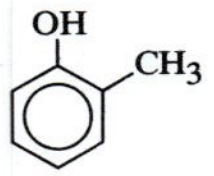	30.944	191.004	2.60E+004	1.95	2.99E-001	1.20E-006
pentachlorophenol	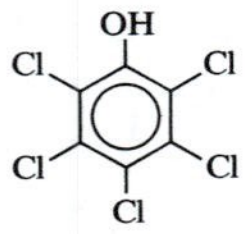	174	309-310	1.95E+003	5.12	1.10E-004	2.45E-008

FIGURE 6–28 Structures and properties of aromatic alcohols (values from Howard and Meylan, 1997). mp: melting point. bp: boiling point. P°: vapor pressure. K_H: Henry's constant.

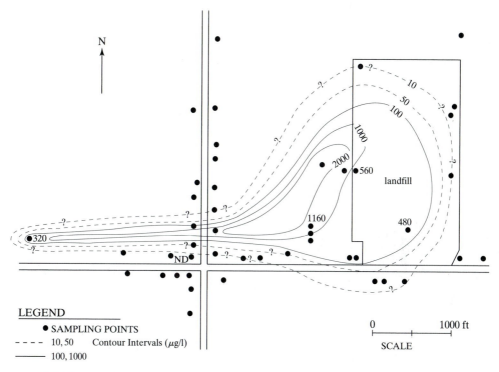

FIGURE 6–29 Isoconcentration map of phenols (μg/l) in ground water at the KL Avenue Landfill, Kalamazoo, Michigan (from Broede, 1990).

Ethers

The *ethers* are derivatives of water formed by replacing both hydrogens with carbon, instead of just one hydrogen as in the alcohols. The general formula for these compounds is $R-O-R'$. Ethers are used as solvents, in pesticides, and in many other manufacturing processes. Consequently, they are not uncommon in contaminant plumes. Simple ethers are less mobile than the alcohols, but some cyclic ethers are highly mobile compounds.

    Names of ethers are formed by naming the groups attached to the oxygen atom in alphabetical order and adding "ether." The ethers include aliphatic, aromatic, and heterocyclic subtypes; examples are shown in Fig. 6–30. In structurally complicated compounds, the ether group is named as a *methoxy* group ($-O-CH_3$), an *ethoxy* group ($-O-C_2H_5$), or a *phenoxy* group ($-O-C_6H_5$). Ethers occur in many compounds used daily in our society. Diethyl ether is one of several ethers used as anesthetics. These compounds produce a state of unconsciousness without affecting the pulse rate or blood pressure. Three-membered heterocyclic rings are known as epoxides and are used to produce synthetic fibers, resins, paints, adhesives, and cosmetics, among other products. Tetrahydrofuran and 1,4-dioxane are solvents that are exceedingly mobile in ground water as a result of their low $\log K_{ow}$ values. Figure 6–31 shows cross sections of plumes

Compound (IUPAC Name)	Structure	mp (°C)	bp (°C)	Water Solubility (mg/l at 25°C)	log K_{ow}	P° (mm Hg)	K_H (atm – m³/mol)
dimethyl ether	$CH_3—O—CH_3$	−141.5	−24.8	3.53E+004	0.10	4.45E+003	1.31E−003
ethyl methyl ether	$CH_3—CH_2—O—CH_3$	−113	7.35	2.05E+005	0.56	1.51E+003	2.93E−004
ethylene oxide (1,2-epoxyethane)	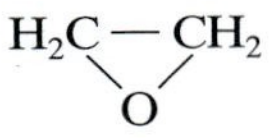 $H_2C—CH_2$ ring with O	−111	10.7	1.00E+006	−0.30	1.31E+003	1.48E−004
tetrahydrofuran	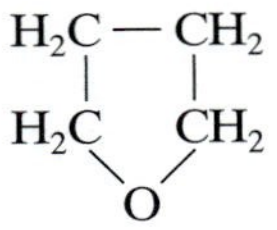 ring $H_2C—CH_2$ / H_2C CH_2 O	−108.5	66	1.00E+006	0.46	1.62E+002	7.05E−005
1,4 dioxane	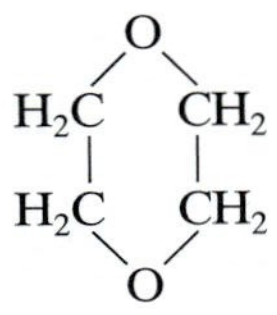 ring with two O, CH_2 groups	11.80	101.32	1.00E+006	−0.27	3.81E+001	4.80E−006
methoxychlor [2.2-bis (4-methoxyphenyl) -1,1,1-trichloroethane)]	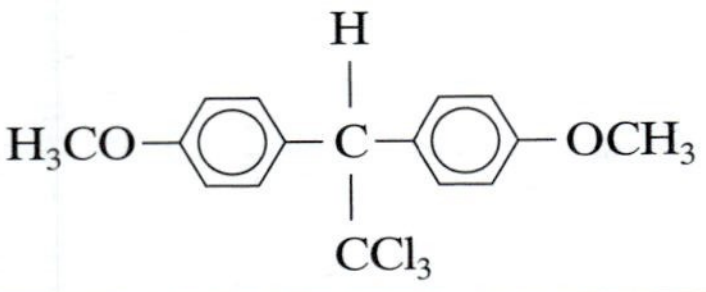 H_3CO—(ring)—$C(H)(CCl_3)$—(ring)—OCH_3	89	346	1.00E−001	5.08	2.58E−006	1.58E−005

FIGURE 6–30 Structures and properties of the ethers (values from Howard and Meylan, 1997). mp: melting point. bp: boiling point. P°: vapor pressure. K_H: Henry's constant.

FIGURE 6–30   *(continued)*

methoxy tert butyl ether	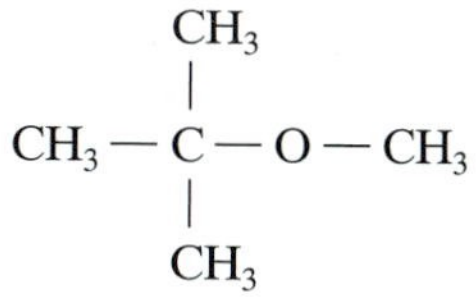
chlorinated dibenzofurans	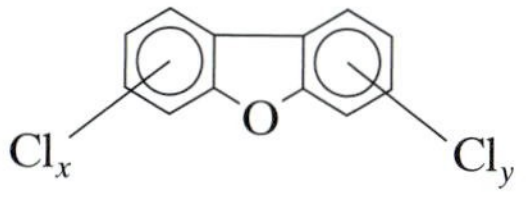
chlorinated dibenzodioxins	

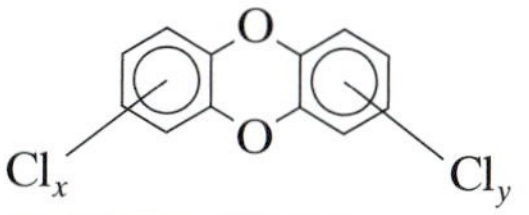

of these compounds emanating from a landfill near Ottawa, Canada (Jackson et. al, 1985). Methoxychlor is a widely used herbicide. Methoxy tert butyl ether (MTBE) is used as an unleaded gasoline additive. Notorious at UST sites for its mobility, MTBE commonly occurs at the leading edge of a plume, with a center of mass ahead of BTEX compounds along the ground water flow path.

Chlorinated ethers include dibenzofurans and dibenzo-*p*-dioxins. The chlorinated dibenzo-*p*-dioxins constitute a family of compounds that are among the most toxic chemicals known. They occur as impurities in herbicides and defoliants such as Agent Orange and as by-products of the incineration of industrial chemicals. Because of their high log K_{ow}, they are very immobile in soil and ground water.

Aldehydes

The *aldehydes* are one of several classes of compounds containing the *carbonyl group*, an important functional group defined by a double bond between carbon and oxygen. In the aldehydes, the carbonyl group is bonded to only one carbon atom. So, in aliphatic compounds, the group lies at one end of the carbon chain. The general formula for the aldehydes can be written

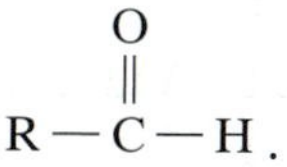

$$R - \overset{\displaystyle O}{\overset{\displaystyle \|}{C}} - H \,.$$

The naming of aldehydes is based on selecting of the parent carbon chain— the longest chain containing the carbonyl group—and replacing the -*e* in the name of the chain by -*al*. In the simplest aldehyde, methanal, the carbonyl group bonds to two hydrogen atoms. This compound is better known by its common name, *formaldehyde*.

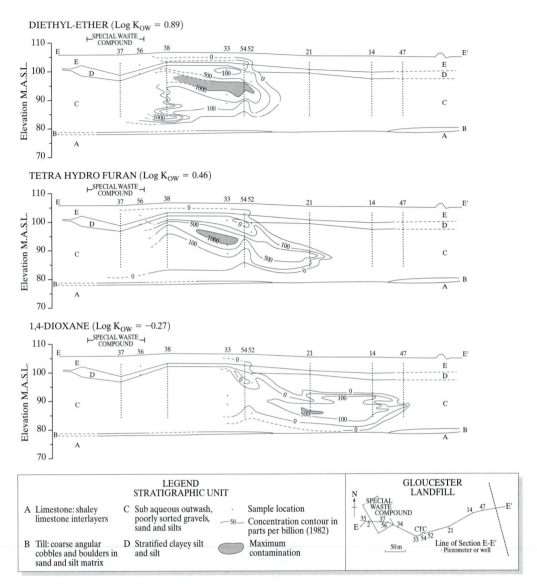

FIGURE 6–31 Cross sections showing plumes of diethyl ether, tetrahydrofuran, and 1,4-dioxane at the Gloucester Landfill, Ontario. Distance of migration is inversely proportional to $\log K_{ow}$ (from Jackson et al., 1985. Contaminant hydrogeology of toxic organic chemical at a disposal site, Gloucester, Ontario, Environment Canada, National Hydrology Research Institute. Paper no. 23. Reproduced with the permission of the Minister of Public Works and Government Services, Canada, 2000).

Representative aldehydes are shown in Fig. 6–32. The electronegativity difference between carbon and oxygen gives the carbonyl group, and therefore the aldehydes, a polar nature. Their solubility in water is high, and they are highly mobile in ground water.

Compound (IUPAC Name)	Structure	mp (°C)	bp (°C)	Water Solubility (mg/l at 25°C)	$\log K_{ow}$	P° (mm Hg)	K_H $\left(\dfrac{atm-}{m^3/mol} \right)$
formaldehyde (methanal)	$H-\overset{\overset{O}{\|\|}}{C}-H$	−92	−21	4.00E+005	0.35	3.89E+003	3.37E−007
acetaldehyde (ethanal)	$CH_3-\overset{\overset{O}{\|\|}}{C}-H$	−123	20.4	1.00E+006	−0.34	9.02E+002	6.67E−005
isobutyraldehyde (2-methylpropanal)	$CH_3-\overset{\overset{CH_3}{\|}}{CH}-\overset{\overset{O}{\|\|}}{C}-H$	−65.9	64	8.90E+004	0.74	1.73E+002	1.80E−004
benzaldehyde	$\bigcirc\!\!-\overset{\overset{O}{\|\|}}{C}-H$	−56.5	179	3.00E+003	1.48	1.27E−001	2.67E−005
cinamaldehyde (3-phenyl-2-propenal)	$\bigcirc\!\!-CH=CH-\overset{\overset{O}{\|\|}}{C}-H$	−7.5	246	1.42E+003	1.90	2.89E−002	1.60E−006

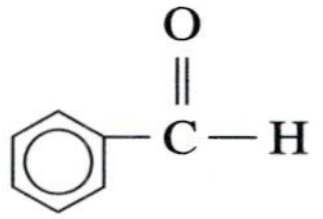

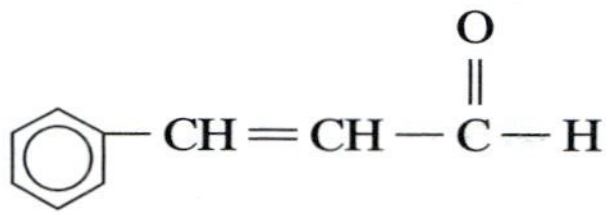

FIGURE 6–32 Structures and properties of the aldehydes (values from Howard and Meylan, 1997). mp: melting point. bp: boiling point. P°: vapor pressure. K_H: Henry's constant.

Ketones

The *ketones* are similar in structure to the aldehydes, with the difference that the carbonyl group is bonded to two carbon atoms. The general formula for the ketones is

$$R - \overset{\displaystyle O}{\overset{\displaystyle \|}{C}} - R'.$$

IUPAC names for the ketones are derived in the same way as for the aldehydes, except that the suffix of the carbon-chain name becomes *-one* and the location of the carbonyl group is designated by a number. In the older terminology, carbon chains on either side of the carbonyl group are named, and the word *ketone* is added. Under this nomenclature, 2-butanone is called methyl ethyl ketone (MEK). This compound is a solvent commonly found in leachate plumes at hazardous waste sites. Like other ketones, MEK is soluble and mobile. The structures and properties of some simple ketones are illustrated in Fig. 6–33.

Carboxylic Acids

When the carbonyl group is bonded to an OH group, the resulting functional group is called the *carboxyl group*, with structure

$$-\overset{\displaystyle O}{\overset{\displaystyle \|}{C}} - OH.$$

The group occurs in both aliphatic and aromatic compounds. The compounds based upon the presence of the carboxyl group are *carboxylic acids*. These compounds are oxidation products of other organics and are therefore important in many contaminant plumes. Compound names follow the usual practice of modifying the name of the carbon chain, in this case by dropping *-e* and adding *-oic acid*. Examples are shown in Fig. 6–34. The common names *formic acid* and *acetic acid* are retained for the C_1 and C_2 members, respectively. The phenoxyacetic acids are herbicide compounds. The defoliant Agent Orange is a mixture of 2,4-D and 2,4,5-T.

Like inorganic acids, the carboxylic acids ionize to varying degrees in aqueous solutions. For acetic acid, the reaction is

$$CH_3COOH \rightleftharpoons CH_3COO^- + H^+. \tag{6–15}$$

The equilibrium constant for the reaction is an acid dissociation constant, written

$$K_a = \frac{a_{CH_3COO^-} a_{H^+}}{a_{CH_3COOH}}. \tag{6–16}$$

Acid dissociation constants are commonly reported as pK_a values, using the negative logarithmic form. Some examples of pK_a values are given in Table 6–7.

Carboxylic acids are significant components of most contaminant plumes originating from organic-rich wastes. In the reducing ground water downgradient from the

Compound (IUPAC Name)	Structure	mp (°C)	bp (°C)	Water Solubility (mg/l at 25°C)	log K_{ow}	P° (mm Hg)	$K_H \left(\frac{atm-m^3/mol}{} \right)$
acetone (2-propanone)	$CH_3-\overset{O}{\overset{\|}{C}}-CH_3$	−94.7	56.07	1.00E+006	−0.24	2.31E+002	3.97E−005
methyl ethyl ketone (2-butanone)	$CH_3-\overset{O}{\overset{\|}{C}}-CH_2-CH_3$	−86	79.6	2.23E+005	0.29	9.53E+001	5.69E−005
methyl phenyl ketone (phenylethanone) (acetophenone)	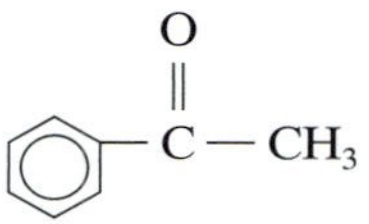	19.62	202.0	6.13E+003	1.58	3.79E−001	1.07E−005
diphenyl ketone (diphenylmethanone) (benzophenone)	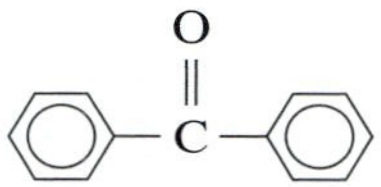	48.5	305.4	1.37E+002	3.18	1.93E−003	1.94E−006

FIGURE 6–33 Structures and properties of the ketones (values from Howard and Meylan, 1997). mp: melting point. bp: boiling point. P°: vapor pressure. K_H: Henry's constant.

Compound (IUPAC Name)	Structure	mp (°C)	bp (°C)	Water Solubility (mg/l at 25°C)	$\log K_{ow}$	P° (mm Hg)	K_H $\left(\text{atm} - \dfrac{}{\text{m}^3/\text{mol}}\right)$
formic acid (methanoic)	$H-\overset{\overset{\displaystyle O}{\|}}{C}-OH$	8.27	100.56	1.00E+006	−0.54	4.26E+001	1.67E−007
acetic acid (ethanoic)	$CH_3-\overset{\overset{\displaystyle O}{\|}}{C}-OH$	16.6	117.9	1.00E+006	−0.17	1.57E+001	1.00E−007
propanoic acid	$CH_3-CH_2-\overset{\overset{\displaystyle O}{\|}}{C}-OH$	−20.8	140.7	1.00E+006	0.33	3.53E+000	4.45E−007
benzoic acid (benzenecarboxylic acid)	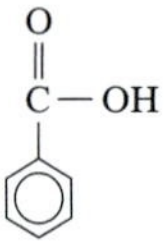	122.4	249.2	3.40E+003	1.87	1.20E−003	1.08E−007
salicylic acid (2-hydroxybenzene-carboxylic acid)	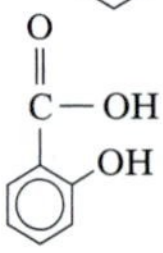	157-159	211	2.06E+003	2.26	8.20E−005	7.34E−009
2-methoxybenzoic acid (o-methoxybenzoic acid)	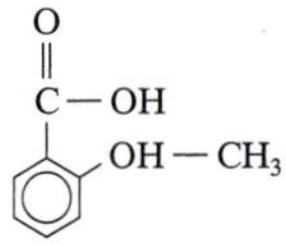	101	200	5.00E+003	1.59	1.59E−005	6.41E−009
2, 4-D (2,4-dichlorophenoxy-acetic acid)	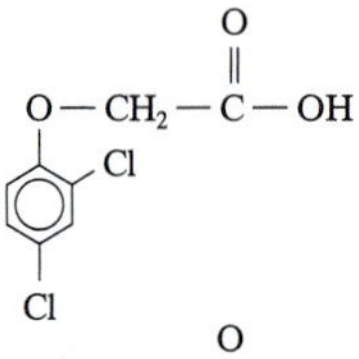	140.5	160	6.77E+002	2.81	6.00E−007	1.02E−008
2,4,5-T (2,4,5-trichlorophenoxy-acetic acid)	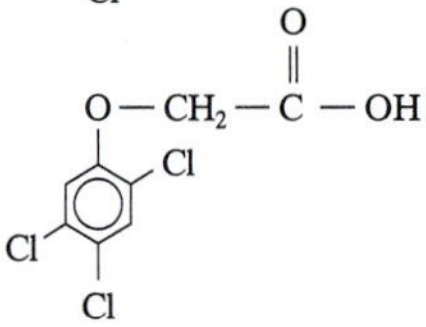	154-155	–	2.78E+002	3.31	3.75E−005	8.68E−009

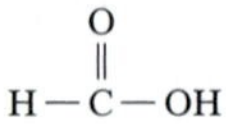

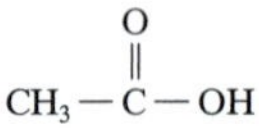

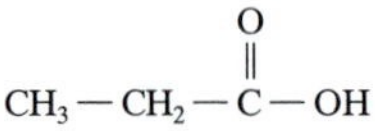

FIGURE 6–34   Structures and properties of carboxylic acids (values from Howard and Meylan, 1997) mp: melting point. bp: boiling point. P°: vapor pressure. K_H: Henry's constant.

TABLE 6–7   Acid dissociation constants for some carboxylic acids.

Acid	pK_a
Formic	3.74
Acetic	4.74
Propanoic	4.85
Benzoic	4.19

oil pool at the Bemidji site, for example, carboxylic-acid concentrations increased by an amount equal to the decrease in primary petroleum compounds. The organic acids include aliphatic, alicyclic, and aromatic structures and are structurally similar to their precursor compounds (Cozzarelli et al., 1990, 1994). A temporally increasing trend in organic-acid concentrations was noted as the aquifer became more reducing (Fig. 6–35).

FIGURE 6–35   Concentrations (μM) of carbon over time (1986–1990) of aliphatic, aromatic, and alicyclic organic acids in ground water in wells installed beneath (421), at the downgradient edge (534b) and just beyond the downgradient edge (533a) of the oil body at the Bemidji, Minnesota, pipeline rupture site (reprinted from *Geochem. Cosmochem.* Acta., v. 58, Cozzarelli, I. M., Baedecker, M. J., Eganhouse, R. P., and Goerlitz, D. F., The geochemical evolution of low-molecular weight organic acids derived from the degradation of petroleum contaminants in groundwater, pp. 863–877, Copyright 1994, with permission from Elsevier Science).

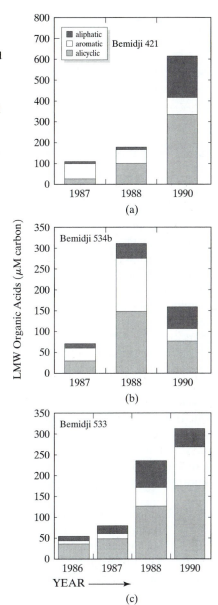

Esters

Functional derivatives of carboxylic acids in which the —OH group of the acid is replaced by an —OR group are called *esters*. Similar acid derivatives, with their replacement groups in parentheses, include the *amides* (—NH$_2$), *anhydrides* (—OCOR), and *acid chlorides* (—Cl). Names of esters include the name of the carbon group attached to oxygen, followed by the name of the acid, with the suffix *-ic* replaced by *-ate*. Examples are given in Fig. 6–36.

Numerous esters occur in hazardous waste and the contaminant plumes derived from it. The phthalates are particularly common. These compounds are used as plasticizers, solvents, and components of insect repellents. In a study of the relative abundances of organic compounds at hazardous waste sites, Plumb (1992) listed bis (2-ethylhexyl) phthalate as the 11th most commonly detected organic compound. Like the organic acids from which they are derived, most esters are polar and mobile in aquifers.

Compounds of Nitrogen, Sulfur, and Phosphorus

Earlier, we discussed compounds containing nitrogen in aliphatic or aromatic rings. Other important classes of compounds, however, contain nitrogen, sulfur, and phosphorus. The *amines* are a group of compounds based on the structure of ammonia in which one or more of the hydrogen atoms are replaced by alkyl or aromatic groups. Names are formed by naming the alkyl groups attached to the nitrogen atom and adding *-amine*. Thus methylamine, dimethylamine, and trimethylamine are formed by replacing one, two, and three hydrogens on the ammonia structure (Fig. 6–37). These amines are known as primary, secondary, and tertiary, respectively. An NH$_2$ group can also be named as an *amino* substituent.

Aromatic amines are called *anilines*. Aniline itself is formed by bonding an amino group to a benzene ring. Other substituents can be added to the ring to form many other compounds. The presence of single methyl groups on the aniline structure is named using *o, m,* and *p* nomenclature to designate the position of the methyl group (Fig. 6–37).

Amino acids are important components of proteins and are common in the dissolved organic fraction of natural waters. Typically, 1 to 3% of DOC is composed of amino acids. The amino acid structure includes an amino group on the carbon adjacent to the carbonyl carbon of a carboxylic acid. If the hydroxyl group of a carboxylic acid is replaced by an amino group, an *amide* is formed. Urea is an amide found in urine and is also an important component of fertilizers. In soils, urea reacts to form carbon dioxide and water.

Double- and triple-bonded C–N compounds include some important members. *Nitriles* are organic cyanides used in the production of synthetic fibers. Compounds containing the nitro (NO$_2$) group include explosives such as 2,4,6-trinitrotoluene (TNT) and nitroglycerine. The nitroso functional group, —N=O, occurs in compounds such as the nitrosamines, which are carcinogenic and are found in foods and beverages such

Compound (IUPAC Name)	Structure	mp (°C)	bp (°C)	Water Solubility (mg/l at 25°C)	log K_{ow}	P° (mm Hg)	K_H $\left(\text{atm}-\text{m}^3/\text{mol}\right)$
methyl formate	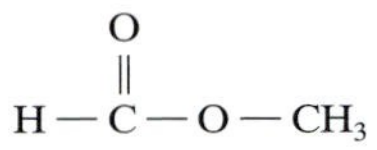 $H-C(=O)-O-CH_3$	−99	31.75	2.30E+005	0.03	5.86E+002	2.23E−004
ethyl acetate	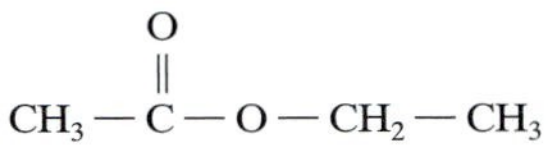 $CH_3-C(=O)-O-CH_2-CH_3$	−83.6	77.06	8.00E+004	0.73	9.37E+001	1.34E−004
phenyl acetate	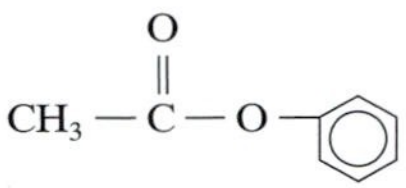 $CH_3-C(=O)-O-$ (phenyl)	−	195-196	4.64E+003	1.49	3.98E−001	6.48E−005
vinyl acetate (ethenyl ethanoate)	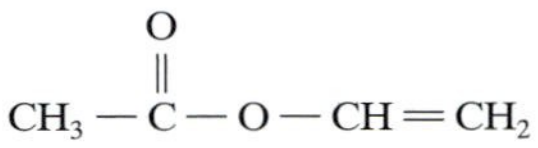 $CH_3-C(=O)-O-CH=CH_2$	−92.8	72.5	2.00E+004	0.73	9.02E+001	5.11E−004
dimethyl phthalate	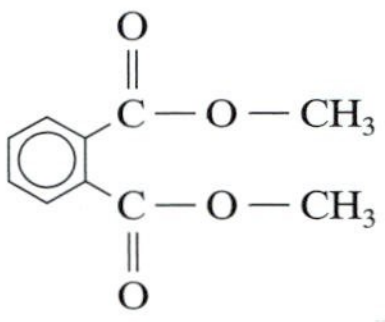 benzene ring with two $C(=O)-O-CH_3$ groups	0-2	282	4.00E+003	1.56	1.65E−003	1.05E−007
bis (2-ethylhexyl) phthalate	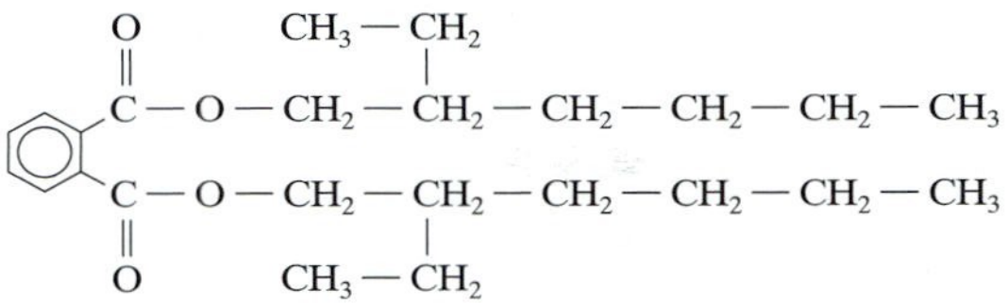 benzene ring with two $C(=O)-O-CH_2-CH(CH_2-CH_3)-CH_2-CH_2-CH_2-CH_2-CH_3$ groups	−50	384	3.40E−001	7.60	9.75E−006	1.47E−005

FIGURE 6-36 Structures and properties of the esters (values from Howard and Meylan, 1997). mp: melting point. bp: boiling point. P°: vapor pressure. K_H: Henry's constant.

Compound (IUPAC Name)	Structure	mp (°C)	bp (°C)	Water Solubility (mg/l at 25°C)	log K_{ow}	P° (mm Hg)	K_H ($\frac{atm-}{m^3/mol}$)
ammonia	H—N—H with H on top						
methylamine (primary amine)	CH_3—N—H with H on top	−93.5	−6.3	1.08E+006	−0.57	2.65E+003	1.11E−005
dimethylamine (secondary amine)	CH_3—N—CH_3 with H on top	−92.2	6.89	1.63E+006	−0.38	1.52E+003	1.77E−005
triethyl amine (tertiary amine)	CH_3—CH_2—N—CH_2—CH_3 with CH_2—CH_3 branch	−115	89–90	7.37E+004	1.45	5.71E+001	1.49E−004
ethanolamine (2-amino ethanol)	HO—CH_2—CH_2—NH_2	10.3	170.8	1.00E+006	−1.31	4.04E−001	3.25E−008
aniline	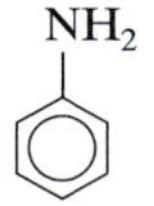 NH_2-phenyl	−5.98	184.40	3.60E+004	0.90	4.90E−001	1.90E−006
m-methylaniline (m-toluidine)	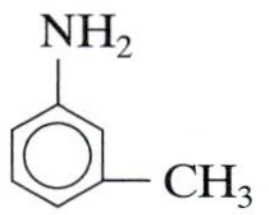 NH_2-phenyl-CH_3	−50	203–204	1.50E+004	1.40	3.03E−001	2.85E−006

FIGURE 6-37 Structures and properties of compounds containing nitrogen (values from Howard and Meylan, 1997). mp: melting point. bp: boiling point. P°: vapor pressure. K_H: Henry's constant.

Name	Structure						
amino acid	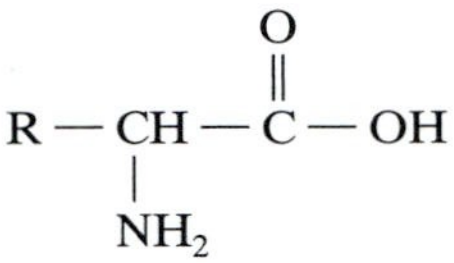 $R-CH(NH_2)-C(=O)-OH$						
urea (i-aminomethanamide)	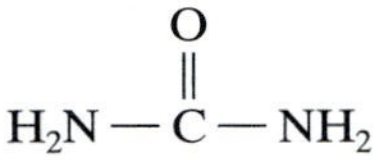 $H_2N-C(=O)-NH_2$	-95	91.5	5.45E+005	-2.11	1.20E-005	1.74E-012
hydrogen cyanide	$H-C\equiv N$						
acetonitrile (methyl cyanide)	$CH_3-C\equiv N$	-45	81.6	1.00E+006	-0.34	8.88E+001	3.45E-005
2,4,6-trinitrotoluene	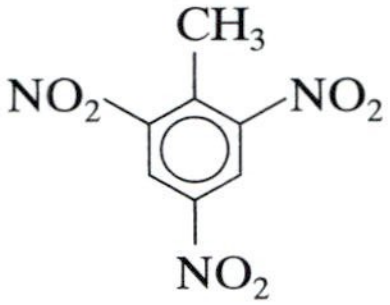 (structure)	80.1	186.6	1.30E+002	1.60	8.02E-006	4.57E-007
N-nitrosodimethylamine	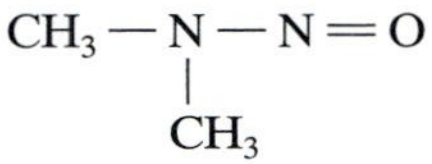 $CH_3-N(CH_3)-N=O$	–	154	1.00E+006	-0.57	2.70E+000	1.82E-006
aldicarb	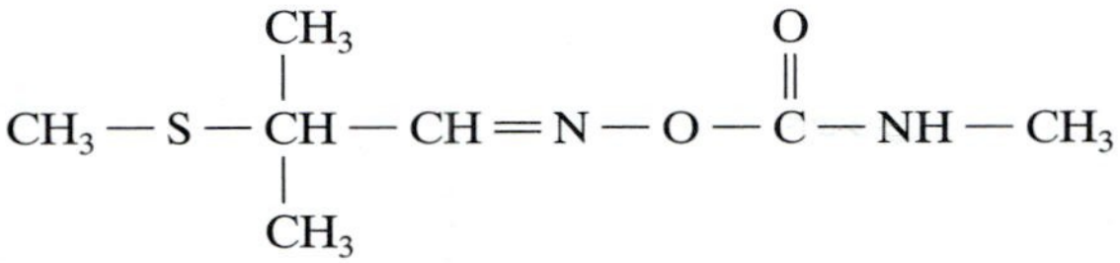 (structure)	98-100	–	6.03E+003	1.13	3.47E-005	1.44E-009

FIGURE 6-37 (continued)

as beer, whiskey, and bacon. *Carbamate pesticides* and *herbicides* are based on the carbamic acid structure,

$$
\begin{array}{c}
\text{O} \\
\parallel \\
\text{HO} - \text{C} - \text{OH} .
\end{array}
$$

Aldicarb, a carbamate pesticide, has been detected in several shallow unconfined aquifers, such as the Central Sand Plain of Wisconsin, and in Long Island, New York.

Sulfur occurs in many organic compounds, often replacing other elements, including oxygen or carbon. *Thiols*, or *mercaptans* (Fig. 6–38), for example, are compounds in which sulfur replaces the oxygen in the alcohol structure. Similarly, *sulfides*, $R-S-R'$, are sulfur analogues to the ethers. The general formula for thiols is $R-S-H$. Thiols are notorious for their odor, which is a major component of skunk spray. The odor is put to beneficial use in the addition of thiols to natural gas and liquified petroleum gas to provide a warning for the detection of these otherwise odorless explosive gases in a house. The thiols differ significantly from the alcohols because of the properties of the S—H bond, which is nonpolar covalent, in contrast to the polar covalent O—H bond. Thiols are therefore less soluble in water and have lower boiling points than alcohols.

The organophosphorus compounds form a large and structurally complex group that includes nerve gases and pesticides. Malathion and diazanon are some of the more common examples of these compounds. Pesticides will be discussed further in Chapter 11.

Miscellaneous.   In contrast to the specific structures described throughout most of this chapter, a certain component of the dissolved organic carbon fraction of ground water consists of poorly defined natural substances derived from the breakdown of plant material. The terms *humic* and *fulvic* acids, which are applied to these substances, refer to a wide range of structural and chemical characteristics. Humic and fulvic acids are differentiated from each other by the insolubility of humic acids in solutions less than pH = 2, relative to the solubility of fulvic acids in both acid and basic solutions. Humic and fulvic acids are nonvolatile, are resistant to biodegradation, and have molecular weights ranging from 500 to 5,000 g/mol. Structurally, they include carboxylic, phenolic, and aromatic functional groups. In the vadose zone, humic and fulvic acids play an important role in the complexation and leaching of aluminum and iron within the soil horizons.

Abiotic Transformation Reactions of Organic Compounds

In addition to the dissolved-phase transport of organic compounds in ground water and sorptive processes, the fate of organic compounds in the subsurface involves chemical and biological transformation reactions. In the next chapter, we examine biotransformation reactions. Here, the most important abiotic transformation reactions will be

Compound Name	Structure	mp (°C)	bp (°C)	Water Solubility (mg/l at 25°C)	log K_{ow}	P° (mm Hg)	K_H $\left(\dfrac{\text{atm}-}{\text{m}^3/\text{mol}}\right)$
methanethiol	CH_3—S—H	−123	5.95	1.54E+004	0.78	1.51E+003	3.12E−003
dimethyl sulfide	CH_3—S—CH_3	−83	36.2	2.20E+004	0.92	5.02E+002	1.61E−003
malathion	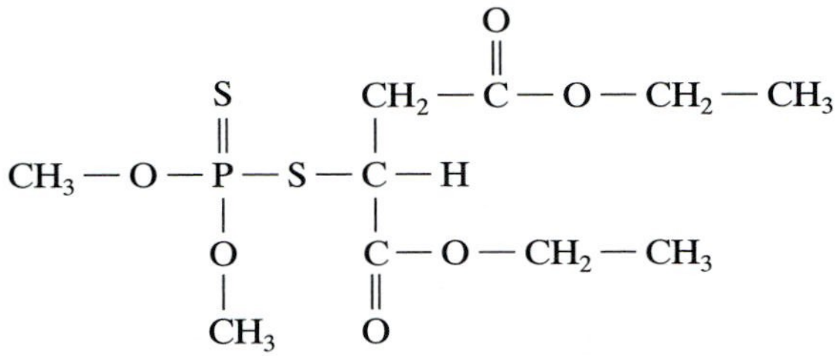 (structure shown)	2.85	156-157	1.43E+002	2.36	7.90E−006	2.40E−008

FIGURE 6–38 Structures and properties of sulfur and phosphorus compounds (values from Howard and Meylan, 1997). mp: melting point. bp: boiling point. P°: vapor pressure. K_H: Henry's constant.

introduced. The difference between the two types of processes is that biotransformation requires the involvement of metabolically active microorganisms.

Hydrolysis describes the reaction of an organic compound with the water molecule or one of its dissociation products, H^+ or OH^-. In these reactions, the water molecule attacks an organic bond and the compound is broken down into smaller portions. For example, halogenated aliphatics hydrolyze via the reaction

$$R-CH_2X + H_2O \longrightarrow R-CH_2OH + H^+ + X^-, \tag{6–17}$$

where X represents a halogen atom. These reactions are pH sensitive, whereby at low pH values the water component, or *nucleophile*, is H^+, and at high pH values the nucleophile is OH^-. The reaction rate can be written as

$$k_h = k_a a_{H^+} + k_n + k_b a_{OH^-}, \tag{6–18}$$

where k_h is the overall rate constant, k_a is the acid-catalyzed rate constant, k_n is the neutral rate constant, and k_b is the base-catalyzed rate constant. Typically, only one of the components dominates the rate, depending on the pH value, and the rate can be assumed to be first order or pseudo first order. Half-lives for reported values of several halogenated compounds are given in Table 6–8 and are generally much longer than biotransformation half-lives.

Reactions in which a hydrogen and a halogen are removed from the organic compound are termed *elimination,* or *dehydrohalogenation*, reactions. The structure of the organic compound is internally reordered with the formation of a double bond. No change in oxidation state is involved. An example is the transformation of 1,1, 1-trichloroethane to 1,1-dichloroethene:

$$\underset{\underset{Cl}{|}}{\overset{\overset{Cl}{|}}{Cl-C}}-\underset{\underset{H}{|}}{\overset{\overset{H}{|}}{C}}-H \longrightarrow \overset{\overset{Cl}{|}}{Cl-C}=\overset{\overset{H}{|}}{C}-H + HCl . \tag{6–19}$$

TABLE 6–8   Hydrolysis half-lives of selected chlorinated solvent compounds (from Sawyer et al., 1994).

Compound	$T_{1/2}$ (years)
Chloroform	1849
Carbon tetrachloride	40
1.1.1-Trichloroethane	1.1
Pentachloroethane	0.01 (3.7 days)
Trichloroethylene	1.23×10^6
Tetrachloroethylene	9.62×10^8

Oxidation and reduction reactions involving organics are just like those involving inorganics, in that the transformed compound either loses an electron to an electron acceptor or gains an electron from an electron donor. Although abiotic oxidations are rare, abiotic reductions are quite common. Many of the halogenated aliphatics and aromatics undergo *reductive dehalogenation* as follows:

$$\underset{\text{Trichloroethene}}{\overset{\displaystyle \overset{\text{Cl}}{\underset{|}{}} \quad \overset{\text{Cl}}{\underset{|}{}}}{\text{Cl}-\text{C}=\text{C}-\text{H}}} + 2e^- + H^+ \longrightarrow \underset{\text{1,2-dichloroethene}}{\overset{\displaystyle \overset{\text{Cl}}{\underset{|}{}} \quad \overset{\text{Cl}}{\underset{|}{}}}{\text{H}-\text{C}=\text{C}-\text{H}}} + Cl^- . \tag{6-20}$$

Analysis of Organics. A thorough discussion of analytical procedures is well beyond the scope of this book. While hydrogeologists are not usually involved in the laboratory analysis of organic compounds, they should be familiar with the instruments and procedures used to obtain the concentrations that will be reported back from the laboratory. The following minimal comments are meant to provide a starting point for hydrogeologists taking part in contaminant investigations involving organic compounds.

The most common technique for analyzing trace organic compounds is the combination of *gas chromatography* with *mass spectrometry* (GC/MS). Analytical methods are specific to the sample *matrix*, which, for most hydrogeological investigations, will be either water or soil. When a sample is brought to the laboratory, it is first separated from the matrix by a method determined by the volatility of the compounds in question. Volatile compounds can be extracted by a method called *purge and trap*, in which the sample is sparged with an inert gas such as helium, which in turn transports the compounds to the trapping material that adsorbs them. The compounds are then desorbed thermally into the gas chromatograph, which is composed of a capillary column as much as 20 to 30 meters in length, coiled inside the instrument. The inner wall of the capillary column is coated with an absorbent material, and as the sample is forced through the column by a transporting gas, individual compounds sorb to the wall of the column according to their sorptive properties. This tends to separate the sample into individual compounds that pass through the column at different rates, much like ions passing through an aquifer at different rates determined by the distribution coefficient. The least sorbed compounds pass through the column first and arrive at the end, where they pass into the detector. Each compound passes through the column in this fashion, from the least to most highly sorbed. The arrival time at the detector is the critical factor in identifyng the compound.

Several different types of detectors are used to analyze organic compounds: the *electron capture* detector, which is useful for analyzing chlorinated pesticides and PCBs, the *flame ionization detector* (FID) for analyzing hydrocarbons, the *thermal conductivity detector* (TC) for analyzing gases, the *photoionization detector* (PID) for aromatic hydrocarbons, and the *mass spectrometer* (MS), which bombards the organic molecule with electrons and separates the resulting ions by their mass-to-charge ratio. The spectra of molecular fragments for each compound, along with the relative amounts of the fragments, are compared against known spectra, and the compounds emerging from the gas chromatograph can be both identified and quantified. The MS is the most useful detector for a broad range of compounds.

Nonvolatile and thermally unstable compounds are analyzed using a method known as high-pressure liquid chromatography (HPLC). Several types of detectors, including ultraviolet and fluorescent detectors, as well as detectors involving electroconductivity, are used with HPLC.

Another aspect of the laboratory analysis of ground water samples with which the hydrogeologist needs to be familar is the system of procedures known as *quality assurance/quality control* (QA/QC). QA/QC programs specify a series of steps taken to ensure that any data produced in the laboratory are both *accurate* and *precise*. Accuracy is a measure of the nearness of the measured value to the true or expected value, and precision is an indication of how closely repeated measurements of the same sample approach one another. In the laboratory, QA/QC involves a regular plan for the periodic calibration of instruments, analysis of blanks, and analysis of replicates. Blanks are samples of purified water that supposedly do not contain organic compounds which may be present at the site. *Trip* blanks are filled sample bottles that are taken into the field from the laboratory during sampling. Blanks can also be prepared in the field by running purified water through a pump or bailer in order to demonstrate that the equipment has been properly decontaminated between samples. Blanks can be extremely important to the conclusions drawn about the ground water being sampled. For example, methylene chloride is an extremely volatile compound that is commonly used in the laboratory. If methylene chloride is detected in trip blanks and samples at roughly the same concentration, it would indicate that its source was the laboratory rather than the aquifer. Even the most carefully controlled laboratory conditions cannot always prevent the presence of compounds that could be attributed to the samples being analyzed. A complete QA/QC program is a long and extensively documented set of procedures. While these programs are expensive, they are essential to the collection of data that are representative of the aquifer being sampled.

PROBLEMS

1. A certain soil has a bulk density of 1.28 g/cm^3 and an organic carbon content of 1%. The porosity of the soil is 30%, of which 50% is occupied by water and 50% by air. Using values from Table 6–4, find the percentages of benzene and 2-butanone contained within the solid, liquid, and gas phases.

2. How do Henry's constant and vapor pressure differ? What characteristics of a compound make it most likely to volatilize from an aqueous phase?

3. Draw the structures of the following compounds:
   a. methylene chloride;
   b. methyl ethyl ketone;

    c. m-xylene;

    d. 1,1,1 trichloroethane;

    e. ethanol.

4. What is the difference between saturated and unsaturated hydrocarbons?

5. Describe the geochemical conditions in each of the zones in the Bemidji plume. How do the classes of hydrocarbon compounds vary among these zones?

6. Are PCBs significant threats to ground water quality? Why or why not?

7. Which of the halogenated hydrocarbons are most commonly detected in ground water? What are the most important sources of these compounds, and what properties control their mobility?

8. Describe the similarities of the structures of alcohols and ethers. Which representatives of these groups are important ground water contaminants and why?

9. Which groups contain the carbonyl group? What types of waste contain these compounds?

C H A P T E R   7

Biotransformation of Organic Compounds

Research into the fate and transport of organic contaminant compounds in ground water flow systems over the past several decades has produced a growing body of evidence supporting the importance of microbial transformations of these substances under a wide range of geochemical conditions. Simultaneously, our knowledge of the role of microorganisms in controlling natural geochemical conditions in aquifers is steadily growing. In this chapter, these processes will be examined in light of basic microbial ecology.

The realization by geologists that microorganisms play an important role in subsurface geochemical processes is a relatively recent phenomenon. With the exception of a few pioneering studies, most subsurface geochemical processes, particularly those occurring beneath the soil zone, were assumed by most geoscientists to be mostly inorganic (Chapelle, 1993). It was not until the 1970s and 1980s that evidence for active subsurface microbial communities began to grow. The prevailing assumption that drilling was responsible for the introduction of bacteria into the subsurface was gradually replaced by the recognition that indigenous microorganisms are extremely important in subsurface processes. Microbiologists began to turn their attention to the subsurface, and aseptic methods of drilling and coring were developed to collect uncontaminated samples for laboratory analysis and experimentation. A growing consensus that bioremediation is the most likely long-term solution to many types of organic contamination is currently driving the integration of hydrogeology and microbiology.

CHARACTERISTICS, GROWTH, AND METABOLISM OF SUBSURFACE MICROORGANISMS

Microorganisms comprise a huge number of organisms that can be divided into four groups: *procaryotes*, *eucaryotes*, *archaebacteria*, and *viruses*. Procaryotes, which include bacteria and *cyanobacteria* (blue-green algae), have a simple cellular structure that lacks a nucleus. They are widely distributed throughout surface and subsurface environments. Eucaryotes have a more complex structure including a nucleus at the center of each cell. Members of the eucaryote group include *algae*, which produce oxygen through photosynthesis, *fungi*, which decompose existing carbon for growth and energy, and *protozoa*. Protozoa include the organisms that form the familiar microfossils foraminifera and diatoms. Eucaryotes are less abundant in the subsurface relative to procaryotes, although they do exist.

The archaebacteria inhabit the type of inhospitable environments that may have been present in the early history of the earth—hence the prefix "archae," which means ancient. Their members include the *methanogens*, a very important type of microorganism that metabolizes hydrogen and carbon dioxide and gives off methane. The methanogens are limited to anaerobic environments, which are common in the subsurface. The occurrence of methane gas in landfills is the result of their activity. Other archaebacteria live in extreme environments ranging from brines to hot, acid solutions.

Viruses are extremely small entities that differ significantly from the previously described microorganisms in that, although they contain genetic information in the form of DNA, they are parasites and cannot reproduce by themselves. Instead, in order to reproduce, they must inject their DNA into bacteria or other microorganisms. Because of their extremely small size they are potentially mobile in aquifers. Their persistence in ground water environments, however, is uncertain, and pathogenic viruses may not pose a serious health threat via ground water transport except under very favorable circumstances.

Classification of Microorganisms

Like all other organisms, microbes are grouped into a hierarchical ordering system ranging from individual species to more inclusive classes such as genera, families, and orders. Names are formed by combining the genus name and species name, as in the intestinal bacteria *Escherichia coli*, which is shortened to *E. coli*. The characteristics used in classification include several different categories, such as the shape of the bacterial cells (Fig. 7–1). Equally important, however, are other characteristics such as growth, physiological, and metabolic differences. Microbial metabolism is particularly important in hydrogeology because the metabolic processes of microorganisms can control the geochemical environment of an aquifer and the ground water chemistry.

Metabolic Processes

Metabolism of microorganisms must satisfy two basic needs: the need for carbon for cell growth and the need for energy to synthesize organic compounds from the carbon source and to maintain cell functions. The two basic types of microorganisms in the subsurface are *heterotrophs* and *chemolithotrophs*. The heterotrophs obtain both carbon

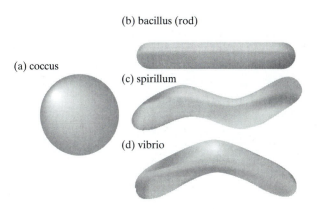

FIGURE 7–1   Common shapes of bacterial cells (from Chappelle, F. H. *Ground Water Microbiology and Geochemistry*, Copyright © 1993. John Wiley & Sons, Inc. Reprinted by permission of John Wiley & Sons, Inc.).

(a) coccus

(b) bacillus (rod)

(c) spirillum

(d) vibrio

and energy from existing organic carbon. By contrast, the chemolithotrophs obtain energy by oxidizing inorganic compounds in the environment. To obtain carbon, the chemolithotrophs must reduce carbon dioxide to synthesize organic compounds. In this regard, the chemolithotrophs are similar to photosynthetic microorganisms, which obtain their energy from the sun and their carbon from atmospheric carbon dioxide. The chemolithotrophs and the photosynthetic organisms are also known as *autotrophs* because they must manufacture their own organic carbon for cell growth. The energy needed for cell maintenance and growth is obtained from the transfer of electrons from electron donors, either organic or inorganic, to electron acceptors. A schematic illustration of the two strategies is shown in Fig. 7–2.

Heterotrophic bacteria are able to oxidize decaying organic matter for both their carbon and energy sources. The ways in which they perform these transformations, however, differ significantly. The two basic biochemical processes utilized by heterotrophic bacteria are *fermentation* and *respiration*. Fermentation is an anaerobic process in which the substrate material is partially oxidized to other organic compounds including carboxylic acids and alcohols. The souring of milk by fermentation, for example, produces lactic acid. A much more pleasant example to some of us is the fermentation of sugars in various fruits and grains by yeast to form ethanol.

An alternative process occurs when electron acceptors are available to the bacteria, and electrons from the electron donor can be transferred to these compounds. This metabolic process is known as *respiration*. It leads to the complete oxidation of organic substrates by heterotrophic bacteria to CO_2. The production of carbon dioxide in the soil by this process was discussed in a general way in Chapter 3.

Both fermentation and respiration are complex pathways with many intermediate steps leading to the final products. The ultimate benefit of these processes to the microorganisms is that energy used for growth can be produced by the transfer of electrons from the electron donors to the electron acceptors. Energy is stored in the cell prior to

FIGURE 7–2   Schematic diagram of heterotrophic and chemolithotrophic metabolic pathways.

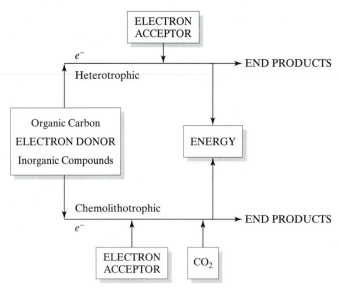

use in the form of a compound such as adenosine triphosphate (ATP), which the cells maintain in a sort of energy savings account until the stored energy is spent in the synthesis of organic compounds needed by the cell.

Enzymes

In Chapter 2 we discussed the release, or loss, of Gibbs free energy that occurs in a spontaneous reaction. Biochemical oxidations, in which organic compounds (electron donors) are oxidized by transferral of electrons to electron acceptors, are useful to microorganisms precisely because they result in a release of energy that the organism can use to live. For example, Eq. (7–1) illustrates that the oxidation of a carbohydrate by free oxygen will release 120 kJ of energy.

$$1/4 \, CH_2O + 1/4 \, O_2 \longrightarrow 1/4 \, CO_2 + 1/4 \, H_2O \qquad \Delta G^0 = -119.98 \, kJ \qquad (7\text{–}1)$$

Although this substantial energy yield is available to heterotrophic organisms in aerobic environments, the reaction rate at low temperatures is too slow to sustain life. Living cells use catalyzing compounds called *enzymes* to overcome this kinetic problem. Enzymes perform their function by forming a complex with the reactants so that they can be brought into close proximity, allowing the reaction to occur.

Enzymes are generally specific to a certain substrate or closely related substrates. Of the many types of enzymes, the most important for our purposes are those that facilitate oxidation and reduction reactions. These are known as *oxygenases* and *reductases*, respectively. Thus, *methane monooxygenase* is an enzyme that catalyzes the oxidation of methane by certain types of aerobic bacteria. Enzymes, which are proteins, often require the presence of nonprotein compounds called *cofactors*. Cofactors include metal ions necessary for metabolism as well as organic compounds known as *coenzymes*. NAD, or nicotinamide adenine dinucleotide, is an important coenzyme that accepts and donates electrons in intermediate steps of complex metabolic pathways.

One of the most remarkable characteristics of microorganisms is the ability to produce new enzymes that can utilize a previously unavailable substrate encountered by the organism. This process is called *enzyme induction*. It is particularly important in the consideration of biodegradation of *xenobiotic* chemicals (those that are not naturally found in the environment). Many of the contaminant compounds described in the previous chapter fall into this category. When a microbial community encounters a new natural or xenobiotic compound, a period of time called the *acclimation period* transpires during which enzyme induction is taking place in the microbial cells. If successful, the concentration of the new substrate will then begin to decline as the appropriate enzymes are produced. The acclimation period can vary greatly, depending on the composition of the xenobiotic compound, the microbes present in the environment, and the geochemical conditions in the environment.

Microbial Habitats

The range of habitats that can sustain microbial life are truly amazing. Microorganisms can survive in almost any near-surface environment, near surface in this case including deep petroleum reservoirs. The wide range in conditions, however, has produced a wide range in microbial adaptions to extreme environments.

The most basic requirement for microorganisms, with the exception of growth substrates, is water. Although bacteria need water for survival, some types can tolerate very dry conditions because their cell walls resist loss of water. The salinity of the water is also very important. Bacteria that are adapted to salinities of brines in sedimentary basins are called *halophiles*. Salinity is fundamental to cell survival because differences in salinity between the cell interior and the environmental conditions outside the cell create osmotic pressure, the tendency of ions to move from an area of higher electrolyte concentration to lower concentration. The cell wall, however, more easily permits the movement of water in the direction tending to equalize concentrations. So if the salinity of the environment outside the cell is greater than the salinity inside the cell, water will tend to move from the cell to the exterior in an attempt to bring the salinity inside the cell closer to external values. If the contrasts in salinity are too great *osmostic pressure* builds up against the cell wall and can be stressful to the organism.

Microbes can survive in temperatures that range from 0°C to 100°C and perhaps higher. They can even be frozen, provided that ice crystals do not damage the cell. In this condition, cells can exist in a dormant state for long periods of time but can reactivate later if the temperature rises. Microbial activity generally increases as the temperature increases, to a point. In fact, the reaction rate commonly doubles for each 10°C rise in temperature. An optimum temperature is eventually reached, however (Fig. 7–3), above which the growth rate declines until a maximum temperature is reached at which survival is no longer possible. This maximum may approach 100°C in deep subsurface environments or even higher if high pressure prevents boiling at temperatures above 100°C.

Another important environmental parameter is pH. Microorganisms have become adapted to the extreme levels of pH occurring in nature. These conditions may range from values of 2–4 in aquifers containing oxidizing sulfide minerals or surface waters produced by acid mine drainage to pH values greater than 10 in alkaline lakes. Despite the overall range, individual strains of bacteria can tolerate a fairly limited range in pH. Enzyme activity is the critical factor affected by pH. Figure 7–4 shows the optimum levels for several enzymes.

Molecular oxygen is an essential requirement for microorganisms known as *obligate aerobes*. Oxygen is used as an electron acceptor for these species, and they cannot

FIGURE 7–3   Relationship between temperature and growth rate for microorganisms (from Chappelle, F. H. *Ground Water Microbiology and Geochemistry*, Copyright © 1993. John Wiley & Sons, Inc. Reprinted by permission of John Wiley & Sons, Inc.).

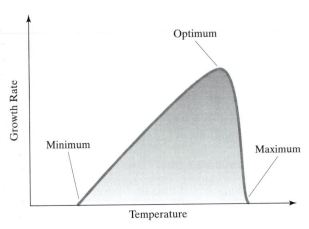

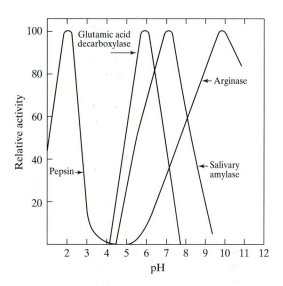

function in its absence. Certain other microbes are able to use oxygen if it is available but can switch to alternative electron acceptors if oxygen becomes depleted. These types are called *facultative anaerobes*. Finally, *obligate anaerobes* use alternative electron acceptors exclusively and find oxygen to be toxic to their survival. Although a particular aquifer may be dominantly aerobic or anaerobic, it is not uncommon to find both aerobes and anaerobes present. For example, an aerobic aquifer may have isolated microenvironments in which oxygen is excluded and anaerobes can survive.

HETEROTROPHIC MICROBIAL METABOLISM AND GROUND WATER GEOCHEMISTRY

Concentrations of many common dissolved species in ground water are controlled by the dominant metabolic processes occurring in the aquifer. Chief among these species are the electron acceptors and the metabolic by-products. As mentioned in Chapter 5, reactions that control the redox potential occur in a specific sequence defined by the relative energy yield to microorganisms. These reactions are summarized in Fig. 7–5. The relationships between microbial metabolism and redox potential are the same whether the electron donor is a natural component of the system or whether it is introduced into the environment as a contaminant. For example, heterotrophic bacteria can adapt themselves to metabolize most organic compounds present in petroleum, including the more soluble compounds of most environmental concern: benzene, toluene, ethylbenzene, and xylene (BTEX). Thus when these compounds are introduced into an aquifer, the enhanced microbial activity begins to change the redox potential and the concentrations of many species in the system. The most common metabolic pathways involved in these transformations will be discussed later in order from high to low redox potential.

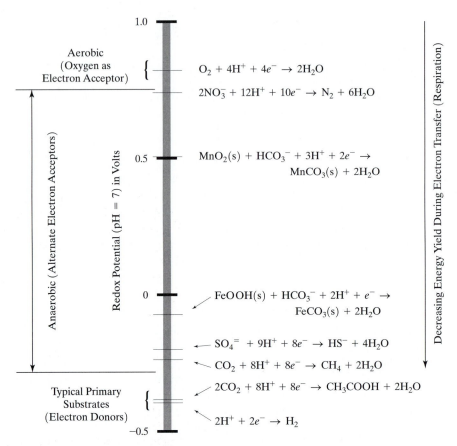

FIGURE 7–5 Important electron donors and acceptors in biotransformation processes (from Norris et al., 1994, reprinted by permission of CRC & Press, LLC).

Aerobic Respiration

In the presence of dissolved oxygen, aerobic respiration will be the dominant form of microbial metabolism. The reason for this is that the energy yield from the use of oxygen as an electron acceptor is the highest of any metabolic process. Table 7–1 compares the standard Gibbs free-energy values for the metabolism of carbohydrate by various possible metabolic pathways. Aerobic metabolism yields the most energy and is therefore favored until the oxygen is consumed. Because the concentration of dissolved oxygen in ground water has a saturation limit of around 13 mg/l, aerobic metabolism is limited by the amount of O_2 present. In shallow aquifers with high hydraulic conductivity and relatively rapid recharge, oxygen is commonly at or near the saturation value and therefore will serve as the dominant electron acceptor. The amount of a xenobiotic compound that can be metabolized by aerobic metabolism, the *assimilative capacity*, can be estimated (Wiedemeier et al., 1994). For example, if benzene is metabolized to carbon dioxide without the production of additional microbial cells, the reaction can be expressed as

$$C_6H_6 + 7.5O_2 \longrightarrow 6CO_2 + 3H_2O. \tag{7-2}$$

According to this reaction, 7.5 moles (240 g) of oxygen are required to metabolize 1 mole (78 g) of benzene for a mass ratio of $240/78 = 3.08:1$. Thus the assimilative capacity of an aquifer with 10 mg/l dissolved oxygen is 3.25 mg/l of benzene. For total BTEX, an average value of the mass ratio can be determined for the four compounds to estimate assimilative capacity.

Anaerobic Respiration with Nitrate Reduction

If the assimilative capacity of dissolved oxygen in an aquifer is exceeded by the abundance of an electron donor, the oxygen is consumed and the system becomes anaerobic. As illustrated in Figs. 7–5 and 5–17, the redox potential is still high at this point. The first anaerobic electron acceptor that is used, if available, is nitrate. In agricultural regions of rapid recharge and shallow ground water, nitrate concentrations in ground water can be significant.

Nitrate reduction can be performed by obligate anaerobes or facultative anaerobes. An example reaction is shown in Table 7–1. The reduction of nitrate to nitrogen gas (N_2) is known as denitrification. The overall reaction shown actually occurs in a series of steps, with unstable intermediate products nitrite (NO_2^-) and nitrous oxide (N_2O) (Chapelle, 1993). Each intermediate step requires a specific enzyme to catalyze the reaction.

Nitrate reduction can be an important metabolic pathway for xenobiotic compounds such as benzene. The overall reaction is

$$C_6H_6 + 6NO_3^- + 6H^+ \longrightarrow 6CO_2 + 3N_2 + 6H_2O. \tag{7-3}$$

With 6 moles of nitrate (372 g) needed to assimilate 1 mole of benzene, assuming no cell production, a nitrate-to-benzene mass ratio of $372/78 = 4.77:1$ is obtained. Thus it requires nearly 5 mg/l nitrate to metabolize 1 mg/l of benzene. In terms of nitrate-N, the mass ratio becomes $84/78 = 1.08$, so that metabolism of 1 mg/l benzene consumes just over 1 mg/l of nitrate-N. With concentrations of nitrate-N in the range of 20–30 mg/l and more in many agricultural area, nitrate reduction becomes an important electron acceptor in the biodegradation of petroleum compounds.

TABLE 7–1   Energy yields from various potential metabolic pathways.

Reaction	ΔG^0 (kJ per Electron Equivalent)
$1/4\,CH_2O + 1/4\,O_2 \longrightarrow 1/4\,CO_2 + 1/4\,H_2O$	−119.98
$1/4\,CH_2O + 1/5\,NO_3^- + 1/5\,H^+ \longrightarrow 1/4\,CO_2 + 1/10\,N_2 + 7/20\,H_2O$	−113.51
$1/4\,CH_2O + Fe^{3+} + 1/4\,H_2O \longrightarrow 1/4\,CO_2 + Fe^{2+} + H^+$	−116.24
$1/4\,CH_2O + 1/8\,SO_4^{2-} + 3/16\,H^+ \longrightarrow 1/16\,H_2S + 1/16\,HS^- + 1/4\,CO_2 + 1/4\,H_2O$	−17.73

Half reactions from Sawyer et al., 1994. ΔG^0 of H^+ assumed to be −40.46 kJ/electron-equivalent.

Anaerobic Metabolism with Ferric Iron Reduction

At lower values of redox potential (Fig. 7–5), nitrate is no longer present and ferric iron is the next major electron acceptor to be used. Ferric hydroxide coatings of grains constitute a large reservoir of assimilative capacity for anaerobic metabolism. The occurrence of this metabolic pathway in natural environments, however, was not demonstrated until recently (Lovley and Phillips, 1988; Lovley et al., 1989). Soon afterward, iron-reducing bacteria were also shown to be able to metabolize petroleum contaminant compounds and their breakdown products (Lovley et al., 1989; Lovley and Lonergan, 1990). In contrast to the reaction illustrated in Table 7–1, the metabolism of toluene by iron-reducing bacteria requires a large amount of ferric iron (Lovley and Lonergan, 1990):

$$C_7H_8 + 36Fe^{3+} + 21H_2O \longrightarrow 36Fe^{2+} + 7HCO_3^- + 43H^+. \qquad (7\text{--}4)$$

This reaction requires 36 moles of ferric iron to metabolize one mole of toluene for a mass ratio of 21.9:1. Because of the higher solubility of ferrous iron under anaerobic conditions relative to aerobic conditions, the concentration of dissolved iron can be greatly increased. Much of the iron dissolved during toluene metabolism, however, may be precipitated as sulfide, oxide, or clay minerals or sorbed by other aquifer solids (Heron and Christensen, 1995). Thus, the amount of dissolved iron in ground water is not a completely reliable indicator of the amount of toluene metabolized. Because the electron acceptor for iron reduction is a solid-phase constituent (ferric iron), the assimilative capacity of the aquifer is harder to measure than for reactions in which the electron acceptor is a dissolved-phase species. To do this, it is necessary to obtain core samples of aquifer materials and conduct lab experiments to determine the amount of available ferric iron that can serve as an electron acceptor (Heron and Christensen, 1995).

The ferrous iron that does remain in solution is mobile until aerobic conditions are encountered (Fig. 7–6). If this occurs, the ferrous iron can serve as an electron donor for chemolithotropic bacteria, which oxidize the ferrous iron to ferric iron. The subsequent precipitation of hydroxides by these "iron bacteria" causes tremendous clogging problems in well screens, pump intakes, and treatment systems. In addition to precipitation of hydroxides, certain strains of iron bacteria secrete slimy, extracellular polysaccharide coatings that contribute to clogging. In addition, they create favorable habitats for other bacteria, such as sulfur-reducing bacteria, that cause odor and corrosion problems. Wells that have iron bacteria can be rehabilitated, usually by adding sterilizing compounds such as chlorine, by adding acid to lower the pH below the tolerance level of the bacteria and dissolve the hydroxides, or by treating the wells with heat. Despite these measures, bacterial growth and clogging usually resume after treatment. This situation creates a continual problem for well owners plagued by iron bacteria.

Anaerobic Respiration with Sulfate Reduction

Under highly reducing conditions, sulfate can be used as an electron acceptor (Table 7–1) by sulfate-reducing bacteria. This reaction is extremely important in ground water geochemistry because of the potential lowering of sulfate concentrations, the increase in bicarbonate concentrations by reaction of CO_2 with water, and the increase in sulfide concentrations. Sulfide, the reduced form of sulfur, can exist in several forms (Fig. 5–13).

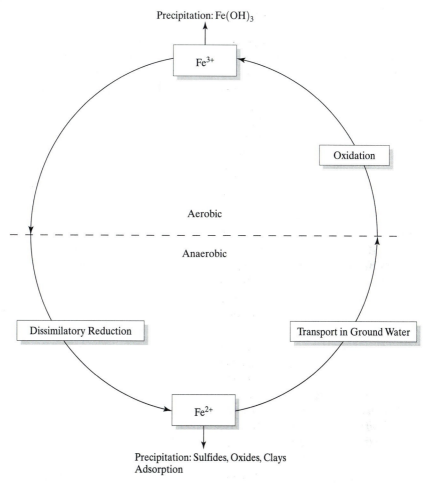

FIGURE 7–6   Metabolic processes (dissimilatory reduction and oxidation) controlling fate of iron in subsurface.

In solution, hydrogen sulfide yields a characteristic "rotten-egg" odor. If ferrous iron is present from the iron-reduction process discussed previously, it will react with hydrogen sulfide to form iron-sulfide precipitants. Sulfide ore bodies can be produced by this process under the right conditions.

The presence of dissolved sulfate in an aquifer also adds significantly to the assimilative capacity of the aquifer for metabolism of organic contaminants. For example, benzene can be degraded by sulfate-reducing bacteria according to the reaction

$$C_6H_6 + 3.75\ SO_4^{2-} + 7.5H^+ \longrightarrow 6CO_2 + 3.75H_2S + 3H_2O. \tag{7–5}$$

With a sulfate-to-benzene mass ratio of 4.6:1, 4.6 mg/l of sulfate accounts for the metabolism of 1 mg/l of benzene. With the significant sulfate concentrations present in many aquifers, this mechanism provides a large assimilative capacity for the degradation of xenobiotic compounds.

Methanogenesis

When the supplies of nitrate, ferric iron, and sulfate have been consumed as electron acceptors by microbial metabolism, one final respiratory process is possible. This process, *methanogenesis*, is well known in many natural and waste-related environments. For example, the generation of methane in landfills occurs by the microbial transformation of organic wastes under very reducing conditions. Methane production in natural ground water systems is also well known. For example, Simpkins and Parkin (1993) found that late Wisconsinan till and loess in a thick aquitard sequence in Iowa actively produces methane using organic matter incorporated in the sediment at the time of glacial deposition.

The bacteria that perform methanogenic processes are the archaebacteria, organisms that are thought to be very ancient. Before the earth had an oxidizing atmosphere, these bacteria probably occupied a much wider range of environments than their restriction today to habitats isolated from the atmosphere or any aerobic aqueous environment. Not all organic compounds are able to be metabolized by methanogenic bacteria. In fact, there are two main reactions that account for most of the methane produced. The first is called the CO_2 reduction pathway. The stoichiometry of this reaction is shown in Eq. (7–6).

$$4H_2 + CO_2 \longrightarrow CH_4 + 2H_2O \tag{7–6}$$

Molecular hydrogen is an essential component for methanogenic bacteria, which must rely on a symbiotic relationship with fermentative bacteria that produce hydrogen. The hydrogen produced is then used by anaerobic respiratory bacteria as an electron donor. Nitrate reducers, iron reducers, and sulfate reducers also use hydrogen produced by fermentative bacteria. Because of the greater energy yield of these processes relative to methanogenesis, the bacterial colonies are able to maintain hydrogen concentrations at levels that are too low for utilization by methanogens (Lovley and Goodwin, 1988; Chapelle et al., 1995). This is the main reason that sulfate reducers and methanogens do not coexist in the same environment.

The other main methanogenic pathway is known as *acetate fermentation* or *acetate dissimilation* [Eq. (7–7)]. In this process, acetate or acetic acid produced as a by-product by fermentative bacteria is degraded to methane by the reaction,

$$CH_3COOH \longrightarrow CH_4 + CO_2. \tag{7–7}$$

Despite the dominance of the reactions given in Eqs. (7–6) and (7–7), other substrates can be used in methanogenesis. For example, benzene can be biodegraded according to the reaction (Wiedemeier et al., 1994)

$$C_6H_6 + 4.5\ H_2O \longrightarrow 2.25\ CO_2 + 3.75\ CH_4. \tag{7–8}$$

The mass ratio of methane to benzene is 0.77:1, indicating that approximately 1 mg/l of methane will be produced for each mg/l of benzene degraded.

CHEMOLITHOTROPHIC PROCESSES

The heterotrophic processes discussed above occur through the transfer of electrons from organic compounds to inorganic electron acceptors. The reduced inorganic species, in turn, provide an electron source for chemolithotrophs. These bacteria obtain their energy from electron transfers, but must obtain carbon for cell growth from other sources—either through the assimilation of organic compounds or the fixation of CO_2.

Electron donors include most of the reduced electron acceptors described previously: hydrogen, hydrogen sulfide, ferrous iron, and ammonia. The reactions are shown in Table 7–2. Notice that these reactions are aerobic and that there must be a source of oxygen present. Several of these reactions have important environmental implications. The oxidation of sulfide contained in pyrite or marcasite to sulfuric acid is the major cause of acid mine drainage. Sulfide minerals occur in both metallic ore bodies and coal, and when these minerals are brought into contact with oxygen and water during mining or reclamation, extremely acidic solutions can be produced. The oxidation of ferrous iron has already been mentioned as a troublesome process leading to the clogging of wells, pump intakes, and treatment systems. The oxidation of ammonium, known as nitrification, causes a widespread non-point-source ground water contamination problem.

KINETIC FACTORS

The rate of metabolism of organic compounds is of enormous practical importance when the compounds of interest are contaminants that we wish to remove from the aquifer. The kinetics of metabolic processes are very complex under controlled laboratory conditions and even more so in natural systems. A detailed discussion of microbial growth kinetics is beyond the scope of this text. Simple kinetic models that can be fit to declining concentrations were touched upon in Chapter 2. Alexander (1994) presents a useful summary of kinetic relationships. The purpose of this section is to qualitatively discuss some factors that influence the rate of decline of contaminant concentrations by biotransformation.

Acclimation

When an organic contaminant in significant concentration is first introduced into a subsurface environment in which microorganisms exist that are capable of utilizing the substrate as a growth substrate, the population experiences an interval of exponential growth during which the compound is rapidly biodegraded. Before the growth phase occurs, however, a period of time passes during which metabolism of the compound

TABLE 7–2   Chemolithotrophic reactions.

$$2H_2 + O_2 \longrightarrow 2H_2O$$
$$2H_2S + O_2 \longrightarrow 2H_2O + 2S$$
$$2S + O_2 + 2H_2O \longrightarrow 2H_2SO_4$$
$$4Fe^{2+} + 4H^+ + O_2 \longrightarrow 4Fe^{3+} + 2H_2O$$
$$NH_4^+ + O_2 + H_2O \longrightarrow NO_3^- + 6H^+$$

does not take place or may occur at such a low rate that it is not observed. The length of this period, known as the *acclimation period*, varies tremendously among different compounds and environments. It is considered to be complete when detectable biodegradation is observed. Once a microbial community has become acclimated to a specific compound, biodegradation commences immediately upon addition of more of the compound. In addition, biodegradation rates are higher for the second influx of contaminant relative to the rates observed after the initial acclimation period. This higher rate is the result of the growth and proliferation of the population that has become adapted to using the compound as a source of carbon and energy. The factors that influence the length of the acclimation period include temperature, pH, nutrient levels, and the concentration of the compound that will be used as a growth substrate.

Alexander (1994) discusses the following possible explanations for the acclimation period.

- Proliferation of small populations
- Presence of toxins
- Predation by protozoa
- Appearance of new genotypes
- Diauxie
- Enzyme induction

When a contaminant is introduced into an environment, the population of microorganisms capable of metabolizing that compound may be very small. The exponential growth of these bacteria, which is characteristic of an abundant substrate, may require many doublings before a change in the contaminant concentration is detected.

If the concentration of the substrate is too high, it may actually be toxic to organisms that can degrade it at lower concentrations. Also, if the contaminant includes multiple compounds, toxicity of one compound to a bacterial population may prevent biodegradation of another compound that normally could be used as a growth substrate by the population.

Predation by protozoa is less common in the subsurface than in surface waters because of the relatively low numbers of protozoa that are present relative to bacteria. Because protozoa prey on bacteria, if sufficiently abundant they could inhibit biodegradation by reducing the number of bacteria present. The random appearance of new genotypes through mutation could produce organisms that are better able to metabolize a specific compound. Thus, acclimation could involve the occurrence of mutants which then begin to multiply more rapidly than others in the population because of their ability to metabolize the substrates available. *Diauxie* refers to a sequential utilization of two or more carbon substrates. The microbial population will utilize the substrate with the greatest energy yield first, before the less favored substrate is used. Acclimation to a specific compound in this case would represent the period during which the microbial population was utilizing a more favored substrate.

Enzyme induction, the synthesis of enzymes able to metabolize a substrate new or increased in concentration to the environment, was mentioned earlier as a process that would delay biodegradation of the compound. Whereas induction may contribute to the acclimation period, the time required for induction is much less than acclimation periods observed. One or more of the other acclimation factors must, therefore, be invoked.

Growth-Linked vs. Non-Growth-Linked Metabolism

It has been mentioned that microorganisms metabolize substrate compounds for energy and growth. When the concentration of the substrate is low, all of the substrate metabolized must be used for cell maintenance and growth or reproduction of the cell does not occur. It is only when the substrate concentration is high relative to the cell density of a microbial population that growth will occur. The boundary between the states of maintenance (nongrowth) and growth is called a *threshold*. Above the threshold, the population expands and the substrate compound is metabolized more rapidly. The kinetics of disappearance of the substrate compound, a critical factor in aquifer cleanup, is dependent upon whether the microbial population is in the growth or nongrowth phase. When declining substrate concentrations are plotted against time (Fig. 7–7), the curves fit several mathematical models. When the substrate concentration is high enough to allow growth, the three curves in the lower part of the diagram are applicable. Substrate concentrations determine which of the three is followed. The substrate concentrations must be considered in terms of the Monod equation,

$$\mu = \frac{\mu_{max} S}{K_s + S},\qquad(7\text{–}9)$$

FIGURE 7–7   Kinetic models for declines in concentration of compounds metabolized by microorganisms under growth-linked conditions (bottom) and non-growth-linked conditions (top) (from Alexander, 1994, reprinted by permission of Academic Press, Inc.).

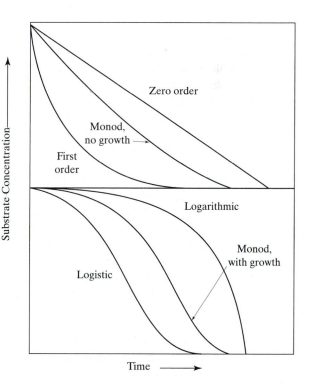

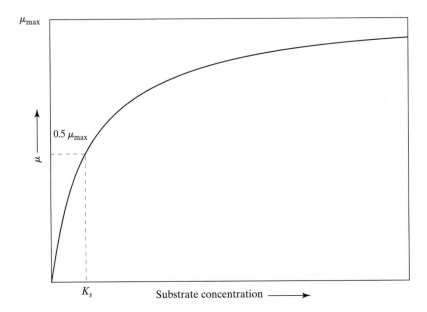

FIGURE 7–8 Growth rate of a population (μ) vs. substrate concentration. This hyperbolic relationship is described by the Monod equation (from Alexander, 1994, reprinted by permission of Academic Press, Inc.).

which produces the hyperbolic relationship illustrated in Fig. 7–8. In this equation, μ is the growth rate of a population, μ_{max} is the maximum growth rate, S is the substrate concentration, and K_s is a constant representing the substrate concentration at one half the maximum growth rate. Both S and K_s are expressed in mg/l or similar units. When the substrate concentration is very high, near the right-hand side of the curve, it may require many doublings of the population by growth before any effect is seen in the substrate concentration. Finally, when the population becomes large, large declines in the substrate concentration begin to occur. This constitutes a *logarithmic* relationship (Fig. 7–7). If, on the other hand, the substrate concentration is on the low end—below the value of K_s—*logistic* declines result. Here, growth rate is linearly proportional to substrate concentration (Fig. 7–8), and each doubling of the population takes longer as the concentration is falling. Thus, the curve begins to flatten out with time, relative to the logarithmic case. The intermediate condition, in which the substrate concentration is neither in the very high or very low range, follows the curve labeled Monod, with growth.

    If the substrate concentrations are below the growth threshold, different substrate decline patterns are observed. In this case, an equation similar to the Monod equation governs the kinetics. This equation, known as the *Michaelis–Menten* equation, has the same form as the Monod equation, except that it is formulated in terms of reaction rate rather than growth rate. Otherwise, it is identical to Eq. (7–9). In the nongrowth phase, the three curves shown (Fig. 7–7, upper part) are also based on the level of substrate concentration relative to the value of K_m (analagous to K_s in Eq. (7–9)—the substrate concentration at one half the maximum reaction rate). *Zero-order* kinetics occur when the substrate concentration is higher than K_m because the reaction rate is basically constant in this range and the substrate is thus metabolized at a constant rate. *First-order* kinetics occur when the substrate concentration is much less than K_m. In this range, the reaction rate is slowing in proportion to the drop in substrate. As a result, the substrate persists for long periods of time in the environment at low concentration because it is being metabolized at such a low rate. When the log of concentration is plotted against time for a first-order process, points fall on a straight line and a half life can be easily calculated (Chapter 2). The simplicity of this model has led to its use, and perhaps overuse, in biodegradation computer models. The intermediate model for the nongrowth case is called Monod-no growth (Fig. 7–7). It applies when the substrate concentration is in the mid range of the reaction rate curve, approximately equal to K_m.

The kinetic models previously described have been mainly derived from tightly controlled lab experiments. The application of these models to dynamic ground water environments at the field scale can only be done with a much lower level of confidence. In the subsurface, declines in concentration due to biodegradation must be separated from declines due to changes in source concentrations, advection, dispersion, sorption, and other processes.

Cometabolism

So far, we have focused on compounds that microorganisms can use for metabolic purposes; that is, they use the compound as a source of carbon, energy, or both. Some organic compounds, however, do not serve as either carbon or energy sources for microbial communities, but are still metabolized to other compounds. This process is known as *cometabolism*, and when it occurs, the kinetics will be much different than those situations already discussed. This process is also referred to as fortuitous metabolism because the organism derives no benefit whatsoever from the reaction.

There are several possible explanations for cometabolism. Most metabolic pathways require a series of steps with different enzymes catalyzing successive reactions. Some enzymes are nonspecific; that is, they can act on a variety of substrates. In the case of cometabolism, a nonspecific enzyme may perform the first step in breakdown of the compound, but other enzymes in the pathway necessary to convert the first product to other intermediates that yield energy or carbon for growth or maintenance are not present. Also, an intermediate product might be toxic to bacteria that could potentially transform it further. In either case, the product of the cometabolized compound accumulates in the solution. Cometabolism can thus be a beneficial process or, if the product is environmentally more harmful than the parent compound, an undesirable process.

The kinetics of cometabolism are compared to a compound that supports the growth of a particular microorganism in Fig. 7–9. The cometabolizing compound declines only slowly from its maximum concentration, whereas the growth substrate experiences a rapid drop-off resulting from the exponential growth of the population.

BIODEGRADATION OF PETROLEUM HYDROCARBONS IN THE SUBSURFACE

Petroleum fuels constitute the largest group of compounds released to the subsurface through spills and leaks from buried tanks, gas stations, pipelines, and transportation accidents. The potential biodegradation of these compounds is therefore a critically important issue. Fortunately, it appears that most compounds derived from petroleum do

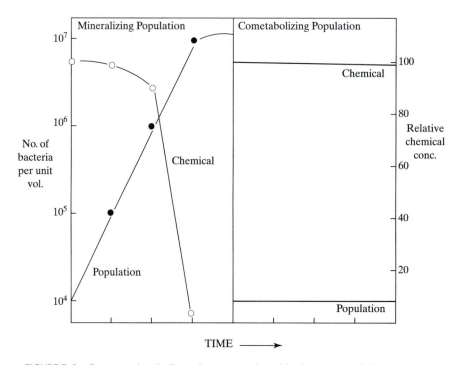

FIGURE 7–9   Concentration declines of a compound resulting from growth-linked metabolism (left) vs. cometabolism (right) (from Alexander, 1994, reprinted by permission of Academic Press, Inc.).

biodegrade under a range of conditions. The growing understanding of these processes suggests that, in many circumstances, natural or enhanced *bioremediation* is the best and most cost-effective solution for subsurface cleanups. Subsurface microbial processes are commonly studied using *microcosms*, which are small quantities of aquifer sediment enclosed in vials or test tubes in the laboratory under *in-situ* redox conditions. Organic compounds are added to the microcosms to see if they can be biodegraded by indigenous microorganisms. Biodegradation is assessed by measuring concentration changes in the target compounds, electron acceptors, or metabolic by-products. For example, biodegradation of a particular compound could be documented by decreases in its concentration along with increases in CO_2 gas relative to a control microcosm which is sterilized to eliminate microbial processes.

Aerobic Biodegradation

Aerobic biodegradation has long been known to be an effective process for almost all compounds contained in petroleum. Heterotrophic communities of aerobic bacteria use petroleum compounds for carbon and energy sources, and degradation to CO_2 can be expected, although compounds of heavier molecular weight are more resistant to metabolism. A wide range of field and laboratory investigations have shown the effectiveness of aerobic metabolism of aromatic petroleum compounds. One of the first studies

FIGURE 7–10 Configuration of plumes produced by injection of chloride and BTX at three points in time after injection (from Barker et al., 1987, reprinted by permission of the National Ground Water Association).

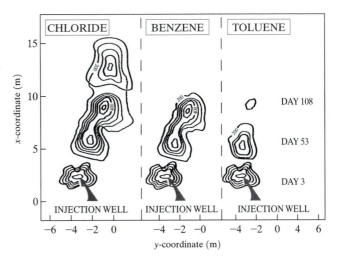

to document BTX biodegradation (ethylbenzene was not included) in an aquifer was an injection experiment by Barker et al. (1987), in which BTX was introduced into a shallow aquifer and monitored over time and distance as a plume formed and moved downgradient (Fig. 7–10). Under aerobic conditions, all injected compounds biodegraded, with o-xylene disappearing most rapidly followed by toluene and benzene. Davis et al. (1993) used field evidence, aerobic microcosms, and modeling to demonstrate the mineralization of benzene and other compounds in a shallow aquifer impacted by petroleum contamination. Because dissolved oxygen levels remained above detection limits throughout the plume, biodegradation was attributed to aerobic metabolism, although biodegradation also occurred in anaerobic microcosms.

In the subsurface, significant spills of petroleum products lead to nonaqueous phase accumulations at the water table surface. Water table fluctuations smear these LNAPL compounds over a thicker vertical range of the aquifer. In pools of free product, oxygen is quickly depleted and aerobic biodegradation ceases, except for the edges of the free product lens, where oxygen can diffuse through the vadose zone atmosphere to sustain some activity. The small amount of the product that dissolves into ground water is dominated by BTEX and other low-molecular-weight aromatics. These make up only 2–8% of the total weight of the original product, but have solubilities of hundreds of mg/l. The supply of dissolved oxygen is quickly consumed, and aerobic biodegradation ceases. Sustaining aerobic biodegradation is a problem encountered in attempts to stimulate or enhance biodegradation of petroleum compounds because of the low solubility of molecular oxygen in water. Once oxygen is consumed, fresh ground water containing dissolved oxygen has to flow into the contaminated section of the aquifer. Because of the low velocities of ground water flow and the high concentrations of contaminants, most subsurface NAPL accumulations quickly reach anaerobic conditions, in which they remain for long periods of time. Hydrogen peroxide, H_2O_2, also a strong oxidant, is frequently mentioned as an alternative to oxygen because of its much higher solubility. If ground water containing hydrogen peroxide is artificially recharged at a site, it could represent a way to get oxygen to aerobic bacteria to enhance biodegradation of the contaminant. Problems have been encountered, however, with injections of hydrogen peroxide in aquifers. Barcelona and Holm (1991) experimentally determined the reduction capacities of aquifer solids and found that a large proportion of hydrogen peroxide was lost very rapidly to chemical reduction by aquifer solids and/or decomposition to oxygen and formation of oxygen bubbles. Both processes represent a loss of potential oxygen to serve as microbial electron acceptors. Although other techniques exist for enhancing aerobic biodegradation, anaerobic biodegradation will be prevalent at many sites.

Anaerobic Biodegradation Using Nitrate As an Electron Acceptor

Anaerobic biodegradation of the common aromatic petroleum compounds under denitrifying conditons has been demonstrated under both laboratory and field conditions, although the results vary significantly for individual compounds and from site to site. Biodegradation of all BTEX compounds was observed in microcosm experiments under denitrifying conditions by Major et al. (1988). Hutchins et al. (1991) noted similar results in lab experiments using both uncontaminated and contaminated aquifer material, with the exception that benzene did not appear to be biodegraded. A creosote contaminated aquifer in Denmark was studied by Flyvbjerg et al. (1993), with the result that toluene was the only BTEX compound that was biodegraded, although other types of compounds were also metabolized. Benzene and the xylenes proved to be recalcitrant to biodegradation. Borden et al. (1997a) conducted a comprehensive field and lab investigation at a site contaminated by BTEX and MTBE in North Carolina. The conditions at the site were determined to be mixed aerobic–anaerobic (denitrifying). All compounds biodegraded at the site, although the first-order biodegradation rates varied significantly. These differences are apparent by the distributions of individual contaminants. Toluene and ethylbenzene degrade very rapidly and, as a result, their distribution was restricted to the source area. Other studies support the rapid breakdown of these compounds, particularly toluene, under denitrifying conditions. The biodegradation of benzene, MTBE, and m- and p-xylene was much slower. The biodegradation of MTBE was surprising as it had not been observed previously under aerobic or denitrifying conditions (Borden et al., 1997a). Microcosm experiments using aquifer materials showed that benzene did not biodegrade under denitrifying conditions, suggesting that the field biodegradation must have been aerobic.

Anaerobic Biodegradation Using Ferric Iron and Sulfate As Electron Acceptors

The use of ferric iron as an electron acceptor in anaerobic biodegradation of aromatic compounds has only been demonstrated in recent studies (Lovley and Phillips, 1988; Lovley and Lonergan, 1990). Lovley et al. (1989) showed that toluene was degraded with a pure culture of iron-reducing bacteria and at the Bemidji, Minnesota, crude oil spill, Baedecker et al. (1993) demonstrated metabolism of BTEX coupled with the reduction of ferric iron in microcosms using anaerobic aquifer sediments and ground water from the site.

FIGURE 7–11   Concentrations of BTEX parameters (a), electron acceptors (b), and indicator parameters (c) along a longitudinal profile along the centerline of a plume in Rocky Point, North Carolina. Electron acceptors are reported in meq/l for easy comparison. Conversion factors are shown in legend. Well U10, at a distance of 0 on the horizontal scale, is in the source area.(from Borden et al., 1995, reprinted by permission of the Natural Ground Water Association).

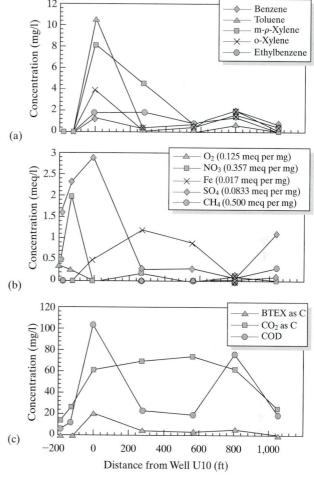

Borden et al. (1997b, 1995) studied the sequential mineralization of BTEX compounds in a plume in Rocky Point, North Carolina, using field and microcosm evidence. One of the most interesting conclusions to result from this work was the differences in biodegradation that were shown to occur in various parts of the plume. The rapid decline of O_2 and NO_3 between −200 and 0 feet (Fig. 7–11) indicate that these compounds serve as dominant electron acceptors at the upgradient margin of the source. With distance downgradient from the source, increases in dissolved iron and decreases in sulfate suggest that Fe^{3+} and SO_4^{2-} become the dominant electron acceptors. In microcosms from the plume source area amended with BTEX, no degradation over the study period occurred. By contrast, in microcosms of sediment taken from the mid-plume area, where iron-reducing conditions prevailed, the BTEX compounds were metabolized in a distinct order (Fig. 7–12). Toluene, m-xylene, and o-xylene declined rapidly after introduction to the microcosm, whereas benzene persisted much longer prior to biodegradation. Ethylbenzene displayed minimal reduction in concentration, on average, over the

FIGURE 7–12   Changes in concentration in BTEX compounds (a) and dissolved iron (b) in microcosms of sediment taken from mid-plume area of plume in Rocky Point, North Carolina. Abiotic microcosms indicate behavior without biodegradation (from Borden et al., 1997b).

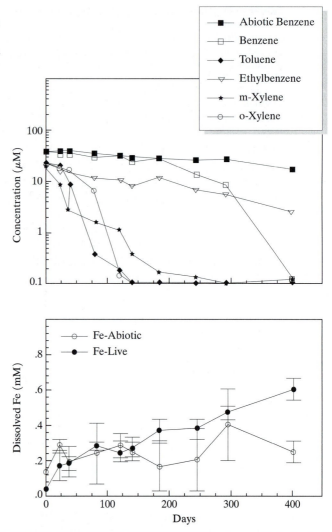

course of the study. Sulfate concentrations did not decline in the mid-plume microcosms, suggesting that it was not an active electron acceptor, whereas iron concentrations did increase (Fig. 7–12). Sulfate depletion was observed closer to the source of the plume, indicating that it was utilized as an electron acceptor, but methane was not detected in any of the microcosms. Sulfate was linked to the degradation of p-xylene and toluene in microcosm studies by Haag et al. (1991).

Borden et al. (1997b) concluded that individual compounds did not follow a first-order decay pattern. Each compound experienced an initial variable acclimation period, followed by an interval of very rapid decline, and finally, a period of very slow decrease. Overall, first-order decay does approximate the decline in total BTEX under site conditions.

Anaerobic Biodegradation via Methanogenesis

Direct evidence for biodegradation of aromatic hydrocarbons has been obtained from a number of sites. Borden et al. (1997b) conducted microcosm studies using aquifer material from a spill site in Michigan in which Wilson et al. (1994) had previously conducted field studies. Here again, toluene proved to be the most rapidly biodegraded compound, whereas no benzene metabolism was detected. The field and lab studies differed with respect to xylene and ethylbenzene. Although the field study showed slower biodegradation of both constituents, xylene did not biodegrade in the microcosms. Some biodegradation of ethylbenzene was recorded in microcosms from locations near the end of the plume. The relatively high natural organic carbon content in the aquifer may have contributed to an unusually long lag time for the onset of toluene biodegradation, suggesting that the microbial community utilized natural organic carbon as an electron donor prior to toluene.

Another field site in which methanogenic conditions prevailed was the Bemidji, Minnesota, crude oil spill. Both benzene and toluene were degraded in microcosms utilizing aquifer sediments from the anaerobic portion of the plume (Baedecker et al., 1993). Methane was not generated in the microcosms but was abundant in the plume. Cozzarelli et al. (1990) analyzed the plume ground water for possible intermediate

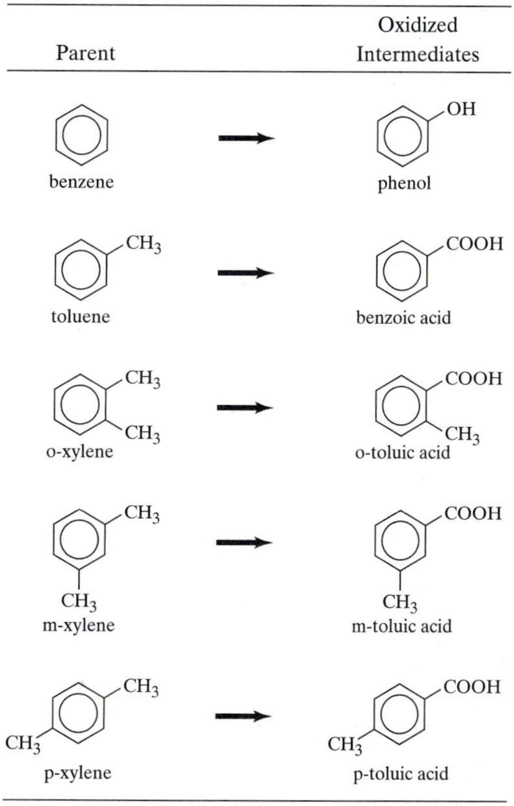

FIGURE 7–13 Parent compounds and oxidized intermediates produced by degradation under methanogenic conditions in the Bemidji, Minnesota, pipeline rupture site (from Cozzarelli et al., Transformation of monoaromatic hydrocarbons to organic acids in anoxic ground water environment, *Environ. Geol. Water Sc.* 16, pp. 135–141, Copyright 1990, reprinted by permission of Springer-Verlag GmbH & Co. KG).

products of BTX biodegradation and found one potential oxidized intermediate product for each primary compound (Fig. 7–13). These compounds, which included phenol and various carboxylic acids, were most abundant in the part of the plume where the primary compounds were disappearing. Further downgradient, the intermediate compounds disappeared as well. Cozzarelli et al. (1990) suggested that such intermediate products temporarily accumulate under very reducing conditions because of slower biodegradation rates, but that the intermediate products are themselves metabolized too rapidly for significant detection under aerobic conditions.

The studies cited previously, and others in the literature, document a wide variation in biodegradation processes and rates in petroleum-contaminated aquifers. The causes of these differences include differences in terminal electron acceptor, BTEX concentrations, organic carbon content of the aquifer, aquifer pH, and possibly others. It is therefore difficult to predict field biodegradation rates and processes without site specific investigations. It is also a problematic task to extrapolate laboratory biodegradation rates to field situations.

BIODEGRADATION OF CHLORINATED SOLVENTS IN THE SUBSURFACE

Compared to the aromatic petroleum hydrocarbons, the common chlorinated solvents are much more resistant to biodegradation in many aquifers. The fundamental problem with most of these compounds is that bacteria cannot use them as sources of carbon and energy. As a result, biodegradation must occur through cometabolism, which is slower than heterotrophic metabolism and, in addition, requires the presence in the aquifer of suitable electron donors, acceptors, and nutrients to sustain growth of the microbial populations that cometabolize the solvents (Table 7–3).

TABLE 7–3 Potential for biodegradation of chlorinated aliphatic hydrocarbons (CAH) as a primary substrate or by cometabolism (from McCarty and Semprini, 1994).

Compound	Primary Substrate		Cometabolism		
	Aerobic Potential	Anaerobic Potential	Aerobic Potential[a]	Anaerobic Potential[a]	CAH Product
CCl_4			0	XXXX	$CHCl_3$
$CHCl_3$			X	XX	CH_2Cl_2
CH_2Cl_3	Yes	Yes	XXX		
CH_3CCl_3			X	XXXX	CH_3CHCl_2
CH_3CHCl_2			X	XX	CH_3CH_2Cl
CH_2ClCH_2Cl	Yes		X	X	CH_3CH_2Cl
CH_3CH_2Cl	Yes		XX	[b]	
$CCl_2=CCl_2$			0	XXX	$CHCl=CCl_2$
$CHCl=CCl_2$			XX	XXX	$CHCl=CHCl$
$CHCl=CHCl$			XXX	XX	$CH_2=CHCl$
$CH_2=CCl_2$			X	XX	$CH_2=CHCl$
$CH_2=CHCl$	Yes		XXXX	X	

[a] 0-very small if any potential; X-some potential; XX-fair potential; XXX-good potential; XXXX-excellent potential.

[b] Readily hydrolyzed abiotically, with half-life on the order of one month.

Although many classes of chlorinated solvents occur in ground water, this chapter will emphasize the common chlorinated ethanes and ethenes, which constitute the vast majority of contaminants. The most common primary solvent compounds include perchloroethene (PCE), trichloroethene (TCE), 1,1,1-trichloroethane (1,1,1-TCA), and carbon tetrachloride (CT). The characteristics of these compounds were discussed in the previous chapter; they are fairly soluble, denser than water, and not strongly sorbed to aquifer solids in aquifers with low organic carbon contents. As a result, they can contaminate large aquifer volumes, and because most of these compounds and their daughter products are known or suspected carcinogens, their drinking water standards are in the low parts-per-billion range in some cases. Because of their mobility and tendency to occupy large aquifer volumes, bioremediation of these compounds is an extremely desirable prospect.

Aerobic Biodegradation

The biodegradation mechanisms of chlorinated solvents depend to a great degree on the oxidation state of the compound and the redox potential of the aquifer in which biodegradation occurs. Highly chlorinated compounds such as PCE, in which carbon has a $+II$ oxidation number, and $CT (C = +IV)$ contain carbon in oxidized form, and are most susceptible to bidegradation by reduction in strongly reducing environments. Reduced compounds such as vinyl chloride $(C = -I)$ can be more easily biodegraded under aerobic conditions. Aerobic bacteria that cometablize solvent compounds include methane oxidizers, or *methanotrophs*. These bacteria use methane as growth substrate and oxidize it to methanol with the help of the enzyme methane monooxygenase (MMO). MMO is a nonspecific enzyme which will cometabolize TCE (Wilson and Wilson, 1985) to TCE

epoxide, $\underset{Cl_2C - CHCl}{\overset{O}{\triangle}}$, which further degrades through a series of steps to CO_2, Cl^-,

and water. Other aerobic bacteria that have the ability to cometabolize TCE use propane, ethylene, and other compounds as primary substrates (McCarty and Semprini, 1994).

The use of methanotrophs for biodegradation has been tested at several sites. At Moffet Naval Air Base in California, ground water was pumped to the surface, where oxygen and methane were added, and then injected back into the aquifer (Semprini and McCarty, 1991). After 430 hours, methane and oxygen were added in alternate pulses to avoid excessive growth of the bacteria near the well screen which could lead to clogging. The declines of oxygen and methane in the well (Fig. 7–14) indicated that bacterial growth was stimulated. The accompanying declines in concentrations of TCE, cis-DCE, trans-DCE, and vinyl chloride (VC) (Fig. 7–15) showed effective cometabolism of the solvent compounds, although the rates of decline in concentration varied among the compounds. 1,1,1-TCA, which was also present in the aquifer, did not biodegrade in the experiment. This study indicated that methanotrophs can be used to degrade various solvent compounds, but both oxygen and methane must be added to the aquifer.

Anaerobic Biodegradation

Anaerobic biodegradation of chlorinated solvents occurs over the entire range of anaerobic redox conditions, from denitrification to methanogenesis. The most common process

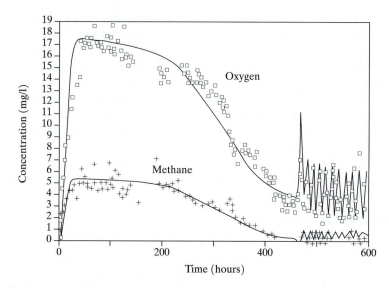

FIGURE 7–14 Methane and oxygen utilization by methanotrophs at Moffett test facility, California. After 430 hours, oxygen and methane were alternately pulsed into the aquifer (from Semprini and McCarty, 1991, reprinted by permission of the National Ground Water Association).

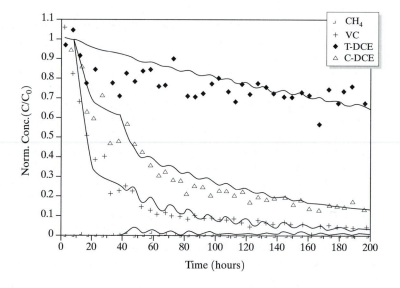

FIGURE 7–15 Cometabolism of chlorinated solvents by methanotrophic bacteria at Moffett test facility, California (from Semprini and McCarty, 1991, reprinted by permission of the National Ground Water Association).

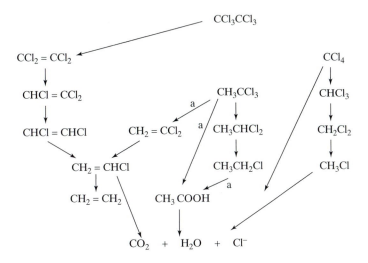

FIGURE 7–16   Anaerobic biodegradation pathways of chlorinated aliphatic hydrocarbons (reprinted with permission from Vogel et. al., Transformations of the halogenated aliphatic compounds. a = abiotic pathway. *Environ. Sci. Technol.* v. 21, pp. 722–736, Copyright 1987 with American Chemical Society).

is *reductive dechlorination*, a cometabolic process in which the solvent is reduced by the replacement of a chlorine atom with a hydrogen atom. The solvent compound serves as an electron acceptor in these reactions. For compounds that contain multiple chlorines, reduction occurs in a sequence of steps, each involving one chlorine atom (Fig. 7–16). Reaction rates for compounds such as PCE and TCE—the more highly chlorinated compounds—are highest under methanogenic conditions and decrease as the compound becomes more reduced. A transition to mildly reducing or aerobic conditions is then required to further transform the compounds rapidly.

Carbon tetrachloride has been observed to biodegrade under the entire range of anaerobic redox conditions (Bouwer, 1994). This reaction produces chloroform (CF) as an intermediate product (Fig. 7–16), which, except under methanogenic conditions, is more resistant to anaerobic biodegradation and therefore tends to accumulate in the flow system. Criddle et al. (1990) isolated a strain of *Pseudomonas* that was found to transform CT to carbon dioxide without formation of CF. This reaction, which occurs under denitrifying conditions, has the potential for enhanced biodegradation with the addition of nitrate as an electron acceptor, because of the high solubility of nitrate.

Considerable evidence exists to document the reductive dechlorination of PCE and TCE (McCarty and Semprini, 1994; Bouwer, 1994). This process occurs sequentially (Fig. 7–16), with the formation of intermediate products cis-1,2-DCE and VC. The final reduction of vinyl chloride to ethene is slow and incomplete as long as anaerobic conditions persist in the aquifer. The subsequent increase in VC concentrations in anaerobic aquifers is considered to be a major problem owing to the mobility and toxicity of vinyl chloride. Model results, showing the relative concentrations of cis-1,2-DCE and VC as TCE is reductively dechlorinated (Fig. 7–17), were produced by Vogel (1994).

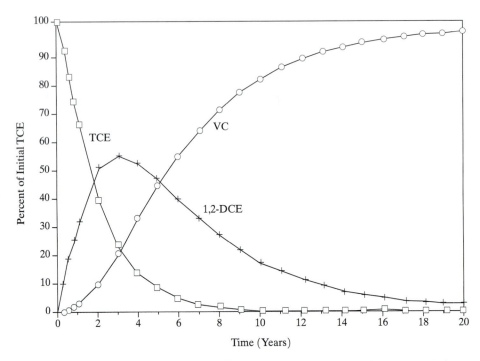

FIGURE 7–17 Reductive dechlorination of TCE under anaerobic conditions (from Vogel, 1994).

Oxidation of VC to CO_2, however, can occur under both aerobic and anaerobic conditions. If the VC plume moves from an area of anaerobic to aerobic conditions, this transformation could occur naturally. Otherwise, various technologies could be used to introduce aerobic conditions with the objective of VC removal. VC has also been shown to oxidize under iron-reducing conditions (Bradley and Chapelle, 1996). This transformation was observed in microcosm studies in which soluble ferric iron chelated with the organic acid EDTA was introduced as an electron acceptor. Bacteria present in the aquifer sediment were capable of oxidizing VC directly to CO_2 without the intermediate production of ethene.

Field examples of solvent biodegradation include a large plume of solvents in St. Joseph, Michigan (Fig. 7–18). At this site, solvent releases occurred over a ground water divide, and plumes developed in two directions. One of the plumes extended under Lake Michigan and discharged small concentrations into the lake. Investigations by McCarty and Wilson (1992) and Semprini et al. (1995) indicated that natural reductive dechlorination of TCE to ethylene was occurring in the plume through intermediate compounds including DCE isomers and VC. Semprini et al. (1995) concluded that reduction of TCE to DCE occurred under sulfate-reducing conditions in the aquifer and that further reductions to VC and ethene required more reducing methanogenic conditions. McCarty and Wilson (1992) measured declines in bulk dissolved organic content of the plume measured as chemical oxygen demand (COD) (Fig. 7–19) and showed that the declines

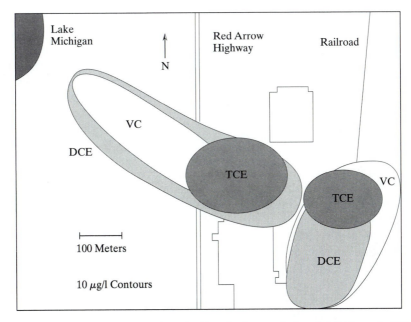

FIGURE 7–18   Distribution of chlorinated ethenes in plume at St. Joseph, Michigan, site. DCE and VC produced by reductive dechlorination (from McCarty and Wilson, 1992).

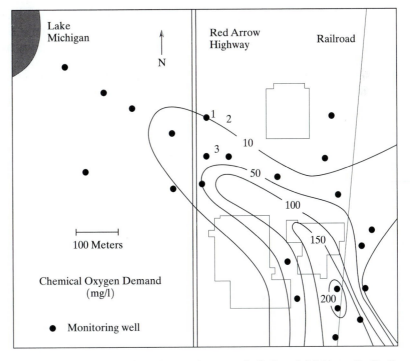

FIGURE 7–19   Distribution of COD in ground water at the St. Joseph, Michigan, site. Declines in COD concentration correspond to declines in chlorinated solvent compounds (from McCarty and Wilson, 1992).

in COD, which represents the electron donor in the biodegradation reactions, correspond to the decreases in solvent concentrations. About 20% of the TCE was biotransformed to ethene. The slow biodegradation rates that occur in plumes such as the St. Joseph plume may be sufficient to gradually remediate the site over time if the plume does not extend to water supply wells or other receptors. This process, known as *natural attenuation*, may not be acceptable when plumes extend into areas of higher risk.

BIOREMEDIATION

Natural attenuation is an increasingly attractive method of site remediation for both hydrocarbon and solvent plumes. Not a no-action alternative, natural attenuation requires extensive delineation and monitoring of the plume and the collection of evidence that biodegradation is occurring (Wiedemeier et al., 1994). The types of evidence that can be used to document natural attenuation include declining concentration of the contaminant and the presence of breakdown products, changes in concentrations of electron acceptors and metabolic by-products, and other indicators of redox potential that may indicate the terminal electron-accepting process. The most direct evidence is the use of microcosms containing aquifer sediment at the *in-situ* redox potential to demonstrate metabolism of the contaminant of interest.

Other, more active bioremediation techniques abound in the environmental industry (Norris et al., 1994). Aerobic methods include *bioventing*, in which oxygen is added to the vadose zone to enhance biodegradation, and *air sparging*, in which air is injected below the water table at high enough pressure to displace the fluid in the aquifer pore spaces. The injected air is used both to volatilize contaminants and drive them to the water table as well as to provide oxygen for use as an electron acceptor. Electron acceptors other than oxygen, as well as electron donors, can be injected into aquifers if biodegradation occurs by cometabolism. The alternate injection of methane and oxygen to stimulate methanotrophic populations is an example of this approach. Ultimately, it may be possible to introduce microorganisms into the aquifer to more readily degrade a recalcitrant compound. In most cases, however, difficulties of transport of the organisms to the zone of contamination and questions concerning the survival of foreign populations in a new environment will limit the widespread utilization of introduced microorganisms.

PROBLEMS

1. How do the metabolic functions of heterotrophs differ from those of chemolithotrophs?
2. Contrast the processes of fermentation and respiration.
3. Ground water has the following concentrations: $O_2 = 8$ mg/l; $NO_3—N = 15$ mg/l; $SO_4^{2-} = 35$ mg/l. In a BTEX contaminant plume in the same aquifer, the dissolved iron concentration is 20 mg/l. Neglecting methanogenesis, what is the assimilative capacity of the aquifer for benzene? What assumption does this calculation make about iron?
4. Describe the major pathways involved in methanogenesis.
5. Explain the possible factors that make up acclimation.

6. Draw hypothetical graphs for decreases in substrate under the following kinetic conditions and explain each curve in terms of the Monod or Michaelis-Menton relations.
   a. logistic
   b. zero order
   c. first order
   d. logarithmic

7. How do the kinetics of cometabolism compare to biodegradation in which the substrate can support growth of the microbial population?

8. Discuss the relationship between an electron-accepting process and the biodegradation of BTEX compounds.

9. Describe the geochemical conditions that would be necessary in a plume to naturally biodegrade TCE and PCE completely to CO_2.

10. What *in-situ* bioremediation technology is capable of degrading PCE and TCE aerobically?

CHAPTER 8

Applications of Isotopes in Hydrogeology

The use of isotopes in hydrogeology continues to grow at a rapid pace. Applications include tracing the evolution of a water mass from its origin as precipitation, through its recharge processes and ending at its occurrence in an aquifer. Isotopes can also be used to determine the origin of specific solutes in ground water. Applications of this type commonly involve *stable* isotopes, because the abundances of different isotopes of an element, measured as ratios, change during numerous physical and chemical processes. These changes are referred to as *fractionation*.

The other main class of applications of isotopes is based on the decay of *radio-isotopes*, those that decay spontaneously through time. In some instances, ground waters can be dated by the use of radioisotopes, although the stable isotopes can also be used in some dating applications.

For much more detailed information on the topics covered in this chapter, see the excellent reference, *Environmental Isotopes in Hydrogeology*, by Clark and Fritz.

STABLE ISOTOPES

The list of stable isotopes used in hydrogeology has grown in recent years. Table 8–1 lists them and their relative abundances. For the purposes of this text, stable isotopes of oxygen, hydrogen, carbon, nitrogen, and sulfur will be emphasized.

TABLE 8–1   Stable isotopes used in hydrogeological studies (from Clark and Fritz, 1997).

Isotope	Ratio	% Natural Abundance	Reference (Abundance Ratio)	Commonly Measured Phases
^{2}H	^{2}H/^{1}H	0.015	VSMOW $(1.5575 \cdot 10^{-4})$	H_2O, CH_2O, CH_4, H_2, OH^- minerals
^{3}He	^{3}He/^{4}He	0.000138	Atmospheric He $(1.3 \cdot 10^{-4})$	He in water or gas, crustal fluids, basalt
^{6}Li	^{6}Li/^{7}Li	7.5	L-SVEC $(8.32 \cdot 10^{-2})$	Saline waters, rocks
^{11}B	^{11}B/^{10}B	80.1	NBS 951 (4.04362)	Saline waters, clays, borate, rocks
^{13}C	^{13}C/^{12}C	1.11	VPDB $(1.1237 \cdot 10^{-2})$	CO_2, carbonate, DIC, CH_4, organics
^{15}N	^{15}N/^{14}N	0.366	AIR N_2 $(3.677 \cdot 10^{-3})$	N_2, NH_4^+, NO_3^-, N-organics
^{18}O	^{18}O/^{16}O	0.204	VSMOW $(2.0052 \cdot 10^{-3})$ VPDB $(2.0672 \cdot 10^{-3})$	H_2O, CH_2O, CO_2, sulphates, NO_3^-, carbonates, silicates, OH^- minerals
^{34}S	^{34}S/^{32}S	4.21	CDT $(4.5005 \cdot 10^{-2})$	Sulfates, sulfides, H_2S, S-organics
^{37}Cl	^{37}Cl/^{35}Cl	24.23	SMOC (0.324)	Saline waters, rocks, evaporites, solvents
^{81}Br	^{81}Br/^{79}Br	49.31	SMOB	Developmental for saline waters
^{87}Sr	^{87}Sr/^{86}Sr	^{87}Sr = 7.0 ^{87}Sr = 9.86	Absolute ratio measured	Water, carbonates, sulfates, feldspar

Measurement and Standards

Because of the difficulty of measuring actual amounts or concentrations of isotopes in a sample, the ratios of the most abundant stable isotopes of a particular element are measured relative to a standard, a sample containing the element in which the isotopic ratio is known. Standards for each isotopic ratio have been established and are used on a world-wide basis. The instrument used to measure these ratios is a mass spectrometer. In practice, each laboratory maintains laboratory (working) standards, to which the sample ratios are compared. The difference in ratios between the sample and the working standard are then corrected to the difference relative to the international standard.

The corrected ratios are expressed in delta values (δ) according to the equation

$$\delta = \frac{R_x - R_{std}}{R_{std}} \tag{8–1}$$

where R refers to the isotopic ratio of the sample (R_x) and the standard (R_{std}). For example, for oxygen, R is the ratio of ^{18}O to ^{16}O. Because of the small differences in isotopic ratios, the delta values are commonly multiplied by 1,000 so that the resulting numbers are greater than 1 or less than -1, depending on the sign. Using this multiplication factor, Eq. (8–1) becomes

$$\delta_x = \left(\frac{R_x}{R_{std}} - 1 \right) \times 1000‰. \tag{8–2}$$

The symbol, ‰, means "per mil," or parts per thousand, by analogy to the symbol for percent, %, which means parts per hundred. If a sample yielded an $^{18}O/^{16}O$ delta value of $+10$ per mil, for example, it means that the sample is enriched in ^{18}O by 10 parts per thousand relative to the standard, or 1 per cent. A value of -10 per mil, on the other hand, indicates that the sample is depleted in the heavier isotope by 1 percent.

The standard for ^{18}O and ^2H is known as VSMOW, which stands for Vienna Standard Mean Ocean Water. This standard is maintained by the International Atomic Energy Agency (IAEA). It replaced an earlier standard that was known as SMOW. The standard for carbon was originally established using a fossil belemnite from the Cretaceous Pee Dee Formation in South Carolina and was therefore abbreviated as PDB. Before the belemnite was totally used up, the IAEA created a new standard by calibration with the original, which became known as VPDB.

Atmospheric nitrogen is used as a standard for $^{15}N/^{14}N$ measurements and sulfur isotopes ($\delta^{34}S/\delta^{32}S$) are measured relative to a standard composed of the troilite (FeS) phase of the Canyon Diablo meteorite (CDT). Standards for a variety of other isotopes are also available.

Isotopic Fractionation

The reason that stable isotopes are useful in hydrological studies is that physicochemical reactions between various species cause a partitioning of isotopes of the same element between reactants and products. For example, when water evaporates to form water vapor, the stable isotopes of oxygen, ^{18}O and ^{16}O, evaporate at slightly different rates. This occurs because the bonds between hydrogen and ^{18}O are slightly stronger than the bonds between hydrogen and ^{16}O. As a result, water containing ^{16}O evaporates more rapidly. The same relationship is true for ^{1}H and ^{2}H. The net result is that after evaporation takes place, the water vapor is enriched in ^{16}O (or depleted in ^{18}O) relative to the water, and it is also depleted in ^{2}H. The difference in isotopic composition between the two phases is called *fractionation*.

Fractionation is also observed in reactions between two chemically reacting species, such as when water and carbon dioxide form carbonic acid:

$$CO_2 + H_2O \rightleftharpoons H_2CO_3. \tag{8-3}$$

As the reaction proceeds, the stable isotopes of carbon, oxygen, and hydrogen are fractionated between the reactants and products, although by different amounts for each isotope. The isotopic fractionation in reactions such as Eq. (8–3) can reach an equilibrium condition similar to chemical equilibrium, or it can be present in a state of disequilibrium, in which the partitioning of isotopes between reactants and products is still occurring. Equilibrium fractionation requires a sufficient time for the forward and reverse reactions to occur such that the total reservoirs of the two isotopes in the reactant and product compounds can be involved in the reaction.

The actual amount of fractionation occurring under equilibrium conditions is a function of the temperature of the reaction because the difference in bond strength between the light and heavy isotopes decreases as the temperature rises. For example, at 0°C, water is enriched in ^{18}O by 11.6‰ at equilibrium, relative to water vapor, but only by 5‰ at 100°C. In this respect, isotopic equilibrium is no different than any other thermodynamic reaction at equilibrium.

The *fractionation factor,* α, is defined as the ratio of the isotopic ratios of reactant and product:

$$\alpha = \frac{R_{reactant}}{R_{product}}. \tag{8-4}$$

When delta values are known for an isotope pair contained in two species, X and Y, then, using the relationship Eq. (8–2), we can derive the fractionation factor as follows:

$$\alpha_{X-Y} = \frac{1 + \dfrac{\delta_X}{1,000}}{1 + \dfrac{\delta_Y}{1,000}} = \frac{1,000 + \delta_x}{1,000 + \delta_y}. \tag{8-5}$$

Fractionation factors are often expressed as $10^3 \ln \alpha$ for consistency with per mil units.

Stable Isotopes of Oxygen and Hydrogen in the Hydrologic Cycle

The isotopic composition of ground water depends upon a complex series of processes beginning with evaporation from the oceans, and followed by movement of vapor in air masses over the continents with periodic condensation, precipitation, and movement through the vadose zone during recharge. The first step in this series of events is evaporation, which is dependent on temperature. As a result, most of the water in the atmosphere is derived from evaporation over subtropical seas. As water evaporates in

TABLE 8–2   Values for fractionation for ^{18}O and ^{2}H in water-vapor reactions (Reprinted with Permission from Clark and Fritz, Environmental Isotopes in Hydrogeology, Copyright, 1997, by CRC Press, Boca Raton, Florida.)

T°C	Water Vapor[1]	
	$10^3 \ln \alpha^{18}O_{w-v}$	$10^3 \ln \alpha^2 H_{w-v}$
−10	12.8	122
0	11.6	106
5	11.1	100
10	10.6	93
15	10.2	87
20	9.7	82
25	9.3	76
30	8.9	71
40	8.2	62
50	7.5	55
75	6.1	39
100	5.0	27

[1] Majoube (1971): $10^3 \ln \alpha^{18}O_{w-v} = 1.137(10^6/T^2) - 0.4156(10^3/T) - 2.0667$

$10^3 \ln \alpha^2 H_{w-v} = 24.844(10^6/T^2) - 76.248(10^3/T) + 52.612$

these areas, the vapor pressure of water containing ^{18}O ($H_2{}^{18}O$) differs from the vapor pressure of water containing deuterium (^2HHO) and both are less than the vapor pressure of H_2O. Thus fractionation occurs, but by a different amount for the heavy isotopes of oxygen and hydrogen.

The fractionation factors for water and water vapor vary as a function of temperature (Table 8–2). At 25°C, for example, the water body being evaporated should be enriched by 9.3‰ in ^{18}O and 76‰ in ^2H. The water vapor would then have a $\delta^{18}O = -9.3‰$ and a $\delta^2H = -76‰$. Evaporation from the ocean in high latitudes would produce slightly more fractionation, but the amount of evaporation is less at lower temperatures. Actual measurements of isotopic fractionation in marine air masses yield values that do not agree with those mentioned above. The conclusion drawn from this discrepancy is that evaporation is a nonequilibrium process. One of the factors that inhibits equilibrium is humidity. When the relative humidity is high, the fluxes of water from the liquid to the vapor and from the vapor to the liquid are roughly equal, a condition which promotes the mixing necessary to achieve equilibrium. As the humidity decreases, however, the flux of water from the liquid to vapor state becomes dominant. Under these conditions, equilibrium does not occur.

Once water vapor has become incorporated in an oceanic air mass, the potential for condensation and precipitation begins. In a general way, precipitation is the opposite of evaporation in that the condensed water is enriched in the heavy isotopes, ^{18}O and ^2H. Precipitation leaves the remaining vapor depleted in these isotopes. A major difference between evaporation and precipitation is that precipitation is an equilibrium process. The water that condenses from the vapor in clouds is enriched in the heavy isotopes to a degree based on temperature (Table 8–2), but in this process sufficient isotopic exchange within the reservoirs of water vapor and liquid water permits equilibrium to be achieved.

As the air masses move over continental land masses, they rise and cool and periodically lose more water by precipitation. The isotopic evolution of the vapor remaining in an air mass can be described in terms of a *Rayleigh distillation model* as follows:

$$R = R_0 f^{(\alpha-1)}. \tag{8–6}$$

In this equation, R is the isotopic ratio of the vapor at any point along the path of the air mass. R_0 is the original ratio in the source area of the air mass, α is the fractionation factor at the temperature of the air mass, and f is the residual fraction of water vapor in the cloud. For example, if half the original vapor had been removed by precipitation, f would be 0.5.

The result of the Rayleigh distillation process is a progressive depletion of precipitation, because of its formation from an increasingly depleted vapor as an air mass moves to higher latitudes or altitudes where cooler temperatures prevail. The progressive depletion of precipitation in an air mass that formed over the ocean at a temperature of 25°C is shown in Fig. 8–1.

The line shown in Fig. 8–1 was derived from measurements of ^{18}O and ^{2}H in precipitation from stations around the world. This line demonstrates the correlation of the two isotopes in precipitation, a relationship first recognized by Craig (1961), who sampled surface waters rather that actual precipitation samples. The equation of the line fit to Craig's data is

$$\delta^2H = 8\delta^{18}O + 10 \text{ SMOW}. \tag{8–7}$$

This line became known as the *Global Meteoric Water Line* (GMWL) because it includes samples from a wide range of climatic regions. Over time the equation of the GMWL has changed slightly because of the change in reference to VSMOW as well as the measurement of precipitation rather than surface waters. A more recent expression of the GMWL (Rozanski et al., 1993) is

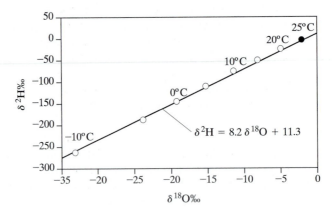

FIGURE 8–1 Progressive depletion in ^{18}O and ^2H during precipitation from an air mass that originally formed over the ocean at 25°C (reprinted with permission from Clark and Fritz, *Environmental Isotopes in Hydrogeology*. Copyright 1997, by CRC Press, Boca Raton, Florida).

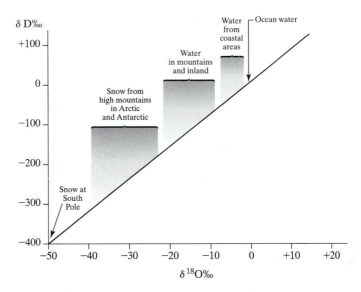

FIGURE 8–2   Variation in isotopic composition by climatic or physiographic region (reprinted with permission from Clark and Fritz, *Environmental Isotopes in Hydrogeology*. Copyright 1997, by CRC Press, Boca Raton, Florida).

$$\delta^2H = 8.17(\pm0.07)\delta^{18}O + 11.27(\pm0.65)\text{‰ VSMOW.} \qquad (8\text{–}8)$$

The position of points on the GMWL can be related in a general way to climatic regions (Fig. 8–2).

The GMWL is really a composite relationship because it includes such a wide variety of climatic regions and because its points are long-term averages for a particular station. Individual points from a specific station can be plotted on the same type of graph as the global data. When this is done, the regression lines commonly diverge slightly from the GMWL because of complicating factors that vary from place to place. Significant deviations from the GMWL can occur through evaporation because of the nonequilibrium effects described earlier. In warm regions where evaporation of precipitation occurs during fall from the cloud to the ground, points often plot on a straight line that diverges from the GMWL at a lower slope.

The factors that control the isotopic composition can be divided into regional and local effects. On a regional basis, latitude and continentality exert important controls on precipitation. Regional patterns of ^{18}O and ^2H content have a close correlation with latitude in many places (Fig. 8–3). This correlation is based on the decreases in temperature from the equator to the poles and also because the general pattern of atmospheric circulation includes poleward movement of air masses, thus enhancing the progressive decrease in water vapor content inherent in the Rayleigh model.

The continentality of a location is important because continents remove more water from air masses than marine areas as a result of elevation. In addition, areas far from the modifying influences of coasts have extreme variations in temperature, which are reflected in the isotopic content of precipitation.

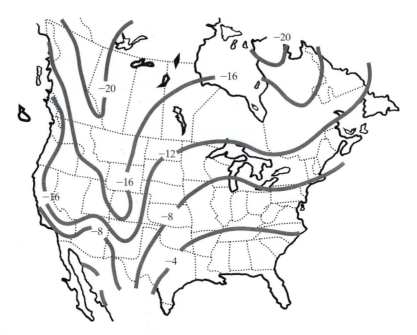

FIGURE 8–3 Weighted mean $\delta^{18}O$ in precipitation in the U.S. and southern Canada (reprinted with permission from Clark and Fritz, *Environmental Isotopes in Hydrogeology*. Copyright 1997, by CRC Press, Boca Raton, Florida).

At the local scale, precipitation patterns and isotopic content are influenced by altitude, seasonal climatic effects and the secondary evaporation of moisture from bodies of freshwater. The altitude effect is manifested in a significant gradient of $\delta^{18}O$ and $\delta^{2}H$ values with elevation because of the corresponding temperature decrease. For example, ^{18}O decreases by -0.15 to $-0.5‰$ per 100 m increase in elevation (Clark and Fritz, 1997). Seasonality results in highly negative $\delta^{18}O$ values for winter precipitation in inland areas and a greater difference in winter and summer precipitation relative to coastal areas. An example of the secondary precipitation effect is provided by Machavaram and Krishnamurthy (1994) who, in a study of the isotopic composition of precipitation downwind from Lake Michigan, showed that evaporation from the lake accounts for a significant amount of the precipitation in downwind areas.

^{18}O and ^{2}H in Ground Water Recharge

The isotopic content of ground water and precipitation can be used to investigate the processes of ground water recharge. Shallow ground water in regions with temperate climates commonly has $\delta^{18}O$ and $\delta^{2}H$ values close to the weighted mean annual values in precipitation (Clark and Fritz, 1997). Because the delta values in precipitation vary significantly over the year, mixing of waters recharged during different times of the year in the vadose zone is quite effective, so that by the time the recharge water arrives at the water table, it is isotopically homogeneous. In other areas, however, the weighted mean annual precipitation values may not match the delta values in shallow ground water.

For example, in western Canada the weighted mean $\delta^{18}O$ value in precipitation is enriched by several per mil relative to the ground water values (Clark and Fritz, 1997). This difference is attributed to the timing of recharge, with a dominance during the cold periods of the year when precipitation is depleted in ^{18}O. Summer precipitation may be significant, but may not contribute equally to recharge because of evapotranspiration.

In areas with high relief, ^{18}O and ^2H contents can be linked to the elevation of recharge. The orographic cooling experienced by air masses moving over mountain ranges results in a progressive depletion of ^{18}O and ^2H in precipitation as the elevation increases. A study in Gran Canaria, a mountainous island located off the coast of Spain (Gonfiantini et al., 1976), yielded a strong correlation between $\delta^{18}O$ and δ^2H values with elevation (Fig. 8–4). The isotopic gradients for this area are $-0.13\ \delta^{18}O$ and $-1.0\ \delta^2H$ per 100 m rise in elevation for the northern flow system and $-0.24\ \delta^{18}O$ and $-1.2\ \delta^2H$ per 100 m in the southern flow system. Thus by obtaining the δ values in a ground water sample, it is possible to estimate the elevation of recharge.

In arid areas, ground water isotopic contents display a strong signature of evaporation. Evaporation prior to recharge can occur during flow across the land surface or from the vadose zone. Differences in the amount of evaporation, as reflected in the isotopic values, can be used to determine the timing and source areas of recharge (Clark and Fritz, 1997). Evaporation is suggested on δ^2H vs. $\delta^{18}O$ plots by points that define a line that diverges from the meteoric water line. The slope of the line of evaporated samples typically ranges from about 2 to 6. This relationship was apparent in a study of ground water in Algeria by Gonfiantini et al. (1974). Ground waters in this area were derived from shallow and deep sources. Some of the shallow waters plotted on an evaporation line below the meteoric water line (Fig. 8–5). These samples were taken in areas where the water table was very close to the surface and evaporation occurred directly from the vadose zone. Regression lines attributed to evaporation intersect the local meteoric water line at the point representing the isotopic composition of the water prior to evaporative enrichment.

Other applications of stable isotopes in recharge studies include the identification of recharge derived from streams, lakes, or wetlands, rather than from infiltration through

FIGURE 8–4 Correlation between oxygen isotope composition and altitude for ground water samples on Gran Canaria. Zone N = north side of island. Zone S = south side of island (from Gonfiantini et al., 1976, reprinted by permission of the IAEA).

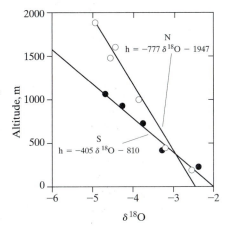

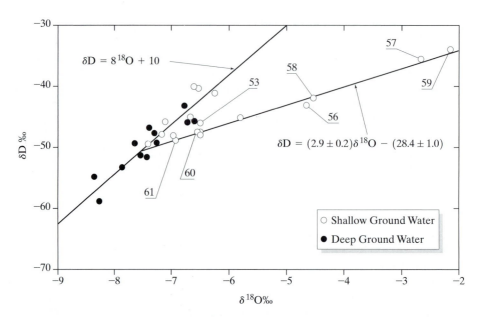

FIGURE 8–5 $\delta^{18}O$–$\delta^{2}H$ plot of shallow and deep ground water samples from a drainage basin in Algeria. The numbers refer to dug wells with high salt content in areas of very shallow water table from which δD and $\delta^{18}O$ values were used to calculate the equation of the line that plots below the meteoric water line. These values were affected by evaporation from the vadose zone (from Gonfiantini et al., 1974, reprinted by permission of the IAEA).

the vadose zone. Wells located near streams commonly obtain a significant fraction of water from the river through induced recharge caused by pumping. This component of river water can be distinguished from other water in the aquifer by its isotopic signature. Rivers reflect the seasonal isotopic composition of precipitation and runoff in contrast to the nonfluctuating ground water delta values that approximate the weighted average of seasonal precipitation variations. The isotopic difference between stream flow and ground water can also be used to separate storm runoff and base flow contributions to storm hydrographs (Fritz et al., 1976).

Lakes and wetlands interact complexly with ground water flow systems, with spatially variable recharge and discharge relationships around the body of water. Ground water recharge from the lake or wetland can be expected to be somewhat enriched in ^{18}O and ^{2}H because of evaporation from the water surface. Krabbenhoft et al. (1990) performed a mass balance analysis on lake–ground water interactions using stable isotopes, and Kehew et al. (1998) showed that both stable isotopes and chemical composition of ground water around and beneath wetlands can be used to determine the recharge-discharge function of the wetland.

In some field settings, highly negative $\delta^{18}O$ values suggest the presence of ground water that is older than the Holocene. Remenda et al. (1994) collected water samples from thick glaciolacustrine clays deposited in Glacial Lake Agassiz in Manitoba and North Dakota. These clay sediments are unweathered and unfractured and have very low hydraulic conductivities. Plots of $\delta^{18}O$ with depth (Fig. 8–6) show a gradual decrease

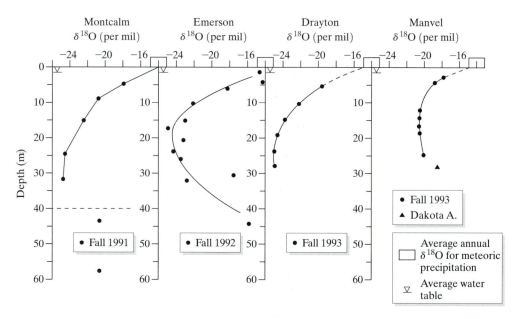

FIGURE 8–6 $\delta^{18}O$ values vs. depth for four monitor sites in Glacial Lake Agassiz plain. The more depleted values at depth represent recharge during colder climatic conditions (reprinted with permission from Remenda et al., Isotopic composition of old ground water from Lake Agassiz: Implications for late Pleistocene climate. *Science*, v. 266, pp. 1975–1978. Copyright 1994 American Association of Science).

from values of ~ 14 per mil in near-surface ground water to -24.5 per mil. The more depleted ^{18}O values are indicative of a much colder climate, conditions that would have prevailed during deposition of the sediments in the late Pleistocene. Similar results have been obtained in glaciolacustrine sediments and tills in other locations in North America (Bradbury, 1984, Desaulniers et al., 1981, and Hendry, 1988). Very low hydraulic conductivities are necessary for retaining ground water from the former climatic environment.

Stable Isotopes of Carbon in Ground Water

As we saw in Chapter 3, reactions among inorganic and organic species in the carbonate system play an important role in the evolution and control of ground water quality. Isotopic fractionation between ^{13}C and ^{12}C during these reactions adds another dimension to the processes that take place in the vadose and saturated zones.

The driving force behind carbonate system reactions is the partial pressure of CO_2 in the soil zone, which is a product of root respiration and organic matter oxidation. The soil gas will retain the $\delta^{13}C$ signature of the plants growing in the soil, which is a function of fractionation during fixation of atmospheric CO_2 during photosynthesis. Organic carbon derived from atmospheric carbon dioxide with a $\delta^{13}C$ value of about $-7‰$ VPDB is depleted in ^{13}C to a degree dependent on which of two different photosynthetic reactions occurs in the plants. These different processes characterize C_3 and C_4

plants. C_3 plants, which constitute the majority of terrestrial plants, are more depleted and have a $\delta^{13}C$ value ranging from $-24‰$ to $-30‰$. C_4 plants are less depleted, falling in the range of $-10‰$ to $-16‰$. C_4 plants include some important agricultural crops such as sugar cane and corn. A third, minor plant type, which occurs in deserts, has an intermediate $\delta^{13}C$.

When the plants die, the organic carbon is oxidized by aerobic bacteria back to carbon dioxide without fractionation so that the soil CO_2 initially has the same $\delta^{13}C$ as the plant. The soil-gas CO_2 is enriched slightly, about 4‰ by diffusion toward the soil surface and downward toward the water table.

The dissolution of CO_2 into vadose water initiates another group of fractionation reactions as CO_2 is hydrated to form H_2CO_3, and then dissociates to form HCO_3^- and CO_3^{2-} (Fig. 8–7). The actual amount of fractionation varies with temperature. A final fractionation occurs if calcite is precipitated. In the absence of other carbon sources, the $\delta^{13}C$ of the dissolved inorganic carbon (DIC) would vary with pH because pH controls the abundances of the individual species (Chapter 3). In water in the neutral vicinity, bicarbonate ion would dominate the DIC and the $\delta^{13}C$ would be approximately -13 VPDB.

In many settings, infiltrating vadose waters encounter calcite and other carbonate minerals. If the calcite has a marine origin, it generally has a $\delta^{13}C$ value close to zero. The enriched DIC produced as calcite dissolves mixes with the depleted DIC from soil-gas sources. The net result is that the overall DIC evolves toward more enriched $\delta^{13}C$ values. The amount of increase, however, is dependent upon whether or not the system is open during dissolution. If open-system conditions prevail, the DIC reaches a final state that is close to the soil-gas bicarbonate value. In closed-system conditions, the final $\delta^{13}C$ values are more affected by the contribution of the dissolved calcite and evolve to heavier values (Clark and Fritz, 1997). The $\delta^{13}C$ values measured in ground water are,

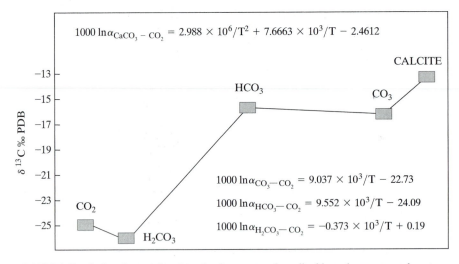

FIGURE 8–7   Carbon isotope fractionation between carbon dioxide and aqueous carbonate species (from Hendry, 1989, reprinted by permission of the National Ground Water Association).

therefore, composite functions of the vadose zone p_{CO_2}, the type of vegetation, the contribution of calcite dissolution, and the openness of the system in which the dissolution occurred. If organic carbon is present in ground water, it is generally more depleted in ^{13}C than the inorganic carbon.

Methanogenesis, the bacterial production of methane, was discussed in Chapter 7. The two reaction pathways mentioned were CO_2 reduction and acetate fermentation. Both of these processes produce a large fractionation in $\delta^{13}C$, in which the methane is highly depleted relative to CO_2. $\delta^{13}C$ values in the methane are typically less than $-50‰$. This fractionation process is useful in distinguishing the source of methane in ground water, because methane produced by methanogenesis, also known as *biogenic* methane, is significantly different in stable isotope composition from methane that is derived from natural gas, which is called *thermogenic* methane, and from *abiogenic* methane, methane that is produced by the reduction of inorganic carbon dioxide in deep crustal or mantle environments. Differentiating types of methane is also aided by measurements of δ^2H, which are plotted against $\delta^{13}C$ as in Fig. 8–8. Biogenic gases plot in one of two fields in the diagram depending on the reaction pathway. The fermentation pathway is more common in shallow environments such as landfills and wetlands, whereas CO_2 reduction is more characteristic of deeper freshwater and marine settings. Samples that do not plot within the fields shown in Fig. 8–8 can be explained by the mixing of different gas types or by other processes. For example, oxidation of biogenic methane, associated with

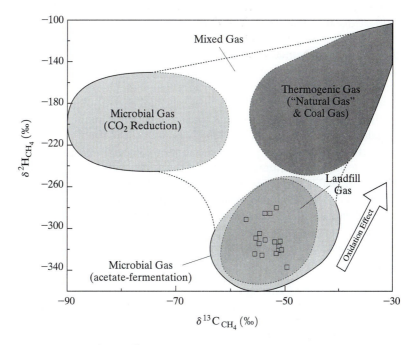

FIGURE 8–8   Plot of δ^2H vs. $\delta^{13}C$ in CH_4. Methane of different origins plots in characteristic fields (from Hackley et al., 1996, reprinted by permission of the National Ground Water Association).

the methanotrophic bacteria, causes $\delta^{13}C$ enrichment of the remaining methane because of the preferential oxidation of ^{12}C isotopes. Thus a partially oxidized methane gas could plot in the thermogenic field.

Other gas characteristics can also be used in determining the origin of methane. Ethane and propane contents of thermogenic gases are higher than biogenic gases and can therefore aid in the analysis. Carbon 14, which is discussed later in the chapter, is useful as well because thermogenic gas is typically very old and lacks detectable ^{14}C. Biogenic methane, on the other hand, may contain moderate to high levels of ^{14}C.

Stable isotopes of carbon and hydrogen were instrumental in the identification of the source of methane in an aquifer in Kingsford, Michigan (Westjohn and Godsey, 1998). The methane was discovered when static electricity from a dryer unfortunately set off a strong explosion in a basement that severely injured the homeowner. Subsequent investigation of soil-gas concentrations around the house indicated that the soil gas contained nearly 100% methane. The natural gas pipeline that served the neighborhood was ruled out as the source of the methane by $\delta^{13}C$ and $\delta^{2}H$ of the gas, which indicated a biogenic rather than thermogenic source. Another potential source, biogenic gas formed from buried organic material in the thick glacial drift in the area, was ruled out by the ^{14}C content of the gas, which suggested a modern carbon source rather than an old source. After an extensive investigation, the most likely source of the methane turned out to be an old manufacturing plant upgradient from the explosion site. The plant was an old auto body manufacturing facility dating from the early days of the auto industry when auto bodies were made from wood. The plant processed huge quantities of wood and used some of the waste products in an innovative chemical plant that produced acetone, acetate, and other chemical products. Later, the process of carbonization of wood by-products to produce charcoal briquets on a commercial scale was developed at the plant. The waste products from these processes were disposed of for years in several natural depressions (glacial kettles) near the plant. Organic carbon was biodegraded, primarily by acetate fermentation, beneath the disposal pits, and high concentrations of methane, along with other compounds, eventually contaminated a large volume of the aquifer over an area of more than 2 square miles (Fig. 8–9). Dissolved methane concentrations in the deeper parts of the aquifer reached levels higher than 60 mg/l and bubbled out of the Menominee River in the discharge area of the aquifer to the degree that the entire water surface was in places covered with gas bubbles. Although most of the shallow parts of the aquifer were protected from upward migration of the methane by thick glaciolacustrine deposits, preferential pathways of some type allowed methane to reach the surface in several locations. Once it accumulated in the vadose zone, the gas was able to migrate into basements.

Stable Isotopes of Nitrogen

The use of nitrogen isotopes, in particular the $^{15}N/^{14}N$ ratio, is an important tool for understanding the source of nitrate contamination of ground water and the transformations of nitrogen that occur in subsurface environments. This technique was pioneered by Kreitler and Jones (1975) in a study in Runnels County, Texas, where nitrate concentrations in shallow ground water were so high that several cattle died from anoxia in 1968

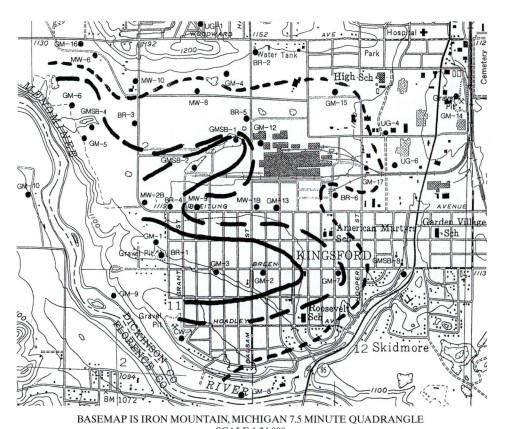

BASEMAP IS IRON MOUNTAIN, MICHIGAN 7.5 MINUTE QUADRANGLE
SCALE 1:24 000

CONTOUR INTERVAL 20 FEET
NATIONAL GEODETIC VERTICAL DATUM OF 1929

EXPLANATION

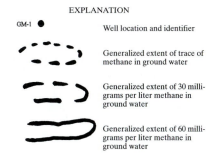

GM-1 ●   Well location and identifier

Generalized extent of trace of
methane in ground water

Generalized extent of 30 milligrams per liter methane in
ground water

Generalized extent of 60 milligrams per liter methane in
ground water

FIGURE 8–9   Distribution of methane in ground water in Kingsford, Michigan. Solid contour = 60 mg/l. Long dashed contour = 30 mg/l. Short dashed contour = trace concentration (from Westjohn and Godsey, 1998).

because they drank this water. Suspecting that animal manures could be the cause, Kreitler and Jones (1975) collected soil samples from a variety of settings including unfertilized fields, fields where cattle grazed, septic tank leach beds, and barnyards, and determined the $\delta^{15}N$ values for the nitrate in these soils. The results clustered in two populations (Fig. 8–10), corresponding to natural soil nitrate and animal-waste nitrate, with animal-waste nitrate having a higher $\delta^{15}N$ range. Ground water nitrate distributions correlated well with soil values in cultivated fields but showed enrichment in wells near barnyards and septic systems. Overall, the high nitrate concentrations in ground water were attributed to natural soil nitrogen. Semi arid grassland regions, such as Runnels County, Texas, have optimum conditions for high soil nitrogen as a result of yearly contributions of nitrogen from decay of grass roots and low rainfall. Cultivation of the soil since the early 1900s led to oxidation of the soil nitrogen in organic form to nitrate.

Although synthetic fertilizers were not added to the fields in Runnels County, Texas, subsequent studies have shown that these compounds range from −5 to +5 in $\delta^{15}N$ (Chapelle, 1993; Clark and Fritz, 1997). Thus, ground water nitrate derived from synthetic fertilizers can often be distinguished from ground water impacted by animal wastes (septic tank effluent or livestock manure), which has a $\delta^{15}N$ range of approximately +10 to +20. This information is useful in implementing changes in land management practices for reduction of ground water nitrate contamination.

Stable isotopes can also be used to discern the causes for changes in nitrate concentration in ground water flow systems. One common change noted in agricultural areas is a decrease in nitrate concentration with depth in the aquifer. Denitrification is a distinct possibility for these declines because of the restriction of nitrate redox stability to oxic conditions near the upper stability limit of water (Fig. 5–12). Decreases in nitrate concentration with depth are commonly accompanied by enrichment in $\delta^{15}N$, as shown by Aravena and Robertson (1998) in a study of ground water beneath a septic tank leach field (Fig. 8–11). At this site, $\delta^{15}N$ values reach as high as +58.3‰. Denitrification, in which ^{14}N is preferentially reduced, is the probable increase in $\delta^{15}N$.

FIGURE 8–10 $\delta^{15}N$ ranges of natural soil nitrate and soils beneath barnyards and septic tank leach fields in Runnels County, Texas (from Kreitler and Jones, 1975, reprinted by permission of the National Ground Water Association).

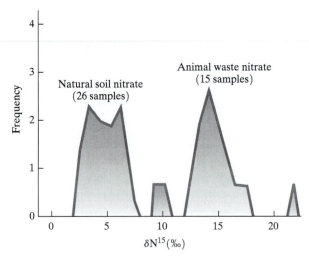

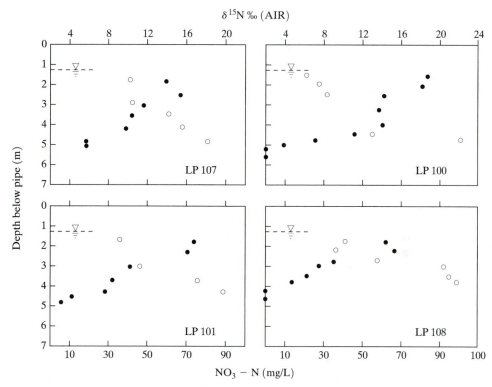

FIGURE 8–11   Depth profiles of $\delta^{15}N$ and nitrate concentration in a shallow sand aquifer beneath a septic tank leach field in southern Ontario. ($\bullet$ = nitrate concentration, $\circ$ = $\delta^{15}N$ values). $\delta^{15}N$ becomes enriched with depth as a result of denitrification (from Aravena and Robertson, 1998, reprinted by permission of the National Ground Water Association).

Hendry et al. (1984) used $\delta^{15}N$ analyses, along with other chemical and microbiological evidence, to conclude that very high nitrate concentrations (up to 300 mg/l) in weathered tills in Saskatchewan were naturally produced by oxidation of ammonium in the tills during a hot, dry period during the mid-Holocene when water tables were low. When water tables rose in the late Holocene, denitrification occurred in the weathered till, in which trace organic carbon in the till was the electron donor. Nitrate remained high in the weathered till only in isolated enclaves surrounded by areas of much lower nitrate concentration. Large enclaves are as much as 1 km in width along the flow path. Enriched $\delta^{15}N$ values where nitrate concentrations were low downgradient of the enclaves relative to nitrate in the enclaves (Fig. 8–12) provide evidence for the denitrification mechanism.

Aravena and Robertson (1998) also showed that enrichment of $\delta^{18}O_{NO_3}$ correlates with the enrichment of $\delta^{15}N$ (Fig. 8–13). The correlation in trends of two stable isotopes, much like $\delta^{18}O$ and δ^2H in water, strengthens the interpretation of the process. Wassenaar (1995) used both $\delta^{15}N$ and $\delta^{18}O_{NO_3}$ to identify the source of nitrate in an agricultural area in British Columbia. The lack of change in $\delta^{18}O_{NO_3}$ with changes in $\delta^{15}N$ suggested that denitrification was not occurring, probably because of the lack of an electron donor. The source of the high nitrate in Wassenaar's study was shown to be poultry manure application, based on the $\delta^{15}N$ values. The $\delta^{18}O_{NO_3}$ was slightly more enriched

FIGURE 8–12 Plot of nitrate concentration vs. $\delta^{15}N$ in weathered till in southern Alberta. High $\delta^{15}N$ downgradient of high-nitrate enclaves suggests enrichment by denitrification (reprinted from *Jour. Hydrol.*, v. 70, Hendry, M. J., McCready, R. G. L., and Gould, W. D., Distribution, source and evolution of nitrate in a glacial till of Southern Alberta, Canada, pp. 177–198, Copyright 1984, with permission of Elsevier Science).

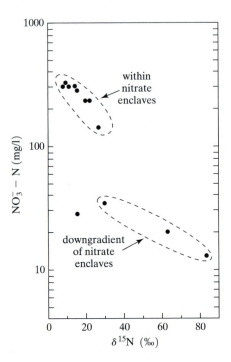

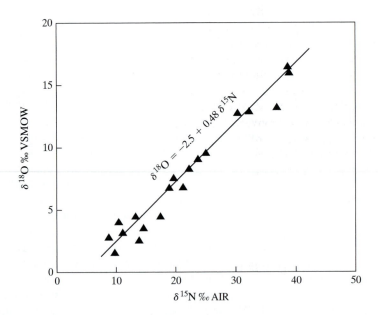

FIGURE 8–13 Correlation of $\delta^{15}N$ and $\delta^{18}O_{NO_3}$ in ground water samples beneath a septic leach field provides further evidence that samples enriched in $\delta^{15}N$ and $\delta^{18}O_{NO_3}$ are produced by denitrification (from Aravena and Robertson, 1998, reprinted by permission of the National Ground Water Association).

than the $\delta^{18}O$ in ground water, pointing to a summer or fall origin of the $\delta^{18}O_{NO_3}$. Wassenaar explained this difference by nitrification of the nitrogen in the poultry manure in the summer and subsequent flushing to the water table by fall rains (Wassenaar, 1995).

Stable Isotopes of Sulfur

The stable isotope ratios of sulfate, including both $\delta^{34}S$ and $\delta^{18}O$, are useful in determining the source of the sulfate and the geochemical processes that have occurred in the aquifer. Sulfate can be derived from a variety of sources in ground water, including dissolution of marine evaporite minerals, mixing with seawater, oxidation of reduced sulfide minerals such as pyrite, and oxidation of aqueous H_2S, which has migrated into the aquifer. The isotopic ratios of sulfur in different types of rocks, minerals, and solutions are shown in Fig. 8–14. It is clear from this diagram that the $\delta^{34}S$ value of marine evaporites has not been constant over geologic time. Because there is no significant fractionation of sulfur isotopes during dissolution of evaporite minerals, the $\delta^{34}S$ content of ground water sulfate derived from evaporite mineral dissolution should be consistent with the value established by the age of the evaporite minerals. These values are mostly enriched to +15‰ or more.

The most significant fractionation of sulfur isotopes takes place during bacterial reduction of sulfate. In this process, a depletion of 40‰ or more is common in H_2S produced by reduction of SO_4^{2-}. Much of the reduced sulfur in sedimentary rocks and sediments originated with bacterial reduction of sulfate under reducing conditions. When

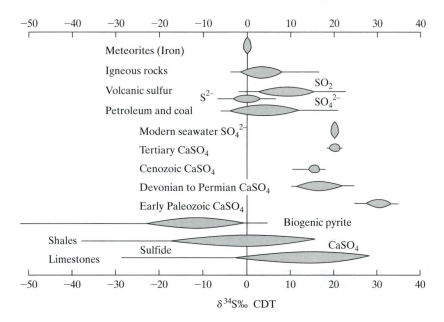

FIGURE 8–14   Ranges in $\delta^{34}S$ contents in various sulfur compounds and environments (reprinted with permission from Clark and Fritz, *Environmental Isotopes in Hydrogeology*. Copyright 1997, by CRC Press, Boca Raton, Florida).

reoxidized to sulfate, fractionation is much less significant. As a result, sulfate derived from oxidation of reduced sulfur is much more negative than sulfate derived from dissolution of evaporite minerals.

The application of these relationships, along with the $\delta^{18}O$ of sulfate, is relevant to a wide range of problems. Rightmire et al. (1974) used sulfur isotopes, along with other evidence, to examine the origin of sulfate in two confined aquifers, the Floridan and Edwards aquifers, that have increasing sulfate concentrations along their flow paths. The Floridan aquifer, in central Florida, has gradually increasing sulfate concentrations from inland recharge areas to discharge areas near the coast. When the sulfate concentrations were plotted against the SO_4^{2-}/Cl^- ratio (Fig. 8–15), it was obvious that the increased sulfate concentrations could be explained by two mechanisms: mixing with ocean waters where the sulfate/chloride ratio was low and dissolution of evaporite minerals where the sulfate/chloride ratio increased along the flow path. Because the $\delta^{34}S$ content of seawater sulfate and sulfate dissolved from evaporites is similar, $\delta^{34}S$ was not particularly useful in separating the waters from different origins. However, the $\delta^{34}S$ values did suggest some sulfate reduction in high-sulfate waters that also contained sulfide (Fig. 8–16) because, where sulfate concentrations were high and sulfide was present,

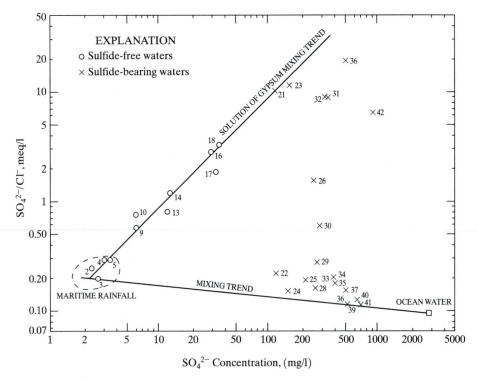

FIGURE 8–15 Plot of sulfate concentrations vs. SO_4^{2-}/Cl^- for the Floridan aquifer. Two mechanisms—mixing with ocean waters and solution of gypsum along the flow path—explain most data points (from Rightmire et al., 1974, reprinted by permission of the IAEA).

FIGURE 8–16   Plot of sulfate concentration vs. $\delta^{34}S$ in sulfate for ground water samples from the Floridan aquifer. High-sulfate, sulfide-bearing waters plot to right of mixing curve, suggesting sulfate reduction. Curve A–B represents mixing of sulfate derived from maritime rainfall with sulfate from ocean water or evaporite minerals. (from Rightmire et al., 1974, reprinted by permission of the IAEA).

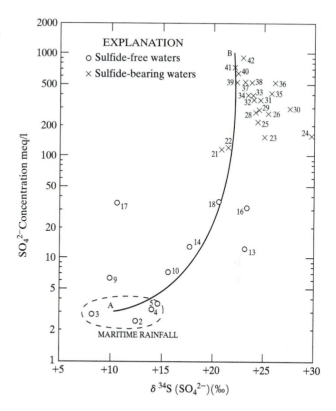

$\delta^{34}S$ values plotted to the right of a mixing curve, implying an enrichment of ^{34}S as a result of preferential bacterial reduction of ^{32}S.

Study of the Edwards aquifer, a confined aquifer that supplies water to the cities of Austin and San Antonio, Texas (Rightmire et al., 1974), using the same methods of analysis, suggested a very different origin for the dissolved sulfate. Rather than a gradual increase in sulfate from recharge to discharge areas, the Edwards aquifer has two distinct zones, a low-sulfate zone with no dissolved sulfide and a downgradient zone with high sulfate and measurable dissolved sulfide. When plotted on a graph of sulfate concentration vs. SO_4^{2-}/Cl^- ratio (Fig. 8–17), the waters also separate into two distinct groups. The low sulfate, sulfide-free waters show a gradual increase in sulfate concentration, which could be explained by dissolution of gypsum, but the high-sulfate, sulfide-bearing waters lie along a trend that is consistent with the composition of brine solutions in nearby oil fields. The sulfate in these waters was explained by the presence of an ancient marine brine that migrated into the aquifer rather than mixing with seawater or dissolution of evaporite minerals. The $\delta^{34}S$ plot (Fig. 8–18) showed no evidence of enrichment in the high-sulfate, sulfide-bearing waters, like the waters of the Floridan aquifer. The sulfide in this zone must have also migrated into the aquifer.

Sulfur isotopes have been applied to help determine the source of sulfate in a variety of hydrogeologic settings. Two additional examples include the studies of Mayo et

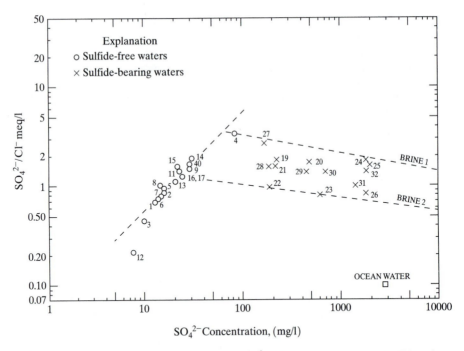

FIGURE 8–17   Plot of sulfate concentrations vs. SO_4^{2-}/Cl^- in ground water for the Edwards aquifer in Texas. The gradual increase of sulfate (along trend of dashed line) in low-sulfate waters in sulfide-free zone of aquifer suggests dissolution of gypsum. The high-sulfate, sulfide-bearing waters suggest mixing with brines in downgradient part of aquifer (from Rightmire et al., 1974, reprinted by permission of the IAEA).

al. (1992) and Siegel (1990). Mayo et al. (1992) used $\delta^{34}S$ values to account for the origin of sulfate in a mining district in Utah. The sulfate in acid mine drainage had a lower $\delta^{34}S$ value than in waters of neutral pH. This relationship implied that the sulfate in the acid mine drainage originated from sulfide mineral oxidation, whereas the sulfate in the waters of neutral pH was derived from dissolution of evaporite minerals. Siegel's study concerned the reversal of the hydraulic gradient in the Cambrian-Ordovician aquifer of eastern Wisconsin and northern Illinois by a Pleistocene glacier in the Lake Michigan basin. Today ground water flows from west to east toward Lake Michigan in the Cambrian-Ordovician aquifer. The high hydraulic head created by the presence of a glacier in the Lake Michigan basin probably reversed the gradient, causing ground water to flow from east to west. Evidence for this hypothesis comes in the form of sulfate with $\delta^{34}S$ values around +20. There is no source of sulfate minerals with this $\delta^{34}S$ value along the current flow path from the recharge area, but Silurian evaoporites beneath Lake Michigan do have this sulfate composition. Movement of this sulfate in the opposite direction of the normal direction of flow could account for this water in the Cambrian-Ordovician aquifer of eastern Wisconsin and northern Illinois.

Aravena and Roberston (1998), in a study of a contaminant plume from a septic system in a shallow aquifer, demonstrate the usefulness of multiple isotopic tracers. The focus of their investigation was to determine the cause of nitrate attenuation in the plume. By determining both the $\delta^{15}N$ and the $\delta^{18}O$ in nitrate they were able to show a steady enrichment of ^{15}N and ^{18}O as the nitrate concentration decreased, a pattern which suggests denitrification. When nitrate is used as an electron acceptor in denitrification, an electron donor must be present. The two possible electron donors considered were organic carbon and reduced sulfur present as pyrite in the aquifer sediments. Both sources were found to be viable. In the case of sulfur, the evidence included an increase in sulfate, accompanied by a depletion in ^{34}S and ^{18}O in sulfate relative to sulfate in the raw wastewater source. These depletions occurred because pyrite, which was oxidized to form sulfate, has a low $\delta^{34}S$, and ground water, which supplied the oxygen to make SO_4^{2-}, had a lower $\delta^{18}O$ than the wastewater source. Thus the sulfate produced was depleted in the heavy isotopes of both sulfur and oxygen.

RADIOISOTOPES

Unlike stable isotope applications that shed light on geochemical processes in aquifers, the radioisotopes are primarily used for determining the relative or absolute age of water in an aquifer. Actually, the dates obtained give some indication of the residence time of water in an aquifer once it has passed through the vadose zone. The decay rate of these isotopes forms the basis for dating the recharge of the water body. Radioactive decay follows first order kinetics, which was discussed in Chapter 2.

FIGURE 8–18   Plot of sulfate concentration vs. $\delta^{34}S$ in sulfate for the Edwards aquifer. Data show two distinct sources of sulfate. Sulfate in sulfide-free waters interpreted to be result of dissolution of gypsum-bearing dust in recharge area. Sulfate in sulfide-bearing waters interpreted to be derived from residual marine brines in downgradient area. Sulfate reduction is not indicated, so sulfide must migrate into aquifer from an outside source (from Rightmire et al., 1974, reprinted by permission of the IAEA).

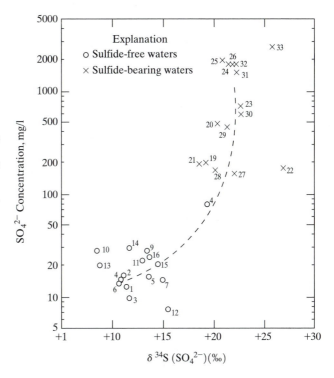

Tritium

Tritium (^{3}H) is a radioactive isotope of hydrogen with a half life of 12.43 years. The use of tritium as a dating tool is based on the huge influx of tritium that entered the atmosphere in the 1950s and 1960s as a result of testing of the hydrogen bomb. The first input of this source occurred in 1952 and levels peaked in 1962, the year before the test-ban treaty between the U.S. and the Soviet Union. Some testing continued after this time, but tritium levels declined steadily. The input of "bomb" tritium into the atmosphere overwhelmed the natural supply of tritium, which is produced by the bombardment of nitrogen by neutrons in cosmic radiation. Tritium is measured in *tritium units* (TUs), each of which consists of one tritium atom per 10^{18} hydrogen atoms. The natural production of tritium is estimated to maintain levels of 5–10 TU. In the early 1960s, atmospheric levels were well above 1,000 TUs in the middle latitudes of the northern hemisphere.

In the atmosphere, tritium reacts with oxgen according to the reaction

$$^3H + O_2 \longrightarrow {}^3HO_2 \longrightarrow {}^1H\,{}^3HO. \tag{8–9}$$

Tritium is then removed from the atmosphere by precipitation. Large seasonal variations characterize the tritium content of the atmosphere, with peak concentrations during a period known as the *spring leak*, when mixing between the stratosphere and troposphere increases the tritium available to precipitation. The Ottawa record (Fig. 8–19), which is the longest continuous tritium record, shows the large seasonal fluctuations superimposed on a rise to a peak in 1962 followed by a decline to the present. Latitudinal variations are also significant, with highest levels in the northern hemisphere occurring in the mid to high latitudes.

Tritium decays by beta emission to helium according to the reaction

$$^3H \longrightarrow {}^3He + \beta^-. \tag{8–10}$$

Peak levels measured in precipitation that fell to the ground in 1962 would, by 2000, have decayed through three half-lives, which, for 1,000 TU, would now represent a level of about 125 TU.

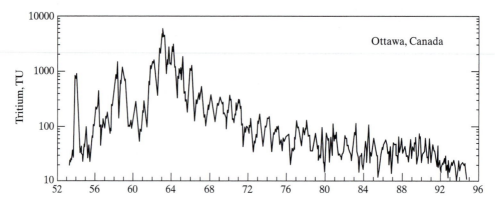

FIGURE 8–19   Tritium content in precipitation at Ottawa, Ontario (reprinted with permission from Clark and Fritz, *Environmental Isotopes in Hydrogeology*. Copyright 1997, by CRC Press, Boca Raton, Florida).

Tritium is measured with liquid scintillation detectors. Direct counting has a precision of ± 7 TU, a value that makes interpretations of waters with low tritium content more difficult. Enrichment techniques improve the precision of the measurement to ± 0.8 TU.

Tritium Applications in Ground Water

The applications of tritium to ground water problems generally fall into three groups: (1) direct dating of the age of the water, (2) use of the tritium distribution in an aquifer to estimate hydrologic parameters such as recharge rates or dispersion, and (3) use of tritium to develop an understanding of flow processes in fractured and mixed fractured and granular aquifer media.

Qualitative estimates of ground water age using tritium content can be estimated as shown in Table 8–3. Absolute age dating is useful for a general understanding of a ground water flow system and also for more practical problems such as the isolation of aquifers by aquitards in the vicinity of a proposed waste disposal site, but determination of an accurate date is complicated by several factors. First, tritium input data for the field site may not be available, requiring that data from the nearest station be used. For example, Bradbury (1990) used regression analysis between measured values at Madison, Wisconsin, and measured values at stations where tritium was measured more frequently. The tritium content of precipitation between times of measurement at Madison could, therefore, be estimated using the best correlated, more continuous record at a distant station. Second, considerable spatial and temporal mixing of infiltrating precipitation occurs in the unsaturated zone prior to recharge (Clark and Fritz, 1997). Thus, weighted averages of input precipitation levels are commonly used. The lack of exact input concentrations precludes the use of tritium alone to calculate an exact date using the first-order decay equation [Eq. (2–52)]. Alternatively, vertical profiles of tritium in vadose zone soils or aquifers where ground water flow is vertically downward in some instances show a characteristic pattern of increasing then decreasing tritium values. If enough samples are available, the peak tritium value is assumed to be a good estimate of the tritium recharge peak of 1962. This approach is becoming less useful as the 1962 peak has moved through the vadose zone in most areas. Bradbury (1990) estimated ages after constructing a weighted tritium input curve for the study area (Fig 8–20) that was corrected for decay. Examination of this curve shows that tritium values between about 20 and 50 TU could have entered the flow system either on the rising limb or declining

TABLE 8–3   Approximate ground water recharge age for continental areas based on tritium values (from Clark and Fritz, 1997).

Tritium Value	Age
<0.8 TU	Submodern—recharged prior to 1952
0.8–~4 TU	Mixture of submodern and recent recharge
5–15 TU	Modern (<5 to 10 yr)
15–30 TU	Some "bomb" tritium present
>30 TU	Considerable component of recharge from 1960s or 1970s
>50 TU	Dominantly 1960s recharge

FIGURE 8–20 Tritium input curve corrected for decay for samples collected in 1984 and 1986. Samples with between 20 and 50 TUs could have two possible ages associated with either rising or falling limb of large spike from 1960s. Other factors must be used to choose between the two dates (from Bradbury, 1990, reprinted by permission of the National Ground Water Association).

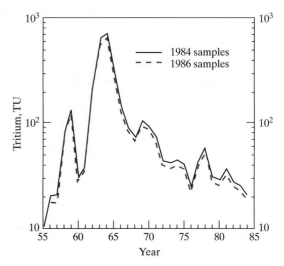

limb of the large spike centered around 1965. Bradbury differentiated between these choices based on other factors known about the sampling point such as well depth and position in the flow system.

A different approach to absolute dating utilizes the daughter product of tritium decay, ^{3}He. The tritium remaining at time t is

$$^3\text{H}_t = {}^3\text{H}_0\, e^{-kt}, \tag{8–11}$$

and the amount of ^{3}He produced by tritium decay $(^3\text{He}_t)$ is the difference between the initial tritium content $^3\text{H}_0$ and the tritium remaining at time t, which is

$$^3\text{He}_t = {}^3\text{H}_0 - {}^3\text{H}_0\, e^{-kt} = {}^3\text{H}_0\!\left(1 - e^{-kt}\right). \tag{8–12}$$

Because, from Eq. (8–11), $^3\text{H}_0 = {}^3\text{H}_t\, e^{kt}$, the later can be substituted into the right-hand side of Eq. (8–12) to obtain

$$^3\text{He}_t = {}^3\text{H}_t\!\left(e^{kt} - 1\right), \tag{8–13}$$

where $^3\text{He}_t$ is expressed in tritium units. The amount of helium in the sample includes both some natural atmospheric helium plus the amount present from the decay of tritium. After a correction is made for the helium not derived from decay of tritium (Clark and Fritz, 1997), the age of the water can be estimated using the equation

$$t = \frac{12.43}{\ln 2}\ln\!\left(1 + \frac{{}^3\text{He}_t}{{}^3\text{H}_t}\right). \tag{8–14}$$

Among applications of this method, Stute et al. (1997) calculated a ground water flow velocity based on the age gradient along flow lines away from a river channel, where bank infiltration recharges a shallow alluvial aquifer.

A second class of applications of tritium data includes using tritium distributions in either vadose or saturated zones in estimates of recharge rates, flow rates, and other parameters. A study of this type was conducted on a till plain in north-central Indiana by Daniels et al. (1991). In the field site chosen, a vertical profile of tritium measurments included a low-high-low pattern at a depth of about 7 m, which was correlated to the 1963 peak. The material in the vadose zone profile was till, a material with relatively low permeability. An assumption in the method is that recharge occurs by piston flow through the vadose zone to the water table. No fractures or other preferential pathways were observed in the core, which, if present, could provide another explanation for the high tritium content beneath lower concentrations. The recharge estimates derived from the tritium measurements, a mean of 4.1 cm/yr, was reasonably close to an estimate made by the traditional water budget method.

Delcore and Larson (1987) calculated the recharge rate for an unconsolidated sand and gravel aquifer using a vertical profile of tritium concentrations in the saturated zone. The recharge rate is determined by an equation developed by Vogel (1967), viz.,

$$R = \frac{(nD)}{t}, \tag{8-15}$$

where R is the recharge rate, n is average porosity of the aqufer, D is the maximum depth of tritium penetration beneath the water table, and t is time, in years since bomb tritium first entered the aquifer. Delcore and Larson chose sites on ground water divides where flow was dominantly in a vertically downward direction. The depth of penetration of bomb tritium ranged from 20 to 32 m in two sites in the study.

Solomon et al. (1995) combined the ^{3}H/^{3}He method with ground water modeling in an unconfined aquifer, but in this case the purpose was to locate the position of a contaminant leak that became the source of a dissolved BTEX plume (Fig. 8–21). This study was successful in determining (1) recharge rates to the aquifer, (2) horizontal ground water flow velocities, (3) the date when the contaminant first reached the water table, (4) the location of the contaminant source, and (5) the average hydraulic conductivity of the aquifer.

Using the configuration of a tritium plume contained within a landfill leachate plume, Egboka et al. (1983) calibrated a ground water flow model to the tritium distribution in order to estimate the dispersivity of the site (Fig. 8–22). The much lower values of tritium in the aquifer, relative to the predicted decay-corrected levels, suggested strong dispersion in the aquifer.

A third category of tritium studies focuses on the preferential flow of water through fractured rocks or sediments of low hydraulic conductivity. Tritium provides an ideal tracer that gives unambiguous evidence that young water has penetrated to a particular depth. For example, Hendry (1983) measured tritium concentrations in a vertical profile through a section of low permeability lacustrine sediment and till in the Canadian prairies. If recharge was assumed to follow a piston flow model downward through these materials, tritiated water would have penetrated to a depth of about 3 m. Instead, the tritium profile (Fig. 8–23) shows an irregular distribution of tritium to a depth of 16 m. In this profile, water containing bomb tritium is overlain by water lacking bomb tritium. The explanation for this distribution is that water has rapidly penetrated to

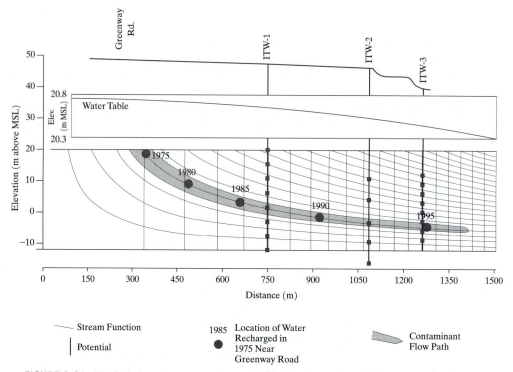

FIGURE 8–21 Vertical cross section of aquifer on Cape Cod, Massachusetts. Benzene concentrations and $^3H/^3He$ ages were determined in three vertical profiles (black squares). Ages and simulation of flow system were used to reconstruct date (1975) and location of spill that produced contaminant plume (from Solomon et al., 1995. Reprinted by permission of the National Ground Water Association).

these depths along fractures that surround blocks of material containing much older water. This type of study is very useful in evaluating potential waste disposal sites. If a site is chosen in an area of low permeability materials to restrict leachate movement, vertical profiles of tritium concentration can provide important evidence relative to whether the material is uniformly low in hydraulic conductivity or whether preferential pathways are present along which contaminants could move.

Radiocarbon

The use of ^{14}C as a dating tool for ground water is an offshoot of its more traditional application in dating solid organic carbon that is collected from Pleistocene or Holocene sediments. In the ground water environment, however, there are numerous complicating factors that make the dates obtained less precise than dates obtained on wood and other forms of old organic carbon.

Like tritium, the formation of ^{14}C occurs in the upper atmosphere by the bombardment of N by neutrons derived from cosmic rays. The reaction produces carbon because nitrogen releases a proton, which in this case has an atomic mass of 1 rather than 3, as in the case of tritium. The reaction is

$$^{14}_{7}N + ^{1}_{0}n \longrightarrow ^{14}_{6}C + ^{1}_{1}p. \tag{8–16}$$

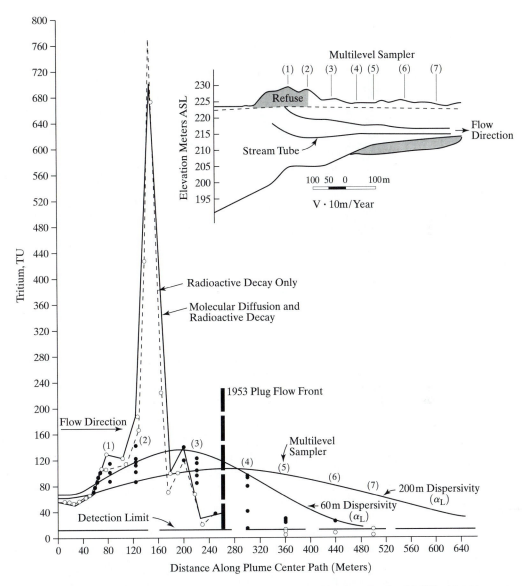

FIGURE 8–22 Cross section drawn along flow line downgradient from Borden landfill with tritium measurements in portion of landfill plume and locations of multilevel samplers (upper diagram). Lower diagram shows measured tritium contents along flow line (solid dots) compared to modeled distribution of tritium with radioactive decay and no dispersion (high peaked spike), and two curves with dispersion (dispersivities of 60 m and 200 m) which fit measured values much better (reprinted from *Jour. Hydrol.*, v. 63, Egboka, B. C. E., Cherry, J. A., Farvolden, R. N., and Frind, E. O., Migration of contaminants in ground water at a landfill: A case study. 3. Tritium as an indicator of dispersion and recharge, pp. 51–80, with permission from Elsevier Science).

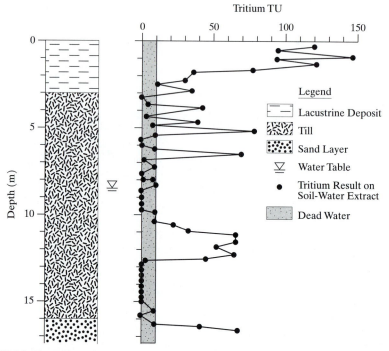

FIGURE 8–23   Tritium profile in low-permeability glacial sediments in Canadian prairies. Spikes of high tritium beneath samples containing no bomb tritium indicates that recharge does not conform to piston flow model, but occurs at variable rates through fractures or other preferential flow paths (reprinted from *Jour. Hydrol.*, v. 63, Hendry, M. J., Ground water recharge through a heavy-textured soil, pp. 201–209, Copyright 1983, with permission of Elsevier Science).

The ^{14}C produced becomes part of the atmospheric reservoir of CO_2 that is fixed by plants as organic carbon. It is then released to the soil, oxidized by microorganisms to CO_2, and dissolved in soil water moving through the vadose zone toward the water table.

As it passes through the hydrogeologic system, radiocarbon decays by beta emission to nitrogen according to the formula

$$^{14}_{6}C \longrightarrow {}^{14}_{7}N + \beta^-. \tag{8–17}$$

The convention for reporting the amount of ^{14}C in a sample is *percent modern carbon*, or *pmC*. "Modern carbon" is defined as 95% of the activity of ^{14}C in the NBS oxalic acid standard in 1950. This amount of radiocarbon is equivalent to the amount contained in wood grown in 1890, resulting in a decay rate of 13.56 dpm/g (disintegrations per minute per gram of carbon). Because of the addition of ^{14}C to the atmosphere by nuclear weapons testing, atmospheric measurements currently run well over 100% pmC.

Determining the age of a ground water sample is an application of first order kinetics. A form of Eq. (2–51) is used, namely,

$$-\ln \frac{A}{A_0} = kt,$$

where A is the activity of radiocarbon at time t, and A_0 is the activity of modern carbon. Because, from Eq. (2–56), k, the decay constant, is equal to $\dfrac{\ln 2}{t_{1/2}}$, it follows that

$$t = \left(\frac{-t_{1/2}}{\ln 2}\right) \ln \frac{A}{A_0} = -8267 \ln \frac{A}{A_0}. \tag{8–18}$$

The half-life of ^{14}C is 5,730 years. If the initial activity of ^{14}C is assumed to be 100 pmC, the resulting date is expressed in "radiocarbon years" before present. Radiocarbon years are not the same as calendar years because A_0, the initial activity of ^{14}C when the plant was alive, has not remained constant over time. Radiocarbon ages must be corrected if an age in calendar years is desired. The variation in ^{14}C over time has been established using tree ring chronologies and corals dated by U/Th methods. For ancient wood and other unaltered organic carbon sources, age determination is a straightforward application of Eq. (8–18). Unfortunately, geochemical processes in ground water flow systems involving the carbonate system complicate the determination of accurate radiocarbon ages.

The most difficult complication in the dating calculation is the estimation of A_0, the initial activity of ^{14}C, because this value is significantly altered by dilution of the ^{14}C contained in the soil gas that dissolves in the infiltrating water by carbon derived from carbonate minerals that dissolve according to the processes discussed in Chapter 3. This carbon is generally old and contains no ^{14}C. One method for correcting A_0 to account for carbonate mineral dissolution was proposed by Tamers (1975). This model is based on the stoichiometry of the carbonate mineral dissolution process, which can be described as

$$H_2CO_3 + CaCO_3 = 2HCO_3 + Ca^{2+}.$$

The carbonic acid is derived from the soil gas which contains ^{14}C, and the calcium carbonate contains no ^{14}C. Thus, the dissolved bicarbonate contains one carbon atom from each source. The correction is calculated as

$$A_0 = \frac{\left(m_{H_2CO_3} + 0.5 m_{HCO_3^-}\right)}{\left(m_{H_2CO_3} + m_{HCO_3^-}\right)} A_g, \tag{8–19}$$

where A_g is the ^{14}C activity in the soil gas. In the near-neutral pH range, the molarity of carbonic acid will be small and A_0 will be approximately $0.5 A_g$, or 0.5 pmC. One assumption of this correction is that the dissolution of calcium carbonate occurs under closed system conditions, which permit no isotopic exchange between the dissolved bicarbonate and the soil gas. Under open system conditions, which is more common in many aquifers, some exchange of ^{14}C will occur and A_0 will be greater than $0.5 A_g$.

The $\delta^{13}C$ mixing model (Clark and Fritz, 1997) can be used when open system conditions in the recharge area were assumed to prevail at the time the ground water was recharged. This method is based upon the changes in $\delta^{13}C$ of the DIC that occur as DIC derived from soil gas CO_2 with a $\delta^{13}C$ value of about -23 is combined with DIC derived from carbonate mineral dissolution with a $\delta^{13}C$ of about 0. For closed-system conditions, the correction can be applied to yield

$$A_0 = \left(\frac{\delta^{13}C_{DIC} - \delta^{13}C_{carb}}{\delta^{13}C_{soil} - \delta^{13}C_{carb}} \right) A_g, \tag{8–20}$$

where $\delta^{13}C_{DIC}$ is the measured $\delta^{13}C$ in the recharging ground water, $\delta^{13}C_{carb}$ is the $\delta^{13}C$ of the carbonate minerals, and $\delta^{13}C_{soil}$ is the $\delta^{13}C$ of the soil gas CO_2. Under open-system conditions, the ^{13}C in the DIC equilibrates with the ^{13}C in the soil gas, leading to an enrichment in the $\delta^{13}C$ of the DIC by a factor of 7 to 10‰, depending on temperature. Equation (8–20) can be modified to account for this enrichment by replacing $\delta^{13}C_{soil}$ with a term called $\delta^{13}C_{rech}$, which reflects the DIC that has been enriched by equilibration with soil gas CO_2. $\delta^{13}C_{rech}$ is calculated as

$$\delta^{13}C_{rech} = \delta^{13}C_{soil} + \varepsilon^{13}C_{DIC\text{-}CO_{2(soil)}}, \tag{8–21}$$

in which $\varepsilon^{13}C_{DIC\text{-}CO_{2(soil)}}$, as mentioned, varies between 7 and 10. The modified $\delta^{13}C$ mixing model correction to A_0 thus becomes

$$A_0 = \left(\frac{\delta^{13}C_{DIC} - \delta^{13}C_{carb}}{\delta^{13}C_{rech} - \delta^{13}C_{carb}} \right) A_g. \tag{8–22}$$

This correction is appropriate at moderate to high pH values. When the pH of equilibration of soil gas CO_2 and DIC is lower than 7, a further correction is needed (Clark and Fritz, 1977).

It should be obvious that estimating the various parameters in the previous equations for recharge waters at some time in the past can be difficult. Although other corrections are available (Clark and Fritz, 1997), the resulting ages are subject to considerable uncertainty. In addition, other geochemical processes can interfere with age estimates. These include oxidation of peat or lignite in the aquifer that is devoid of ^{14}C, sulfate reduction associated with oxidation of organic carbon, ion exchange leading to additional dissolution of carbonate minerals, and migration of CO_2 into the formation from other sources. However, despite all these problems, ages that are within $\pm 50\%$ of the true value can be useful as estimates of ground water age.

Pearson and White (1967) used the $\delta^{13}C$ mixing model to determine the age of water in the Carrizo Sand, an aquifer that becomes confined downgradient of its exposed recharge area in central Texas (Fig. 8–24). Wells along the flow were sampled and ^{14}C activities were measured to yield dates ranging from 0 years in the recharge area to 27,000 years at a downgradient distance of 35 miles (Fig. 8–25).

Other Dating Methods

Tritium and ^{14}C are used to date young and moderately old waters, respectively. Dating of older ground waters has been attempted with a variety of isotopes, including ^{36}Cl and various members of the ^{238}U decay series. These methods are not routine, and the ages obtained are subject to numerous uncertainties. Clark and Fritz (1997) and references therein discuss the uses and limitations of these techniques in detail.

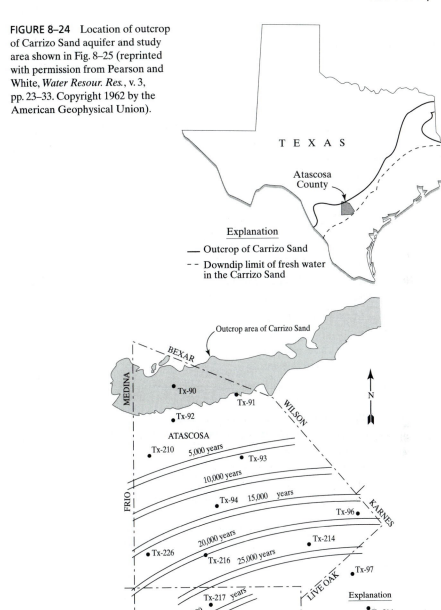

FIGURE 8–24   Location of outcrop of Carrizo Sand aquifer and study area shown in Fig. 8–25 (reprinted with permission from Pearson and White, *Water Resour. Res.*, v. 3, pp. 23–33. Copyright 1962 by the American Geophysical Union).

FIGURE 8–25   Ages of ground water determined by ^{14}C dating in the Carrizo Sand aquifer for area shown in Fig. 8–24 (reprinted with permission from Pearson and White, *Water Resour. Res.*, v. 3, pp. 23–33. Copyright 1962 by the American Geophysical Union).

PROBLEMS

1. The equilibrium fractionation of ^{18}O between CO_2 and water is shown in the equation below:

$$\delta^{18}O_{CO_2-H_2O} = \frac{(^{18}O/^{16}O)_{CO_2}}{(^{18}O/^{16}O)_{H_2O}} \approx 1.04.$$

Express the enrichment of ^{18}O in CO_2 in per mil units.

2. At equilibrium, compound X has a δ^2H of $-5‰$ VSMOW and compound Y has a δ^2H of $-50‰$ VSMOW. What is the fractionation factor?

3. A water sample has a $\delta^{18}O$ of $-15‰$. Using the equation for the Global Meteoric Water Line, what would you estimate δ^2H to be?

4. Explain why precipitation in northern Canada has much lower $\delta^{18}O$ and δ^2H values than precipitation in the tropics.

5. Ground water in a certain area has a $\delta^{18}O$ value that is 2‰ lower than the weighted mean annual precipitation. What does this suggest about ground water recharge in the area?

6. Describe the trend with depth in $\delta^{18}O$ values that could be obtained in an area of fine-grained sediments with extremely low hydraulic conductivities. What is the explanation for this trend?

7. Discuss the fractionation of stable carbon isotopes that occurs during ground water recharge.

8. Methane dissolved in ground water has a $\delta^{13}C$ value of $-50‰$ and a δ^2H value of $-350‰$. What can be said about the source of this gas? What additional isotopic evidence could be used to further pinpoint the source?

9. How can nitrogen isotopes, along with changes in nitrate concentrations, be used to document the occurrences of denitrification?

10. How can sulfate in ground water originating from dissolution of evaporite minerals be distinguished from sulfate originating through the oxidation of bacterially reduced sulfur?

11. What would the tritium concentration in ground water near Ottawa, Ontario, that recharged in 1962 be in the year 2010 if no process other than radioactive decay had affected its concentration?

12. What are the problems involved in age determination of ground water using tritium concentration alone?

13. Ground water at 25°C and pH = 7.46 has a concentration of bicarbonate of 8.6 mmol/l and a radiocarbon activity of 25.9 pmC. Using the Tamers method, what is the age of the water?

Chemical Changes in Ground Water Flow Systems

In previous chapters, we have examined the chemical processes and reactions occurring in aquifers composed of various mineral and rock types. With these principles in mind, the objective of the present chapter is to place these chemical processes in the context of ground water flow systems. Chebotarev (1955) recognized that ground water tends to chemically evolve in long flow systems toward a more concentrated solution similar to the composition of seawater. The systematic changes in anion composition in aquifers of this type have become known as the *Chebotarev sequence*. In the direction of flow, from the recharge area, the sequence can be illustrated as follows:

$$HCO_3^- \longrightarrow HCO_3^- + SO_4^{2-} \longrightarrow SO_4^{2-} + HCO_3^- \longrightarrow SO_4^{2-}$$
$$+ Cl^- \longrightarrow Cl^- + SO_4^{2-} \longrightarrow Cl^-.$$

The sequence can only progress to its conclusion if sulfate- and chloride-bearing minerals are present in the aquifer lithologies. This is generally only true for long flow systems in sedimentary basins containing evaporite minerals. Cation evolutionary sequences are more variable than anion sequences because of ion exchange, precipitation, or other processes.

GROUND WATER FLOW SYSTEMS AND WATER CHEMISTRY

The relationships between ground water chemistry and flow systems can be considered in two ways: (1) chemical differences between types of flow systems and (2) chemical changes within flow systems. The basic classification of ground water flow systems was set forth by Tóth (1963), who recognized local, intermediate, and regional flow systems (Fig. 9–1). The size of these flow systems is governed by the topography in the drainage basin relative to the depth of the flow system. High-relief undulating or hummocky surface topography will increase the depth of local flow systems, and if the depth to a regional aquitard is shallow, the drainage basin may contain only local flow systems. By contrast, in basins of great depth in which the surface relief is small compared to the depth of the base of the flow system, a regional flow system will predominate. Local flow systems recharge at topographic highs on the water table and discharge in adjacent topographic lows. Intermediate flow systems recharge at topographic highs and discharge in topographic lows farther downgradient than the adjacent low. The regional flow system recharges near the drainage divide in the basin and discharges at the topographic

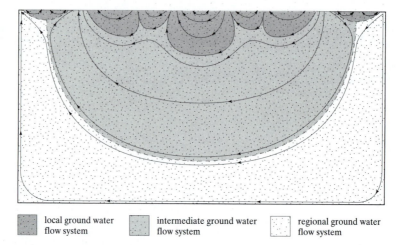

| | local ground water flow system | | intermediate ground water flow system | | regional ground water flow system |

FIGURE 9–1   Distribution of local, intermediate, and regional ground water flow systems in a drainage basin (after Tóth, J. *Geophysical Res.*, v. 68, pp. 4795–4812, 1963, Copyright by the American Geophysical Union).

low point of the basin. The classic numerical modeling study of Freeze and Witherspoon (1966, 1967, 1968) shows the effects of heterogeneity and other variables on the configuration of flow systems in a drainage basin.

The chemical composition of ground water in flow systems is a function of aquifer mineralogy and the rate at which ground water moves through the system. Local flow systems display vigorous flow that flushes soluble mineral salts out of the aquifer in a relatively short period of time. As a result, the ground water may never progress past the bicarbonate hydrochemical facies in the Chebotarev sequence. Flow velocity is progressively more sluggish with depth in a drainage basin, and intermediate and regional systems are therefore more likely to evolve into the sulfate or chloride facies, depending on the availability of evaporite minerals in the drainage basin. Because of the sluggishness of ground water circulation, particularly in regional flow systems, soluble salts are flushed out of the system very slowly and persist for thousands or millions of years. In very deep basins, brines many more times saline than seawater may be present. These concentrated solutions may contain remnants of the original seawater present in the depositional environment, modified by dissolution of evaporite minerals and other processes. A detailed discussion of brine origin and characteristics is beyond the scope of this text.

A hypothetical, regionally unconfined flow system (Tóth, 1984) is shown in Fig. 9–2. Zones of stagnation created by the flow patterns have high total dissolved solids. Near-surface soils and ground water are more saline in the discharge areas of the intermediate and regional flow systems. Sulfate and chloride facies characterize the intermediate and regional flow systems, respectively.

Changes in water chemistry within a flow system are highly variable. In small, local flow systems in permeable aquifers with uniform lithology, very little change in the chemical composition of ground water may take place. If carbonate minerals are present,

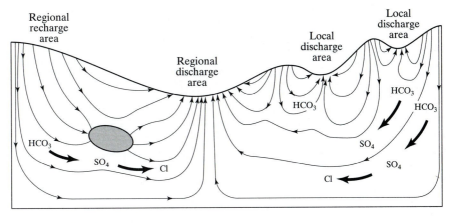

—▶— Direction of ground water flow

FIGURE 9–2   Hydrochemical facies related to type of flow system in hypothetical regionally unconfined basin (from Tóth, 1984. Reprinted with permission of the National Ground Water
Association).

equilibration may occur under open-system conditions in the vadose zone. Few changes
occur after the water reaches the water table and moves through the ground water flow
system to the discharge area. In any flow system in which the aquifer mineralogy changes,
corresponding water chemistry changes will be noticed.

Order of Encounter

The chemical composition of ground water in flow paths that cross multiple rock types
can be highly variable. The greatest variability occurs in sedimentary rock sequences.
Palmer and Cherry (1984) illustrate simulated evolutionary sequences using a mass-
transfer model. The flow paths are shown in Fig. 9–3 and the final chemical compositions
are shown in Table 9–1. The *order of encounter* of various rock types is the critical factor that controls the water quality.

   Each of the flow paths in Fig. 9–3 begins with a soil zone in which the infiltrating
water equilibrates with a specific p_{CO_2}. The underlying rock units include limestone,
sandstone, gypsum, clay shale, and carbonaceous shale arranged in different sequences.
The limestone is assumed to be composed of pure calcite, and the gypsum unit contains
only the pure mineral. The sandstone bed contains the plagioclase albite ($NaAlSi_3O_8$),
and the carbonaceous shale has no reactive mineral phases, but does have labile organic carbon and sulfate reducing bacteria. The clay shale allows the exchange of sodium
and calcium.

   In sequence A, the infiltrating ground water dissolves CO_2 at a partial pressure of
10^{-2}. It then enters the limestone bed and equilibrates with calcite under open-system
conditions in which the p_{CO_2} remains at 10^{-2}. At this point the water is a calcium-
bicarbonate type. The water then enters the gypsum bed in which it dissolves gypsum
to equilibrium and changes the chemical facies to calcium sulfate. This produces

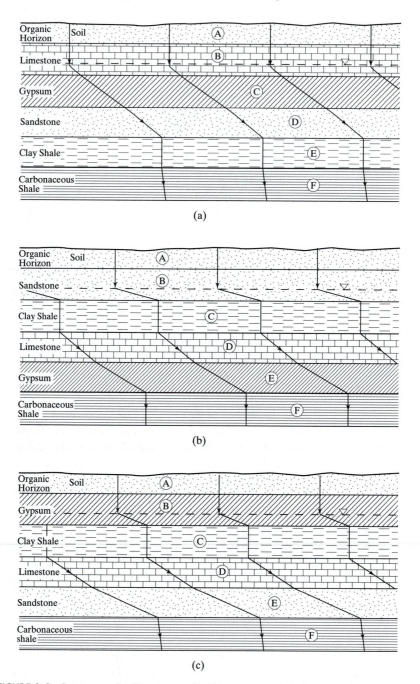

FIGURE 9–3 Sequences of sedimentary rock units encountered produce very different water chemistry. Final chemical composition for each sequence shown in Table 9–1 (reprinted from *Jour. Hydrol*, v. 63, Palmer, C. D. and Cherry, J. A., Chemical evolution of ground water in sequences of sedimentary rocks, pp. 27–65, Copyright 1984, with permission from Elsevier Science).

TABLE 9–1   Final chemical composition of ground water after flow through sequences shown in Fig. 9–3 (from Palmer and Cherry, 1984).

Parameter	Sequence A	Sequence B	Sequence C
pH	6.76	6.28	6.75
$\log p_{CO_2}$	−0.669	−0.422	−0.652
TDS (mg/l)	3,024	2,539	2,981
Na (mM)	25.2	3.3	25.4
Ca (mM)	1.3	6.2	1.4
CO_2 (mM)	29.0	25.6	29.6
SO_4 (mM)	0.0	0.0	0.0
$\log Si_{cal}$	+0.0	+0.0	+0.0
$\log SI_{gyp}$	$-\infty$	$-\infty$	$-\infty$

oversaturation with respect to calcite by the common ion effect and a small amount of calcite precipitates, accompanied by more dissolution of gypsum. In the sandstone below the gypsum, albite is weathered to kaolinite, producing sodium ion. The consumption of hydrogen ions in the precipitation of kaolinite raises the pH and causes more calcite to precipitate. The loss of calcite during calcite precipitation brings the solution below saturation with respect to gypsum. When the water reaches the clay shale, a large amount of the dissolved calcium is exchanged for sodium. The water is now altered to the sodium-sulfate facies. The final unit is the carbonaceous shale. In this unit, all the sulfate is reduced to sulfide. The organic carbon in the shale serves as the electron donor in the sulfate reduction reaction and is oxidized to bicarbonate in the process. The added bicarbonate oversaturates the solution with respect to calcite, which subsequently precipitates. The final chemical facies is sodium bicarbonate, and the concentrations of the solution are shown in Table 9–1. Concentrations at each point in the sequence are given by Palmer and Cherry (1984).

In sequence B, the water enters the sandstone immediately below the soil unit. The weathering of albite produces sodium and kaolinite as before. In the sandstone, however, the exchange reaction is reversed with respect to Sequence A because there is no calcium in solution. Calcium is released from the clay surfaces in exchange reactions with sodium from solution. The water remains in the sodium-bicarbonate facies through both the sandstone and the clay shale. The water then enters the limestone bed, where calcite is dissolved under closed-system conditions. Next comes the gypsum unit where gypsum dissolves and calcite precipitates as before. The chemical facies changes to calcium sulfate at this point. In the final unit, the carbonaceous shale, sulfate is reduced to sulfide and calcite again precipitates. The resulting solution has much less sodium (Table 9–1) and more calcium because the ion-exchange reaction occurred before the water encountered the limestone and gypsum units. The chemical facies is calcium bicarbonate.

The arrangement of formations in Sequence C includes the gypsum unit below the soil zone. When the water enters the evaporite unit, it becomes a calcium-sulfate water with high TDS. In the underlying clay shale, most of the calcium is exchanged for sodium and the water becomes a sodium-sulfate type. Calcite dissolves under closed-system conditions in the limestone and albite dissolves to release sodium in the sandstone, but neither of these reactions alters the overall chemical composition of the water. In the carbonaceous shale, bicarbonate replaces sulfate as the dominant anion in the sulfate-reduction reaction and some calcite precipitates.

Although all three sequences end with bicarbonate as the dominant anion because of the carbonaceous shale, the waters change significantly within each sequence. Most flow systems do not include this many different lithologic units and most real aquifers or aquitards include multiple mineral types that add complexity relative to the hypothetical sequences described. Nevertheless, general conclusions can be drawn about the chemical evolution of water by comparing the concentrations of cations and ions, preferably in milliequivalents. Facies diagrams such as the Piper trilinear plot are very useful. Along with TDS , the ion concentrations suggest the general evolutionary trends of the water. For example, calcium and calcium–magnesium, bicarbonate waters of low to moderate TDS indicate the influence of carbonate rocks. Calcium-sulfate waters of moderate to high TDS are produced by evaporite units containing gypsum or anhydrite. Sodium-bicarbonate and sodium-sulfate facies of moderate to high TDS suggest the presence of carbonates or sulfates along with ion exchange. High chloride contents can be explained by several other conditions such as the presence of halite-bearing evaporites or mixing with seawater or brines. Equilibrium modeling programs such as WATEQF or PHREEQC calculate saturation indices for many minerals. These models are critical in testing hypotheses for the chemical evolution of a particular water.

The evolution of redox conditions along flow paths was discussed in Chapter 5. In general, there is a decline in redox potential from the recharge area, where free oxygen is commonly present. The amount of decline is determined by the abundance of electron donors and acceptors along the flow path. Representative changes in chemical species affected by redox changes are shown in Fig. 5–15.

Examples of Chemical Evolution of Ground Water in Regional Flow Systems

The Floridan Aquifer System The Floridan aquifer system is one of the most productive aquifers in the world and is a major source of potable water in the southeastern United States. In the 1980s, daily production was approximately 3 billion gallons. Although the aquifer has been studied for many years, the hydrochemistry of the entire system was described in a major study by Sprinkle (1989), which summarizes previous research on the aquifer.

The aquifer occurs in a sequence of Tertiary carbonate rock units. Because of a zone of low permeability, it is commonly divided into upper and lower sections. Overlying most of the aquifer is the Miocene Hawthorn Formation, a clayey confining unit of variable thickness. Where the confining unit is absent, the Upper Floridan aquifer receives much more recharge, and circulation through the aquifer to the discharge areas is much more rapid.

The extent of the aquifer and the potentiometric surface of the Upper Floridan aquifer are shown in Fig. 9–4. Recharge occurs over potentiometric highs such as the one in central Florida. The aquifer discharges to springs, streams, and lakes, and in the downgradient part of the aquifer freshwater mixes with seawater along the Atlantic and Gulf of Mexico coastlines. Ground water in the Upper Floridan aquifer evolves chemically

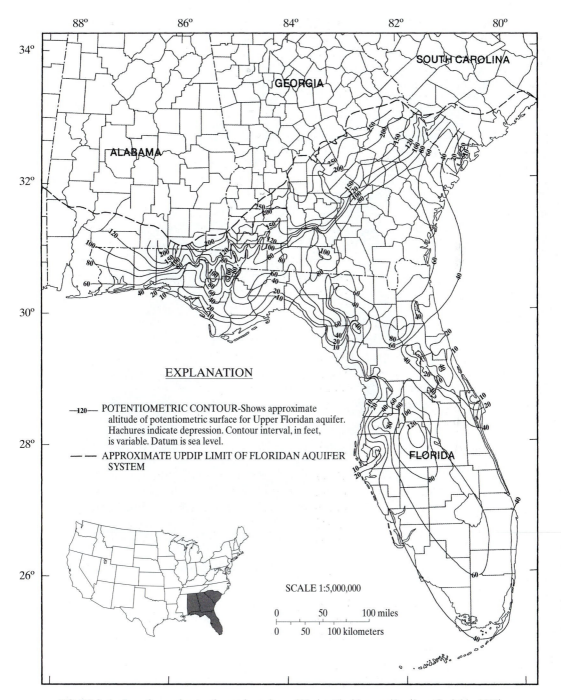

FIGURE 9–4   Location and potentiometric surface of Upper Floridan aquifer (from Sprinkle, 1989).

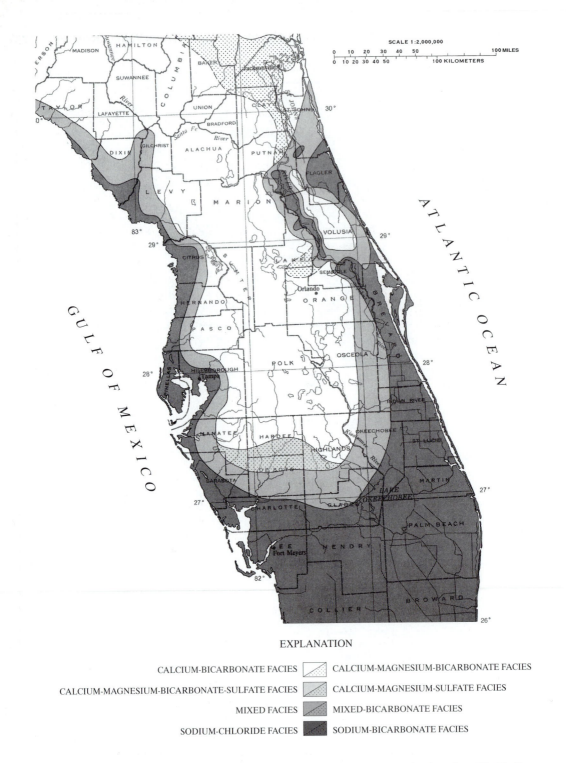

FIGURE 9–5 Hydrochemical facies of part of the Upper Floridan aquifer in central and northern Florida (from Sprinkle, 1989).

from recharge areas to discharge areas under the influence of several major geochemical reactions. In the recharge areas, calcite dissolution dominates, and the water falls into the calcium-bicarbonate facies (Fig. 9–5). Total dissolved solids are low. Carbonate dissolution is assumed to take place under open-system conditions where the aquifer is unconfined and under closed-system conditions beneath the confining unit. In central Florida, ground water flows radially outward from the high on the potentiometric surface (Fig. 9–4). Minor amounts of gypsum in the aquifer dissolve, causing oversaturation with calcite according to the common ion effect. Dolomite is also present in the carbonate units, and as calcite precipitates from the oversaturated waters, dolomite dissolves. The combination of calcite precipitation, gypsum dissolution, and dedolomitization leads to increases in magnesium and sulfate concentrations and the water evolves to the calcium–magnesium–bicarbonate, calcium–magnesium–bicarbonate–sulfate, and calcium–magnesium–sulfate facies (Fig. 9–5). Ion exchange adds sodium to the water, and sulfate reduction is significant in some areas. As the water approaches the coastal discharge areas, it encounters modern seawater or ancient seawater that has not been completely flushed out of the aquifer because of decreases in permeability. In these areas water evolves to the sodium-chloride facies. The evolutionary trends in the aquifer are shown on the Piper trilinear diagram in Fig. 9–6.

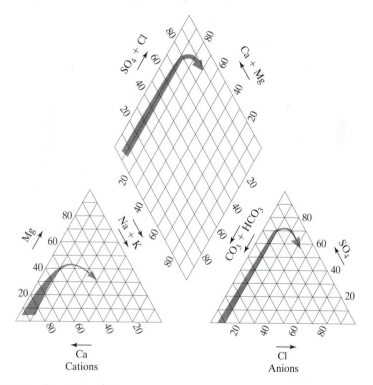

FIGURE 9–6 Chemical evolution of ground water in Floridan aquifer from recharge areas to discharge areas (reprinted from *Jour. Hydrol.*, v. 10, Back, W. and Hanshaw, B. B., Comparison of chemical hydrogeology of the carbonate peninsulas of Florida and Yucatan, pp. 330–368, Copyright 1970, with permission from Elsevier Science).

The Fox Hills–Basal Hell Creek Aquifer System   The chemical evolution of ground water in the Fox Hills–Basal Hell Creek aquifer in North Dakota, South Dakota, and Wyoming, as described by Thorstenson et al. (1979), provides an interesting example of an aquifer with more complex mineralogy than the Floridan aquifer and one that contains organic carbon.

The Fox Hills–Basal Hell Creek aquifer occurs in Upper Cretaceous deposits of the Great Plains of western North America. The Fox Hills Formation is an Upper Cretaceous sandstone that was deposited at the top of a regressive marine deltaic sequence that grades upward into the deltaic Hell Creek Formation. Above the aquifer units lie the coal-bearing formations of the Tertiary Fort Union Group. The aquifer crops out along uplifts and in stream valleys in western North and South Dakota. In North Dakota, it dips eastward where it becomes confined beneath younger rocks. Flowing wells in the aquifer are common, and it serves as the primary supply for municipalities in the region. Lithologic components of the aquifer that are relevant to hydrochemistry include carbonate minerals, iron oxides and sulfides, organic carbon in the form of lignite beds, and bentonitic clays.

The regional flow system of the Fox Hills–Basal Hell Creek aquifer can be divided into recharge, transition, and discharge areas (Fig. 5–18) using both hydrologic and hydrochemical criteria. Downward leakage from overlying confining units occurs in the recharge area; and upward leakage from the aquifer occurs in the discharge area near the Missouri River. Chemically, ground water is very different in the recharge and discharge areas, separated by the narrow transition zone. The characteristics of ground water in the recharge area include high sulfate (200–600 mg/l), high alkalinity (~ 700 mg/l HCO_3^-), and high sodium (400–600 mg/l). Calcium, magnesium, and chloride concentrations are much lower. The reactions postulated by Thorstenson et al. (1979) to account for these concentrations include dissolution of carbonate minerals and exchange of calcium for sodium on clay minerals. High partial pressures of CO_2 generated by oxidation of lignite, along with consumption of calcium and magnesium by clay mineral exchange, are necessary to produce the high alkalinities. The source of the sulfate concentrations is not definitely known. It could be derived either from small amounts of gypsum or oxidation of sulfide minerals.

Within the narrow transition zone of the aquifer, the chemical composition of ground water changes dramatically. The major changes include increases in alkalinity, decreases in sulfate to near zero, increases in chloride concentration by a factor of five, and the presence of methane. These changes are likely driven by decreases in redox potential because of the continued biodegradation of the organic carbon in the aquifer. In the transition zone, the redox potential decreases to the level of sulfate reduction. Alkalinity increases in the sulfate-reduction reaction because the organic carbon is oxidized to CO_2, which subsequently reacts to form bicarbonate. The sulfide produced by sulfate reduction precipitates as sulfide minerals. After consumption of the sulfate, further biodegradation of organic carbon takes place via methanogenesis. The source of the chloride in the discharge area is harder to explain; it may migrate into the aquifer from the underlying shale because of the upward gradients in that part of the flow system.

The Edwards Aquifer.   The Edwards aquifer was mentioned in the discussion of sulfur isotopes in Chapter 8. Water chemistry in the aquifer illustrates another type of chemical evolution—one in which the chemical changes in the direction of flow are the result of mixing with water of a different chemical type. It is somewhat similar to the downgradient part of the Floridan aquifer where freshwater and seawater mix. The Edwards aquifer is a carbonate unit of Cretaceous age that dips southward toward the Gulf of Mexico south of the Balcones fault zone (Fig. 9–7). The aquifer is recharged in the fault zone, which forms a surficial escarpment. The Edwards Formation has been eroded from the upper part of the escarpment. Streams flowing in a large drainage basin to the north bring water to the recharge area, where it infiltrates rapidly in a karst landscape. In the recharge area, much of the flow is contained within a flow system consisting of solutionally enlarged conduits, with natural discharge through springs as the flow system passes into an artesian area. Ground water in the recharge area and spring discharge area is low in dissolved solids, generally less than 350 mg/l and calcium-bicarbonate in type. This part of the aquifer is extensively developed for ground water production. Downdip of the spring discharge areas, movement of water in the aquifer becomes very slow because of sedimentary facies changes and less solutional development. A narrow transition zone separates the freshwater zone from saline water with TDS of greater than 3,000 mg/l (Fig. 9–8). A relatively distinct change in water quality called the "bad water line" occurs within the transition zone. The TDS is approximately 1,000 mg/l at the bad water line. The aquifer is not used for potable water supplies downdip of the bad water line.

In addition to increases in dissolved solids, the ground water changes in chemical facies within the flow system. In the vicinity of the bad water line, the chemical facies becomes sodium sulfate (Fig. 9–9), and within the saline zone, evolves to the sodium-chloride facies (Senger and Kreitler, 1984). The increases in dissolved solids are caused

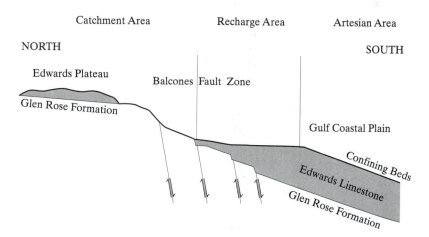

FIGURE 9–7   Cross section of the Edwards aquifer showing the catchment area where water is discharged to the recharge area at the base of the Balcones escarpment and the artesian area of the aquifer (from Burchett et al., 1986).

FIGURE 9–8 Hydrogeochemical cross section of the Edwards aquifer showing fresh, transition, and saline zones. Irregular line in transition zone represents the bad water line (from Burchett et al., 1986).

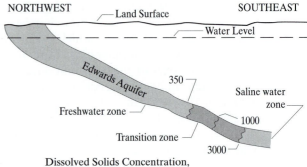

Dissolved Solids Concentration, in Milligrams per Liter

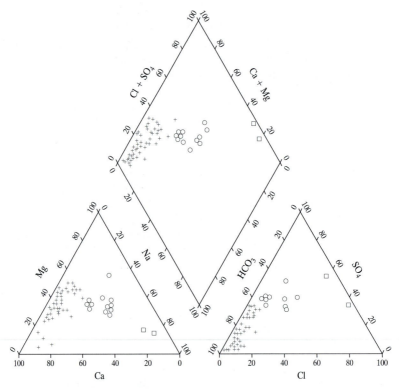

FIGURE 9–9 Piper diagram of water chemical analyses from the Edwards aquifer. Samples from updip of bad water line: +, in vicinity of bad water line: ○, and downdip of bad water line: □ (from Senger and Kreitler, 1984).

by mixing with ancient brines that were never flushed out of the aquifer, as discussed in Chapter 8, and by upward leakage from the underlying Glen Rose formation (Senger and Kreitler, 1984). The concern of resource managers of the Edwards aquifer is that population growth and increasing withdrawals of water from the freshwater zone will cause migration of the transition zone and the bad water line far enough that well fields close to the bad water line will become impacted or unusable.

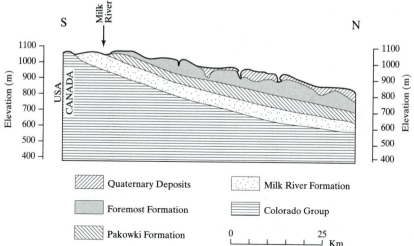

FIGURE 9–10 Location of the Milk River aquifer and stratigraphic cross section of aquifer system (from Hendry and Schwartz, 1990. Reprinted by permission of the National Ground Water Association).

The Milk River Aquifer.   Not to be overlooked in examples of chemical evolution in regional aquifers are geologic changes to the recharge area of the aquifer. The Milk River aquifer of southern Alberta (Hendry and Schwartz, 1990) is an interesting example. Ground water flow in this regional, confined system is downdip from recharge areas at the Canada–U.S. border toward the north (Fig. 9–10). The Milk River aquifer, a Cretaceous sandstone, is confined above and below by shale aquitards. Ground water

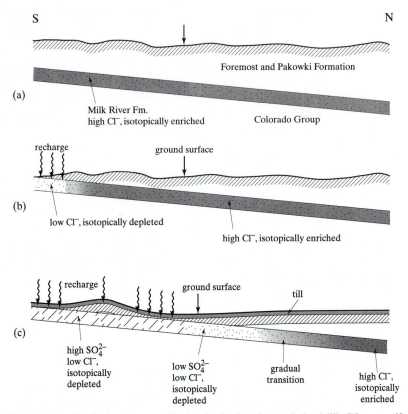

FIGURE 9–11 Geologic changes affecting hydrochemistry of the Milk River aquifer: (a) Recharge area overlain by confining shales, (b) recharge area exposed by erosion about 500,000 years ago, (c) recharge area covered by glacial deposits about 30,000–40,000 years ago (from Hendry and Schwartz, 1990. Reprinted by permission of the National Ground Water Association).

in the recharge area of the aquifer is high in sulfate and low in chloride. Downdip from the recharge area in the confined flow system, sulfate decreases significantly, as do sodium, chloride, and bicarbonate. Farther downgradient, sodium, chloride, and bicarbonate increase, whereas sulfate and calcium remain at low levels. The geologic sequence of events by which Hendry and Schwartz (1990) explain these trends is shown in Fig. 9–11. Prior to about 500,000 years ago, the overlying shale of the Pakowki formation extended over the recharge area (Fig. 9–11a). Ground water in the aquifer was high in sodium and chloride. Hendry and Schwartz postulate that sodium and chloride diffused into the aquifer from the confining shales. Around 500,000 years ago, the recharge area of the aquifer was exposed by erosion (Fig. 9–11b). At that time, low-chloride water entered the aquifer from surface recharge. This water was depleted in ^{18}O and ^{2}H, relative to the water in the aquifer. Water with this chemical signature moved downgradient until 30,000 to 40,000 years ago, when tills were deposited over the recharge area by the Laurentide Ice Sheet. During recharge to the Milk River aquifer through the overlying glacial drift, sulfate became elevated because of the mineral composition of the glacial deposits.

The aquifer has persisted in this condition to the present day (Fig. 9–11c), creating the trends in aquifer chemistry described previously. The decreases in sulfate downgradient are attributed to the presence of low-sulfate recharge that migrated to that part of the aquifer, rather than sulfate reduction, as in the Fox Hills–basal Hell Creek aquifer. The increases in bicarbonate, as well as methane, downgradient in the Milk River aquifer are suggested by Hendry and Schwartz to be the result of methane fermentation that may be taking place in the adjacent shales and diffusing into the aquifer, rather than forming in the aquifer itself. In the Milk River aquifer, the geologic changes affecting the recharge area have been the major controls on the chemical trends in the aquifer.

Ground Water Chemistry Variations in Recharge and Discharge Areas

Ground water interactions with surface water bodies are complex and temporally variable. The hydraulic interactions between ground water flow systems and lakes, wetlands, and ephemeral ponds greatly influence the water chemistry of the surface water bodies and the shallow ground water and soil water around the water bodies. In general, the surface water–ground water interactions can be characterized as recharge, discharge, or flow through (Fig. 9–12). Siegel (1988) showed that the chemical composition of surface water in wetlands is dependent upon the recharge–discharge relationships, or the recharge–discharge *function*. Recharge wetlands are dilute, with a chemical composition similar to precipitation, whereas discharge wetlands reflect the concentrations and chemical species in ground water that discharges into the wetland. The existence of recharge and flow-through lakes and wetlands contradicts the axiom that the water table is a subdued replica of the land surface. If depressions on the landscape contribute to ground water recharge, it must be true that the water table is at least occasionally higher beneath depressions than beneath adjacent downgradient uplands. The hydrochemistry of the surface water body can commonly be related to the recharge–discharge

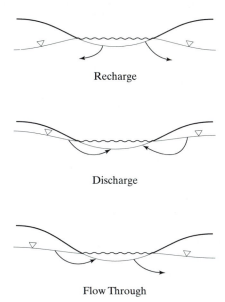

FIGURE 9–12   Generalized recharge/discharge interactions between ground water and surface water bodies.

Recharge

Discharge

Flow Through

interactions of the surface water body. In addition, the chemical and isotopic composition of shallow ground water around the lake or wetland is in places an artifact of the hydrological interactions between the surface and subsurface systems.

Recharge Systems. Recharge from lakes or wetlands can occur on a year-round or seasonal basis. Kasenow (1995) studied a wetland that served as a more-or-less continuous source of ground water recharge. Wetlands of this type are known as *bogs* to distinguish them from *fens* with vegetative communities that are adapted to the high nutrients present in ground water discharge. Bogs are typified by thick accumulations of peat, acidic pH levels, and very dilute major ion concentrations. Government Marsh (Kasenow, 1995) is isolated from the underlying carbonate-rich sand aquifer by a glacio-lacustrine clay of low permeability. Water in the bog has very low TDS but high dissolved organic carbon resulting from organic acids derived from the peat. At depth in the bog, the redox potential is very low because of the organic matter and the low permeability of the peat. As a result of the low redox potential and low pH, very high iron concentrations are present. The dissolved inorganic solids are predominantly derived from rainfall, dry fallout, and limited surface runoff into the bog.

Hayashi et al. (1998) demonstrated that solutes are transported from a recharge wetland in Saskatchewan to the area beneath an adjacent upland area in a shallow ground water flow system (Fig. 9–13). Inorganic salts, including chloride, are drawn upward from the water table in the upland area by evapotranspiration. Chlorides accumulate in the near-surface soils after concentration by evapotranspiration. Part of the near-surface chloride dissolves in winter and spring snowmelt runoff and is returned to the wetland in runoff.

Transient reversals of recharge and discharge conditions around ponds and wetlands are common. Phillips and Shedlock (1993) discussed water-table fluctuations near seasonal ponds on the Atlantic coastal plain in Delaware. During the late summer and fall, water levels in piezometers beneath the ponds were higher than water levels beneath adjacent uplands. The ponds function as recharge systems under these conditions. In

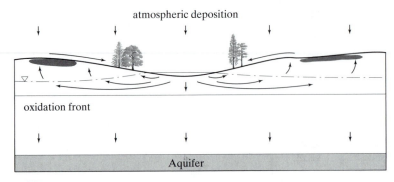

FIGURE 9–13 Transport of chloride and other salts by ground water recharge from a wetland in Saskatchewan to adjacent upland areas, where they are drawn upward by evapotranspiration and accumulate (dark areas) in near-surface soil (reprinted from *Jour. Hydrol.*, v. 207, Hayashi, M., van der Kamp, G., and Rudolf, D. L., Water and solute transfer between a prairie wetland and adjacent uplands, 2. Chloride cycle, pp. 56–57, Copyright 1998, with permission of Elsevier Science).

the winter and spring, however, water-table mounds develop around the wetland margins. The chemical similarities between the ground water in the wetland margins and the pond water, including low alkalinity, low pH, and high aluminum concentrations, indicate that discharge from shallow ground water in marginal areas into the ponds occurs when the mounds form. The flow systems are apparently too short to allow buffering of the recharging waters by mineral weathering.

Transient ground water mounds at the edges of wetlands in North Dakota proved to be important in a study by Arndt and Richardson (1993). In this setting, drawdown of pond elevation in the summer because of evaporation produced high concentrations of salts in the sediments at the pond edges. During subsequent recharge events, transient mounds developed at the pond margins and the accumulated salts were transported away from the ponds to ponds at lower elevations. The studies cited previously illustrate the temporal and spatial complexity of surface water–ground water interactions around lakes and wetlands and the resulting chemical composition of the shallow ground water.

Flow-through Systems.   Flow-through lakes and wetlands receive ground water discharge on the upgradient side and recharge ground water on the downgradient side of the impoundment. Temporal variations in both recharge and discharge are likely to occur. Kehew et al. (1998) delineated plumes of ground water that originate and flow downgradient from flow-through wetlands. Ground water in these plumes maintains its distinct chemical composition for hundreds of meters in the flow system (Fig. 9–14). The ground water recharged from the wetland has elevated dissolved organic carbon, iron, ammonium, and lower concentrations of nitrate and sulfate, relative to shallow

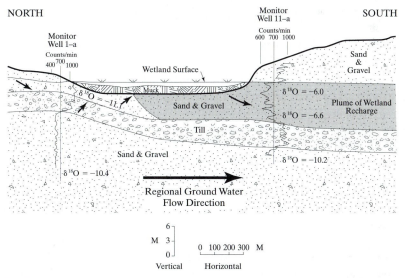

FIGURE 9–14   Plume of ground water recharge from flow-through wetland characterized by elevated dissolved organic carbon, iron, and ammonium and low nitrate and sulfate. High $\delta^{18}O$ values in plume produced by evaporative enrichment in wetland. Gamma-ray logs of boreholes shown for monitor wells 1-a and 11-a (from Kehew et al., 1998. Reprinted by permission of the National Ground Water Association).

ground water in the watershed not recharged from wetlands. Although the water is dilute when it initially passes through the base of the wetland to the aquifer, it quickly equilibrates with carbonate minerals contained in the aquifer solids, and calcium, magnesium, and alkalinity rise to the approximate level of ground water outside the plume. The elevated DOC and iron persist downgradient, imparting a reducing capacity with respect to water infiltrating downward through the vadose zone. Ground water beneath fertilized corn fields downgradient of the wetlands is devoid of nitrate, whereas ground water not impacted by wetland recharge beneath corn fields is commonly above the drinking water standard of 10 mg/l NO_3-N.

Wentz et al. (1995) studied a flow-through lake in Wisconsin for a period of eight years. During that time, ground water discharge into the lake ceased for several years because of drought conditions in the basin. Buffering mechanisms in the lake changed dramatically as a result of this temporal variation in the interaction of the lake with the ground water flow system. During years when ground water discharge was significant, buffering of acid precipitation falling on the lake was accomplished by alkalinity derived from the inflowing ground water. When ground water discharge lapsed in drought years, alkalinity derived from sulfate reduction in the lake sediment buffered the acid precipitation.

Discharge Systems.   The chemistry of discharge lakes and wetlands is controlled by the characteristics of the discharging flow system as well as by the physical and chemical processes occurring in the surface water body. A large wetland adjacent to Lake Michigan in northern Indiana was investigated by Shedlock et al. (1993). Local, intermediate, and regional ground water flow systems were identified in the vicinity of the wetland (Fig. 9–15). Beneath the wetland, an intermediate flow system discharges through its confining layer, a glacial till, into the wetland. In one part of the wetland, the confining layer is absent and the intermediate flow system discharges directly into that part of the wetland. A raised peat bog with an elevated water table developed above the discharge point of the intermediate flow system. Ground water chemical composition within and just beneath the wetland was used to identify and trace the flow systems. Ground water in the intermediate system has low tritium concentrations, high alkalinity, and hardness. These same characteristics are present in the shallow ground water in the raised peat bog. In other parts of the wetland, ground water chemistry is more variable. Water in the local flow systems has high tritium content and lower dissolved solids than the intermediate system, but it is higher in sulfate concentration. The source of the sulfate may be atmospheric deposition from the industrial area that lies just upwind from the wetland. This study shows a strong relationship between the wetland water chemistry and the hydrochemistry of the ground water flow systems that discharge into it.

A close hydrochemical relationship between surface water and ground water chemistry was also noted in evaporative lakes in the northern Great Plains along the border of North Dakota and Montana (Donovan and Rose, 1994). Lakes in this region occur in closed depressions in the glacial drift. Because of the extreme summer evaporation in the semiarid climate of the area, the dissolved solids in lake water increase by evaporation to the point where they become brines and a variety of evaporite minerals are precipitated. Ground water flow systems were separated into shallow, intermediate, and

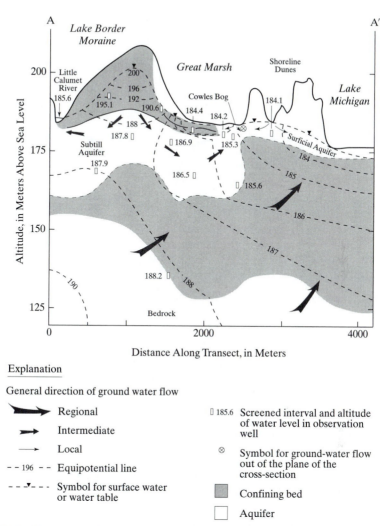

FIGURE 9–15   Cross section of wetland complex in northern Indiana. Intermediate ground water flow system (subtill) discharges through confining layer into wetland. Where confining layer is absent, discharge produces raised bog with water chemistry similar to intermediate flow system (reprinted from *Jour. Hydrol.*, v. 141, Shedlock, R. J., Wilcox, D. A., Thompson, T. A., and Cohen, D. A., Interaction between ground water and wetlands, southern shore of Lake Michigan, pp. 127–155, Copyright 1993, with permission of Elsevier Science).

deep types. Water chemistry differs by flow system, as shown in Fig. 9–16. Water in shallow flow systems is a mixed sodium–calcium–magnesium cation type and a bicarbonate–sulfate anion type. Intermediate-depth water has higher sodium and alkalinity and lower sulfate, calcite, and magnesium. Regional flow system water is from the Fox Hills-basal Hell Creek aquifer system, which, as discussed earlier, evolves toward a sodium–bicarbonate type with low sulfate. The trend of the ground water compositions from shallow to deep flow systems is shown in Fig. 9–16. Lake water chemistry is also

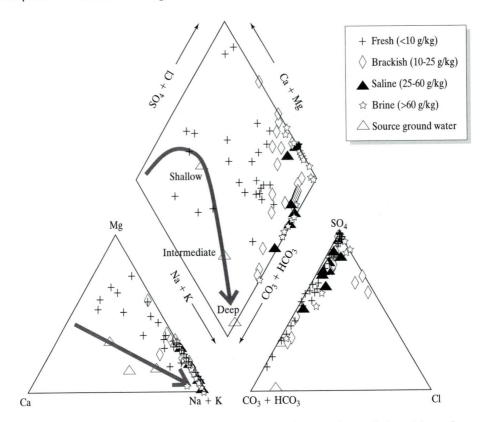

FIGURE 9–16 Piper plot of water composition of evaporative ground water discharge lakes and ground water systems in western North Dakota and eastern Montana. Symbols show different salinity ranges of lake waters. Arrows show trends in ground water chemistry with depth. Open triangles are average compositions of ground water at different depth intervals (reprinted from *Jour. Hydrol.*, v. 154, Donovan, J. J. and Rose, A. W., Geochemical evolution of lacustrine brines from variable–scale ground water circulation, pp. 35–62, Copyright 1994, with permission from Elsevier Science).

shown on Fig. 9–16, which indicates effects of evaporative concentration from ground water discharge. Donovan and Rose used a reaction path model to simulate the evaporative concentration of waters starting at the composition of the shallow, intermediate, and deep points shown on the diagram. The modeling results showed lakes receiving discharge from the shallow flow systems evolved to sodium–magnesium–sulfate brines. Carbonate minerals precipitate early in the reaction sequence. Lakes with discharge from the intermediate flow system evolve to a sodium–carbonate–sulfate brine composition. Ground water from deep flow systems would evolve to a sodium–bicarbonate brine composition because of the lack of sulfate in the downgradient part of the deep flow system. Lake waters of this type were not found in the region, leading to the conclusion that discharge to lakes was not occurring from the deep flow system.

Evaporative salinization of water in ground water discharge areas is not limited to surface water bodies. Large areas of shallow ground water that does not discharge to a

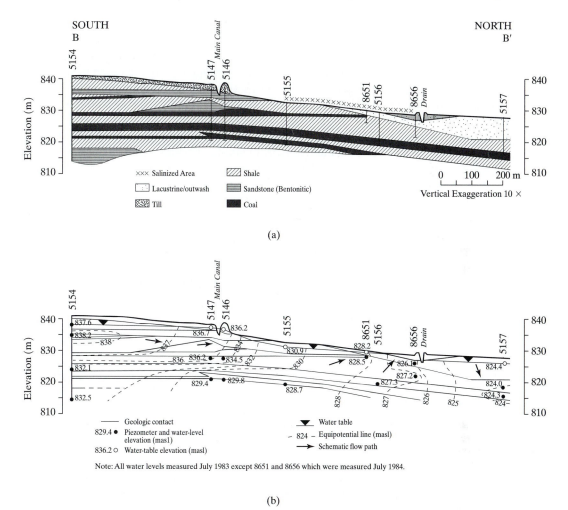

(a)

(b)

FIGURE 9–17 Stratigraphic (a) and hydrogeologic (b) cross sections of area containing salinized soils in southern Alberta. Salinization is caused by evaporation of ground water from water table in discharge area of local flow system developed within bedrock aquifer. Discharge area begins on drainage basin slope (to the right of irrigation canal on cross section a) where glacial deposits pinch out and slope angle increases (from Hendry and Buckland, 1990. Reprinted by permission of the National Ground Water Association).

lake or wetland can become salinized by evaporation. Hendry and Buckland (1990) provide an excellent example of this phenomenon in southern Alberta. Fig. 9–17 includes stratigraphic and hydrogeologic cross sections of part of the study area. Land surface, which slopes to the north (right) in the cross section, is underlain by a discontinuous surficial till and the Tertiary bedrock, consisting of shale, sandstone, and coal. Ground water flow occurs from a drainage divide recharge area to the south to the lowland in the right half of the cross section. Ground water discharge to the surface begins north (right) of the irrigation canal, where the till pinches out and the land surface slope increases. Piezometer data and modeling ruled out leakage from the irrigation canal and deeper bedrock formations as a source of the discharging ground water. Ground water flowing within a local flow system in the bedrock is apparently the cause of soil salinization, which occurs in the discharge area as water evaporates from the shallow water table to concentrate salts in the upper few meters of the soil. Soil salinization prevents the use of the land for agriculture in this and many other parts of the Great Plains of the U.S. and Canada.

The Red River Valley of eastern North Dakota provides a contrasting example to the Alberta study, because shallow ground water and soil salinization in the discharge area is caused by discharge from a deep regional flow system modified by mixing with ground waters of shallow origin, as well as by evaporation. Deep ground waters have dissolved solids concentrations of as much as 50,000 mg/l, and more than 162,000 ha of land surface area have been affected by salinization from ground water discharge. Gerla (1992) combined ground water flow and geochemical modeling to explain the distribution of ground water compositional types in this large regional discharge area. Figure 9–18 shows a stratigraphic cross section and the modeled ground water flow system.

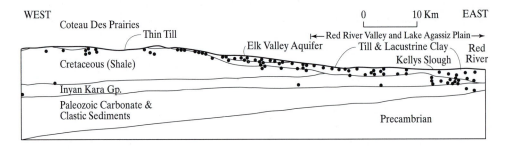

(a)

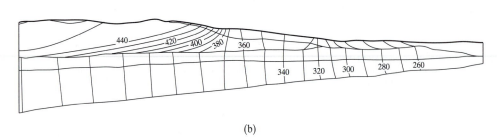

(b)

FIGURE 9–18   Stratigraphic cross section (a) of bedrock aquifer discharge area in Red River Valley of eastern North Dakota. Dots show location of ground water samples. (b) Modeled ground water flow system (equipotentials). Brackish to saline ground water flows from west to east in Inyan Kara and Paleozoic aquifers to eastern end of cross section where mixing and upward discharge through lacustrine sediment occurs. Downward recharge through Cretaceous shale occurs in western part of section. Water mixes with water in Inyan Kara aquifer prior to discharge to surface (from Gerla, 1992. Reprinted by permission of the National Ground Water Association).

Ground water flow in the westward-dipping Cretaceous (Inyan Kara Group) and Paleozoic aquifers is from west to east because of the regional slope of the Great Plains from west to east. These formations continue westward toward the center of the Williston Basin, where they contain highly concentrated brines. The deep bedrock formations are overlain by Cretaceous shale in the western part of the study area (Fig. 9–18). Ground water flow through the shales is downward from surface recharge areas toward the Inyan Kara Group sandstone. Glacial sediments, including an important surficial aquifer, the Elk Valley aquifer, and thick, low permeability lacustrine clays overlie the Cretaceous shales as they pinch out toward the east. Discharge occurs through the lacustrine sediments east of the pinch out of the Cretaceous shales. Mixing of ground water from the Cretaceous shales and the glacial aquifers occurs as these waters move downward toward the contact with the Inyan Kara Group aquifer. In addition, as the deeper aquifers pinch out in the discharge area, waters from the Paleozoic formations mix with the waters of the overlying Inyan Kara Group. Fig. 9–19 shows the ranges in total dissolved solids in the various aquifer units. Gerla (1992) modeled changes in chemical facies occurring because of precipitation, ion exchange, and other processes as the ground waters from different sources mix in this regional discharge area. The chemical evolutionary trends resulting from these processes are shown in Fig. 9–20. Gerla's mass balance calculations indicate that 94% of the ground water discharge reaching land surface comes

FIGURE 9–19 Variations in total dissolved solids concentration in aquifers in Red River Valley regional discharge study area (from Gerla, 1992. Reprinted by permission of the National Ground Water Association).

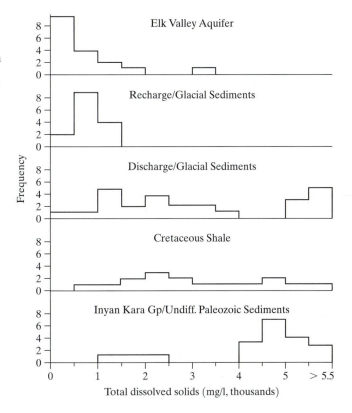

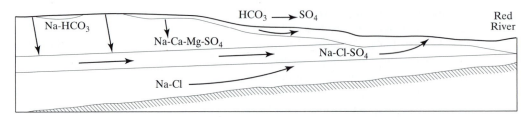

FIGURE 9–21   Chemical evolutionary trends caused by mixing of waters from different sources in Red River Valley regional discharge area (from Gerla, 1992. Reprinted by permission of the National Ground Water Association).

from the Inyan Kara Group and Paleozoic aquifers. Recharge from the Cretaceous shales in the western part of the area contributes the remaining 6% of the discharge.

PROBLEMS

1. How do flow patterns influence ground water chemistry in sedimentary basins?

2. For each of the three sequences described in Fig. 9–3 and Table 9–1, how would the ground water chemistry differ if the clay shale were not present? Is the final composition of sequence C in charge balance? Why or why not?

3. On Fig. 9–6, the arrows show evolutionary ion trends in the Floridan aquifer. Why do the arrows bend sharply at the end after following straight paths to that point?

4. In the recharge area of the Fox Hills–Basal Hell Creek aquifer (Fig. 5–18) the ground water evolves to very high sodium and alkalinity values. How is this possible, considering that the solubility of calcite and dolomite is generally too low to produce alkalinities of this magnitude and sodium this high by ion exchange?

5. What explains the differences in water quality in the Edwards aquifer upgradient from the "bad water line" relative to water downgradient from the line?

6. How did glaciation affect the water quality in the Milk River aquifer in Alberta?

7. How could you distinguish between recharge, discharge, and flow-through wetland systems based on water chemistry alone? How do redox processes fit into this system?

Hydrogeochemistry of Contaminants

Ground water contamination results from widely diverse human activities. Some of these, like solid and liquid waste disposal, are easily identified *point sources*. It has been abundantly clear for several decades that subsurface disposal of hazardous waste can create severe soil and ground water contamination problems. Other sources of contamination result from failure to contain potentially hazardous chemicals. The massive leaking underground storage tank (LUST) cleanup program in the United States in the 1980s and 1990s was a response to the unintended release of fuels, solvents, or other compounds. Today, as a result of environmental laws and regulations, storage and disposal facilities are much safer than in the past.

Still other types of ground water contamination are associated with *non-point-source* activities, agriculture and road salt application being good examples. Exploration for and production of natural resources, including oil and gas, metallic and nonmetallic minerals, and others, have always been potential sources of ground water contamination and will continue to be concerns in the future.

Throughout this text, numerous chemical processes taking place in the subsurface have been described. Many of these have been discussed in the context of contaminants. The purpose of this chapter is to focus on the hydrogeochemical characteristics of some major contaminant sources and activities and to provide well documented examples of contamination that they have caused. The literature on ground water contamination is huge. Any overview of this type can only sample the vast amount of published research and case studies of ground water contamination.

MUNICIPAL SOLID WASTE LANDFILLS

One of the first types of waste disposal to arouse scientific concern about ground water contamination was the landfill. Prior to the 1960s, municipal and industrial solid waste was routinely buried with little regard for the hydrogeology of the site, and few precautions were taken to protect the environment. Landfill design and management have changed tremendously in the ensuing decades. Potential sites are thoroughly investigated, and hydrogeological factors are an important component of site selection. Landfill design, even for nonhazardous municipal facilities, commonly includes multiple liner systems and leachate collection systems to isolate the waste from the hydrologic system. Costs are much higher for these safeguards, and it is likely that most of world's developing countries that are not using these technologies will experience the ground water problems that have been addressed by the developed nations.

The leachate generation processes in landfills are driven by water, and many sites are capped with low-permeability material to limit infiltration of water into the waste. Lee and Jones (1990) have criticized this "dry tomb" model of landfill design on the grounds that caps and liners will eventually fail and leachate generation and migration will only be delayed by trying to keep the waste dry. Instead, Lee and others advocate cycling water and oxygen through the decomposing waste to facilitate biochemical breakdown of the wastes as rapidly as possible. Once this stabilization process is complete, the stable, volumetrically reduced waste products can be capped with less concern for the integrity of the closed cells.

Leachate Generation

Three stages are recognized in the formation of leachate from municipal sanitary landfills (Qasim and Chiang, 1994), which typically have a high content of organic matter. Oxgyen present at the time of initial disposal is quickly consumed during the first, or *aerobic*, stage (Fig. 10–1). Aerobic decomposition reactions are exothermic and the temperature of the landfill increases. As the oxygen is consumed, the aerobic phase is rapidly terminated and the landfill enters the second, *acetogenic*, stage. Facultative anaerobes break down the waste in this stage by fermentation of sugars to acetic and other volatile fatty acids (Owen and Manning, 1997). In the absence of buffering compounds, pH declines in this stage. Carbon dioxide and H_2 gases are produced, and the redox potential declines to low levels. The abundance of volatile fatty acids (Fig. 10–1) produces extremely high alkalinity measurements because when these carboxylic acids dissociate, organic anions such as acetate accept protons in the measurement of alkalinity by titration, just as carbonate and bicarbonate do, to yield very high measurements (Baedecker and Back, 1979a, b; Kehew and Passero, 1990). In the acetogenic stage, biodegradation of the volatile fatty acids by sulfate-reducing bacteria consumes the available sulfate as an electron acceptor. In this stage, the molar concentrations of the organic acids acetate and propionate show a 1:1 proportionality (Manning, 1997) because the oxidation of both acids is coupled with sulfate reduction. Dissolution of inorganic salts in the waste and dissolution of carbonate minerals, if present, in cover soils in response to the formation of organic acids and CO_2, maximizes the total dissolved solids and the specific conductance (Fig. 10–1) of the leachate.

The third stage of decomposition is the *methanogenic* stage. After the sulfate is consumed, methanogens become the dominant microbial community. Volatile fatty acids and other organics are converted to CO_2 and CH_4. The redox potential reaches its minimum level during methanogenesis. Accompanying the formation of microbial methane is fractionation of the carbon and hydrogen isotopes that make up the gas. Fractionation patterns in landfill methane suggest that the dominant CH_4 formation pathway is acetate fermentation rather than CO_2 reduction (Hackley et al., 1996).

Eventually, decomposition rates decrease because of the reduction of organic substrate. The volume of waste is reduced by decomposition processes, which causes physical subsidence of the landfill and potential cracking of the cap. Aerobic conditions

FIGURE 10–1   Changes in gas content and selected chemical parameters during the three stages of landfill leachate generation (modified from Qasim and Chiang, *Sanitary Landfill Leachate. Generation, Control and Treatment*, with permission from Technomic Publishing Co., Inc. Copyright 1994).

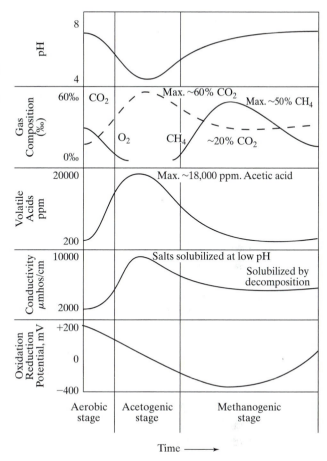

may be re-established in the waste cells as oxygenated recharge water infiltrates into the landfill.

Leachate composition varies tremendously according to the waste type, age of the landfill, precipitation, and other variables. Typical concentrations of various parameters for landfills of different ages are shown in Table 10–1. The high organic content of the leachate is indicated by the BOD (biochemical oxygen demand) and COD (chemical oxygen demand). These nonspecific parameters represent the amount of oxygen needed to oxidize the organic matter, by microbial biodegradation in the case of BOD and by a chemical oxidant for COD. COD is usually high because it includes refractory organic matter that cannot be microbially oxidized in the BOD test. Inorganic ions are high in concentration as well, producing a high specific conductance and total dissolved solids. Sulfate may initially be high, but declines by sulfate reduction in the acetogenic stage of decomposition. Iron and manganese are present in high concentrations in young landfills because of the low redox potential, but significant amounts may precipitate as sulfide and carbonate (siderite) minerals. Nitrogen is present in the form of ammonium because of the instability of nitrate at low redox potential. Concentrations

TABLE 10–1 Composition of landfill leachates over time (From Qasim and Chiang, 1994).

Parameters	Age of Landfill		
	1 Year	5 Year	15 Year
BOD	7,500–28,000	4,000	80
COD	10,000–40,000	8,000	400
pH	5.2–6.4	6.3	
TDS	10,000–14,000	6,794	1200
TSS	100–700		
Specific Conductance	600–9,000	—	
Alkalinity (CaCO$_3$)	800–4,000	5,810	2250
Hardness (CaCO$_3$)	3,500–5,000	2,200	540
Total P	25–35	12	8
Ortho P	23–33	—	
NH$_4$-N	56–482		
Nitrate	0.2–0.8	0.5	1.6
Calcium	900–1,700	308	109
Chloride	600–800	1,330	70
Sodium	450–500	810	34
Potassium	295–310	610	39
Sulfate	400–650	2	2
Manganese	75–125	0.06	0.06
Magnesium	160–250	450	90
Iron	210–325	6.3	0.6
Zinc	10–30	0.4	0.1
Copper	—	<0.5	<0.5
Cadmium	—	<0.05	<0.05
Lead	—	0.5	1.0

Note: All values are mg/l except specific conductance measured as microhms per centimeter pH as pH units.
Source: After Chian and DeWalle (1976, 1977).

of trace metals are commonly not high, although some may be elevated relative to drinking water standards.

Despite the ban of hazardous wastes from municipal waste landfills, some toxic or hazardous chemicals are typically present in landfill leachates, either from past disposal practices or because of the inability to test all waste sent to the site. Table 10–2 gives median concentrations of hazardous inorganic and organic compounds detected in municipal waste leachates. These detections justify the liner systems and leachate collection systems used in municipal landfills, particularly in hydrogeologically vulnerable areas.

Landfill Contaminant Plumes

In the absence of effective liner systems, leachate solutions will migrate out of the waste cells and into the surrounding geologic materials. In some older landfills, excavation of the cells intersected the water table, and the formation of a contaminant plume began as soon as wastes were buried. In landfills confined to the vadose zone, leachate migration

TABLE 10–2   Median concentrations of compounds of environmental concern in municipal landfill leachates.

Substance[a]	Concentration (ppm)	Substance[a]	Concentration (ppm)
Inorganics:		Organics (*continued*):	
Antimony (11)	4.52	Dichlorodifluoromethane (6)	237
Arsenic (72)	0.042	1,1-Dichloroethane (34)	1,715
Barium (72)	0.853		
Beryllium (6)	0.006	1,2-Dichloroethane (6)	1,841
Cadmium (46)	0.022	1,2-Dichloropropane (12)	66.7
Chromium (total) (97)	0.175	1,3-Dichloropropane (2)	24
Copper (68)	0.188	Diethyl phthalate (27)	118
		2,4-Dimethyl phenol (2)	19
Cyanide (21)	0.063	Dimethyl phthalate (2)	42.5
Iron (120)	221	Endrin (3)	16.8
Lead (73)	0.162	Ethyl benzene (41)	274
Manganese (103)	9.59	bis (2-Ethylhexyl)	
Mercury (19)	0.002	phthalate (10)	184
Nickel (98)	0.326	Isophorone (19)	1,168
Nitrate (38)	1.88	Lindane (7)	0.020
Selenium (18)	0.012	Methylene chloride (68)	5,352
Silver (19)	0.021	Methyl ethyl ketone (24)	4,151
Thallium (11)	0.175	Naphthalene (23)	32.4
Zinc (114)	8.32	Nitrobenzene (3)	54.7
		4-Nitrophenol (1)	17
Organics:		Pentachlorophenol (3)	173
Acrolein (1)	270	Phenol (45)	2,456
Benzene (35)	221	1,1,2,2-Tetrachloroethane (1)	210
Bromomethane (1)	170	Tetrachloroethylene (18)	132
Carbon tetrachloride (2)	202	Toluene (69)	1,016
Chlorobenzene (12)	128	Toxaphane (1)	1
Chloroform (8)	195	1,1,1-Trichloroethane (20)	887
bis (Chloromethyl) ether (1)	250		
p-Crestol (10)	2,394	1,1,2-Trichloroethane (4)	378
2.4 D (7)	129	Trichloroethylene (28)	187
4,4-DDT (16)	0.103		
Di-n-butyl phthalate (5)	70.2	Trichlorofluoromethane (10)	56.1
1,2-Dichlorobenzene (8)	11.8	1,2,3-Trichloropropane (1)	230
1,4-Dichlorobenzene (12)	13.2	Vinyl chloride (10)	36.1

[a] Number of samples in parentheses.
Source: Adapted from U.S. EPA 1988.

begins when the waste reaches *field capacity*, the moisture content at which gravity drainage begins. Despite chemical monitoring activities at hundreds of landfills in many countries mandated by governmental regulatory programs, few comprehensive hydrogeochemical studies have been made of contaminant plumes. Among the best documented of these are studies of the Army Creek landfill in Delaware (Baedecker and Back, 1979a; 1979b; Baedecker and Apgar, 1984), the Borden landfill in Ontario (Nicholson et al., 1993), and the Vejen landfill in Denmark (Lyngkilde and Christensen, 1992; Heron and Christensen, 1994).

The Borden landfill study (Nicholson et al., 1983) is notable for its highly detailed sampling network of multilevel piezometers, allowing unprecedented three-dimensional characterization of the plume. The Borden landfill plume differs somewhat from many other plumes in that sulfate concentrations were high and dissolved organic carbon was relatively low. Although methane was present in the plume, the high sulfate levels probably restricted methanogenesis within the landfill. Ground water downgradient from the landfill was saturated with respect to gypsum and supersaturated with respect to calcite and siderite. High iron and manganese concentrations were attributed to dissolution from aquifer solids because of the low redox potentials in the plume. Cation concentrations in the plume were affected by exchange reactions with aquifer solids. Potassium, magnesium, and sodium were adsorbed in these reactions in exchange for calcium (Dance and Reardon, 1983).

Baedecker and Back (1979a, b) recognized anaerobic, transitional, and aerobic redox zones downgradient from a landfill in Delaware. These zones result from a gradual increase in redox potential from its lowest levels near waste cells as organic matter in the plume is gradually biodegraded along the flow path and the ground water returns to near-background levels. Baedecker and Back compared the landfill environment to the redox processes and zonation in organic-rich marine sediments. As a contaminant plume moves away from the landfill in the ground water flow system, redox zonation develops in the transition from highly reducing to aerobic conditions as biodegradation consumes the organic carbon and the plume mixes with aerobic recharge and reacts with aquifer solids. Baedecker and Back (1979b) recognized the oversaturation of carbonate minerals in the plume, but cautioned that precipitation of calcite, for example, could be inhibited by formation of Ca–fatty acid complexes (soaps) and that the contribution of organic acid anions to alkalinity could lead to overestimation of the saturation index for calcite. Kehew and Passero (1990) found that as much as 90 percent of the carbon in a landfill plume in Michigan was organic and that titration alkalinity values highly overestimate the true inorganic alkalinity (bicarbonate). With distance from the source, biodegradation of the fatty acids releases calcium and calcite precipitates. Siderite may also precipitate in the more reducing parts of the plume where dissolved iron concentrations are high. As the redox potential rises from its lows beneath the landfill, ferrous iron oxidizes and precipitates as oxide coatings on aquifer solids. Iron sulfides may also precipitate in the zone of sulfate reduction.

Lyngkilde and Christensen (1992) delineated the three-dimensional distribution of six specific redox zones in a landfill in Denmark. The criteria for the redox zones are given in Table 10–3, and the distribution of the zones is shown in Fig. 10–2. For ease in mapping, the ferrogenic and manganogenic zones were combined within the plume. Redox zones in a landfill plume delineate redox buffers, as defined in Chapter 5, in which a specific redox reaction controls the redox potential within a specific zone. For example, within the ferrogenic zone, the redox potential is controlled by the reduction of ferric iron to ferrous iron. Oxidation of organic matter is the complementary oxidation process. Heron and Christensen (1994) concluded that the reduction of ferric iron solids in the aquifer plays a significant redox buffering role in the biodegradation of the organic carbon of the plume. Only 2% of the ferrous iron in the plume produced by ferric iron reduction was present in dissolved form. The remainder was associated with

TABLE 10–3 Criteria used to delineate redox zones at the Vejen Landfill, Denmark (all values in mg/l). From Lyngkilde and Christensen, 1992.

Parameter	Aerobic	Nitrate-Reducing	Manganogenic	Ferrogenic	Sulfidogenic	Methanogenic
Oxygen	>1.0	<1.0	<1.0	<1.0	<1.0	<1.0
Nitrate	—	—	<0.2	<0.2	<0.2	<0.2
Nitrite	<0.1	—	<0.1	<0.1	<0.1	<0.1
Ammonium	<1.0	—	—	—	—	—
Mn(II)	<0.2	<0.2	>0.2	—	—	—
Fe(II)	<1.5	<1.5	<1.5	>1.5	—	—
Sulfate	—	—	—	—	—	<40
Sulfide	<0.1	<0.1	<0.1	<0.1	>0.2	—
Methane	<1.0	<1.0	<1.0	<1.0	<0.1	>1.0

— = no criterion is applied.

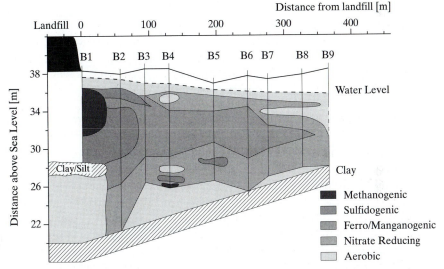

FIGURE 10–2 Distribution of redox zones along a cross section in the direction of ground water flow at the Vejen landfill, Denmark (reprinted from *Jour. Contam. Hydrol.*, v. 10, Lyngkilde, J. and Christensen, T. H., Redox zones of a landfill leachate plume (Vejen, Denmark), pp. 273–289, Copyright 1992, with permission from Elsevier Science).

the solid phase either as exchangeable cations or mineral precipitates including pyrite, siderite, and other phases.

INDUSTRIAL LANDFILLS AND LIQUID WASTE DISPOSAL

Industrial waste disposal encompasses a very large variety of wastes and compounds. Industrial wastes in the past were commonly mixed with municipal wastes or disposed of in the same landfill. The characteristics of the leachate plumes from these compound sites are very similar to those originating from landfills that contain only municipal wastes, except that specific hazardous or toxic compounds are present in addition to the predominantly nonhazardous wastes in municipal landfills. Small to moderate concentrations of hazardous compounds, particularly the volatile organic compounds, are commonly present in plumes from municipal landfills despite attempts to prevent these compounds from disposal.

Occurrences of specific organic compounds detected at hazardous waste landfills were compiled and rank ordered by Plumb (1992), based on the number of detectable events and number of sites out of a total of 500 sites studied (Table 10–4). Volatile organics dominate the database, comprising 75% of the detections. Compound groups include aromatic hydrocarbons, halogenated aliphatics and aromatics, phenols, PAHs, phthalates, and ketones. The mobility of individual compounds is subject to the

TABLE 10–4   Rank-ordered list of 30 most frequently detected organic compounds at hazardous waste sites based on detectable events and number of sites in which compound was detected. From Plumb, 1992.

Compound	Detectable Events	Compound	Number of Sites
Dichloromethane[a]	4558	Dichloromethane[a]	157
Trichloroethene[a]	4001	Trichloroethene[a]	132
Tetrachlorethene[a]	2913	Toluene[a]	131
trans-1,2-Dichloroethene[a]	2357	Benzene[a]	120
Trichloromethane	2137	trans-1,2-Dichloroethene[a]	116
1,1-Dichloroethane[a]	1706	Tetrachloromethane[a]	111
1,1-Dichloroethene[a]	1653	Ethylbenzene[a]	109
1,1,1-Trichloroethane[a]	1609	1,1-Dichloroethane[a]	108
Toluene[a]	1430	1,1,1-Trichloroethane[a]	101
1,2-Dichloroethane[a]	1339	Trichloromethane[a]	89
Benzene[a]	1169	Bis(2-ethylhexyl) phthalate	89
Ethylbenzene[a]	733	Chlorobenzene[a]	86
Phenol	679	1,2-Dichloroethane[a]	82
Chlorobenzene[a]	662	Vinyl chloride[a]	79
Vinyl chloride[a]	580	Phenol	79
Tetrachloromethane[a]	484	1,1-Dichloroethane[a]	76
Bis(2-ethylhexyl) phthalate	383	Chloroethane[a]	61
Naphthalene	369	Naphthalene	61
1,1,2-Trichloroethane[a]	270	Di-N-butyl phthalate	56
Chloroethane[a]	269	Fluorotrichloromethane[a]	45
Acetone[a]	254	Lindane	41
1,2-Dichlorobenzene	240	2,4-Dimethylphenol	38
Isophorone	211	1,1,2-Trichloroethane[a]	37
Fluorotrichloromethane[a]	203	Diethyl phthalate	37
1,4-Dichlorobenzene	191	Isophorone	37
2-Butanone	171	2,4-D	36
1,2,4-Trichlorobenzene	164	1,2-Dichlorobenzene	36
2,4-Dimethylphenol	159	Acetone[a]	34
1,2-Dichloropropane[a]	158	1,4-Dichlorobenzene	34
Dichlorodifluoromethane[a]	154	1,2-Dichloropropane[a]	33

[a] Volatile compound.

partitioning constraints discussed in Chapter 6 and the susceptibility to biodegradation, which was discussed in Chapter 7.

Among the volatile organic compounds, the halogenated aliphatics comprise the most common contaminants in industrial areas. These are the most widely used solvents for oil and grease and, as a result, are utilized in many industries. Although examples of contaminant plumes composed of these compounds were included earlier, one additional case history will be given. In the city of Coventry, UK, an industrial facility had used trichloroethylene (TCE) and related solvents since their development in the 1930s (Bishop et al., 1993). Contamination was apparently the result of sloppy handling techniques around the above-ground storage tanks, rather than by a specific spill. It was common practice in the past, when tanks were filled from tanker trucks, for drivers to empty the remaining liquid from the hoses onto the ground when the tank was filled. Over time, a great deal of solvent makes its way onto the ground by this and other practices. Aquifers beneath the site consist of a series of sandstone beds, interbedded with shales. When a production well became contaminated, test boreholes were drilled at the site. The test wells were open boreholes from which individual sandstone beds could be sampled using packers. The use of open boreholes proved to be a mistake, as concentrations increased in the production well. Bishop et al. (1993) hypothesize that contaminants migrated downward in the open boreholes to lower aquifer units. Because the aquifer units are composed of a dual porosity system, with both granular and fracture porosity, contaminant compounds will diffuse into isolated pores from fractures. If the contaminants are eventually removed from the high-permeability flow paths, a reversal of the diffusion process from isolated pores into the primary flow paths will insure that the aquifer is impacted by solvent compounds for a long time into the future.

Industrial waste sites that contained a more limited assortment of compounds associated with a particular industry abound in the literature. Goerlitz (1992) described six sites studied by the U.S. Geological Survey Organics Program. One of these sites contained chlorinated hydrocarbons from pesticides manufacturing, two contained explosives wastes, and three were wood preservative sites. Most of these sites contained a mixture of solid and liquid wastes. One interesting contrast in the sites was the difference in transport of pentachlorophenol (PCP) at two wood preservative sites. Creosote and PCP are commonly used for treatment of wooden power poles. Creosote is a mixture of more than 200 compounds derived from coal tar. The composition includes 85% PAHs, 12% phenolic compounds, and 3% nitrogen, sulfur, and oxygen heterocyclic compounds (Goerlitz, 1992). Based on the high log K_{ow} of PCP, it would not be expected to migrate significantly in ground water. This did not prove to be the case at a site in Visalia, California, where PCP migrated 500 m in ground water from leaky treatment tanks since its use began in the early 1950s (Fig. 10–3). PAHs, mainly naphthalene and methylnaphthalene, also migrated about the same distance as PCP. The phenolic compounds and heterocyclics were not detected in the ground water samples. A nonaqueous phase liquid was present on the water table near the contamination source.

A wood treatment plant in Pensacola, Florida, used the same types of compounds as the Visalia site, but the fate and transport of contaminant compounds was much different. PCP was relatively immobile at the site, whereas concentrations of phenols and nitrogen herterocyclics were very high. One major difference in the two sites was the pH.

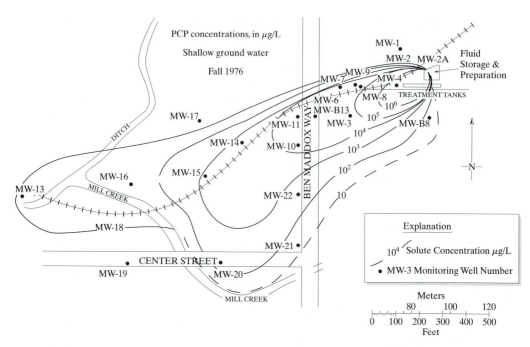

FIGURE 10–3   Concentration of PCP in ground water at the Visalia, California, wood treatment site, fall, 1976 (reprinted from Goerlitz, A review of studies of contaminated ground water conducted by the U.S. Geological Survey Organics Project, Menlo Park, California, 1961–1990, in Lesage, S. and Jackson, R. E., eds., *Ground Water Contamination and Analysis at Hazardous Waste Sites*, 1992, by courtesy of Marcel Dekker, Inc.).

The ground water pH at the Visalia site was 7.9–8.6 in comparison to a pH range of 5.0–6.3 at the Pensacola site. Goerlitz (1992) suggests that at low pH values, PCP is much less soluble and therefore was detected only near the source at Pensacola. Under the high pH conditions at Visalia, PCP was more soluble and migrated without apparent sorption or biodegradation. Differences in biodegradation between the two sites may also have influenced the dissimilar behavior of the contaminant compounds.

Metals are common constituents of industrial wastes. Of primary concern are trace metals that may be elevated many times beyond their natural concentrations in ground water. Trace metals that have drinking water standards include chromium, lead, copper, silver, cadmium, zinc, nickel, and mercury. Prediction of the mobility of these elements in ground water begins with an analysis of their speciation in terms of redox potential. Most metals have several oxidation states, and it is important to understand the geochemistry of the contaminant plume so that redox conditions are known and metal speciation can be predicted.

Chromium is a metal that has caused significant ground water contamination problems. Under oxic conditions, chromium exists in the Cr^{+6} valence state (chromate) in anionic species (Fig. 10–4). Chromium is very mobile under these conditions. One of the earliest documented ground water contamination plumes was the result of disposal of chromium plating wastes on Long Island, New York (Perlmutter and Lieber, 1970).

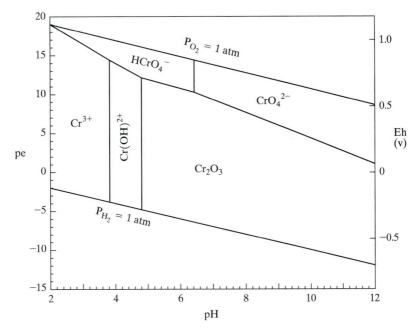

FIGURE 10–4 pe–pH diagram for the system Cr—O—H$_2$O at 25°C and 1 atm. Solubility defined as a dissolved Cr activity of 10^{-6} (from *The Geochemistry of Natural Waters* 3/e by Drever, © 1997).

The surficial materials of Long Island are permeable glacial deposits with shallow and deep aquifers. Ground water is the main source of drinking water and, as urbanization spread to this area, wastewaters were routinely discharged to the subsurface in pits, septic systems, and stormwater retention ponds. A plume of chromium and cadmium from chrome-plating wastes discharged into a surface impoundment by a World War II aircraft factory extended nearly 1,300 m downgradient in the shallow aquifer (Fig. 10–5) and discharged into a creek.

Henderson (1994) studied a chromate plume in a sand aquifer near Odessa, Texas. Maximum concentrations of hexavalent chromium decreased by tenfold in the plume over a six-year period of monitoring. The total mass of dissolved chromium was estimated from the volume of the plume. By measuring the partitioning coefficient (K_d) for adsorption of Cr^{+6}, Henderson was also able to estimate the mass of adsorbed Cr^{+6}. The total mass of hexavalent chromium, which is the sum of the masses in the dissolved and solid phases, followed a first-order decrease in the aquifer (Fig. 10–6). In order to explain the declines in hexavalent chromium mass in the aquifer, Henderson plotted values measured from water samples on an Eh-pH diagram for chromium (Fig. 10–7). Most points plotted within the Cr^{+3} stability fields, in which chromium forms insoluble oxide phases or cationic species that sorb to oxide phases, suggesting that Cr^{+6} was being reduced to Cr^{+3} in the aquifer and precipitating or sorbing to oxide or ferric hydroxide minerals. The proposed reaction is

$$3Fe^{2+} + CrO_4^{2-} + 7H_2O \longrightarrow 3Fe(OH)_{3(am)} + Cr(OH)_2^+ + 3H^+. \qquad (10\text{-}1)$$

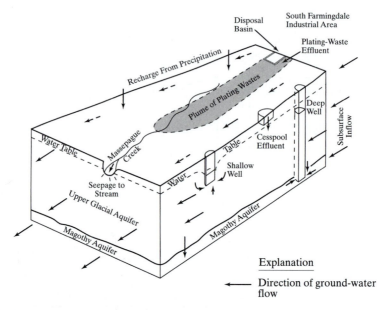

FIGURE 10–5 Chromium plume created by disposal of plating wastes in surface lagoon on Long Island, New York (from Perlmutter and Lieber, 1970).

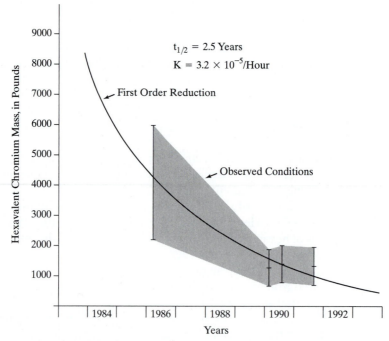

FIGURE 10–6 Reduction in total mass of hexavalent chromium by reduction to trivalent chromium in Trinity Sand aquifer, Odessa, Texas (from Henderson, 1994, reprinted by permission of the National Ground Water Association).

FIGURE 10–7   Eh–pH diagram for chromium. Points from Trinity Sand aquifer plot in trivalent chromium field, supporting hypothesis of reduction of chromium from Cr^{+6} to Cr^{+3} species (from Henderson, 1994, reprinted by permission of the National Ground Water Association).

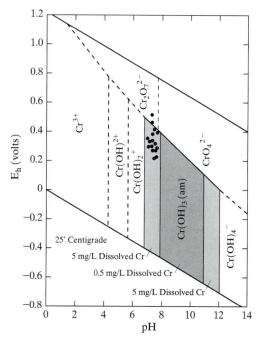

Ferrous iron in the aquifer or natural dissolved organic carbon would serve as the reducing agents, and the trivalent chromium oxide cation would be immobilized in the aquifer by sorption to hydroxide surfaces. By using the half time from the first-order reaction model, Henderson (1994) was able to predict when the Cr^{+6} concentrations would drop below the drinking water standard. This is an example of a contamination problem that may be addressed by natural attenuation rather than by active remediation through an understanding of the geochemical environment of the aquifer.

Some of the most complex geochemical problems occur where sources of both organics and metals interact at industrial waste disposal sites. An excellent example of this type of contaminant setting is described by Davis et al. (1994). The site was the location of a historical hide-tanning and -rendering factory in Woburn, Massachusetts, dating from 1927. Prior to that, it was used for the manufacture of arsenic- and lead-based insecticides, beginning in 1853. The Aberjona watershed, in which the site occurs contains the infamous solvent-contaminated municipal water wells that became the subject of the book and film, *A Civil Action*. One hundred tanneries once operated in this watershed (Davis et al., 1994).

Geochemical reactions and processes proposed for the site are shown in Fig. 10–8. Organic-rich leachate emanated from the hide piles, creating a reducing plume much like landfill plumes. Arsenic and chromium were derived from the insecticide and tanning processes, respectively. Arsenic was present at very high concentrations in the soil. Both arsenic and chromium would be expected to be relatively immobile in their reduced forms in a reducing plume. Instead, arsenic and chromium were surprisingly mobile in the plume. Davis et al. (1994) proposed that organic complexes involving both arsenic

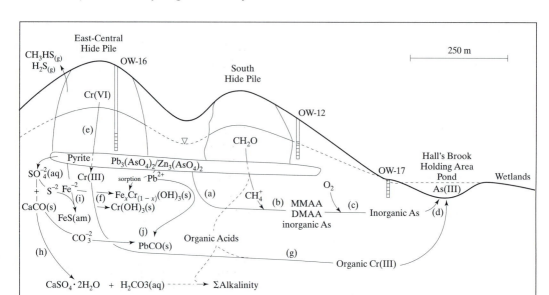

FIGURE 10–8 Cross section, showing ground water flow system and proposed chemical reactions beneath historical tanning site, Woburn, Massachusetts. See text for description of reactions (reprinted from *App. Geochem.*, v. 9, Davis, A., Kempton, J. H., Nicholson, A., and Yare, B., Groundwater transport of arsenic and chromium at a historical tannery, Woburn, Massachusetts, U. S. A., pp. 569–582, Copyright 1994, with permission of Elsevier Science).

and chromium increased the mobility. The hypothesized reactions are shown in Fig. 10–8 by letters. Arsenic (sequence a–b–c–d) reacts with methane in the plume to form the mobile compounds monomethylarsonic acid, MMAA ($CH_3AsO(OH)_2$), and dimethylarsonic acid, DMAA [$(CH_3)_2AsO(OH)$]. Arsenic is in the oxidized (+V) oxidation state in these compounds. As the plume migrates into more oxidizing areas downgradient from the hide piles, MMAA and DMAA are demethylated and arsenic is reduced to As (+III) in the pond discharge area. The reactions affecting chromium are shown by the sequence e–f–g. Hexavalent chromium was used in the tanning process. This is reduced to trivalent chromium in the reducing hide-pile plume, where, as in the studies described previously, it would normally precipitate. However, in the presence of the organic acids in the plume, soluble aqueous complexes are formed and trivalent chromium becomes mobile. It is also transported to the sediments of the pond in the discharge area.

Several other reactions occur in the plume. Gypsum is oversaturated and precipitates (h) because of the contribution of sulfate ions oxidized from sulfide in pyrite and calcium ions from dissolved calcium carbonate. Some of the sulfate may be subsequently reduced to sulfide and precipitate with iron as amorphous iron sulfide (i). The geochemical processes described by Davis et al. (1994) illustrate the complexity of plumes originating from mixed-waste sources. The study also demonstrates the large number of compounds that must be analyzed in order to fully characterize the plume and to understand the geochemical controls on migration of various constituents.

SEWAGE DISPOSAL

Sewage treatment and disposal commonly involve ground water impacts that vary depending upon the methods of treatment and disposal and the hydrogeological and hydrogeochemical conditions in the aquifer. A wide range of chemical and biological contaminants may migrate in ground water.

Practices that result in ground water contamination associated with sewage treatment and disposal include irrigation with treated effluents, spreading of treatment sludges, designed or incidental infiltration of treatment effluents, leakage from sanitary sewers, and surface contamination of wells with sewage.

Pathogenic microorganisms present the most significant threat to consumption of untreated ground water contaminated by sewage disposal. Yates and Yates (1993) review the health effects related to ground water contamination by bacteria and viruses and the fate and transport of pathogenic microorganisms in ground water. For example, almost one half of the approximately 200 waterborne disease outbreaks in the United States in the 1980s are attributed to untreated or inadequately disinfected ground water.

Bacteria and viruses are capable of survival in ground water for months and can be transported hundreds of meters in ground water flow systems (Yates and Yates, 1993). Both types of microorganisms survive longer at low temperatures and may survive indefinitely as temperatures reach 8°C. Physical filtration can retard the movement of bacteria, particularly in finer grained soils, although the small size of viruses precludes significant attenuation by filtration in most soil types. Adsorption is an important attenuating mechanism for both types of microorganisms. Both the Langmuir and Freundlich isotherms are used in modeling adsorption during transport.

Chemical contamination by sewage is more universally recognized than biological contamination. Many of the contaminants in sewage, nitrate for example, are also associated with other types of contamination. Sewage, however, contains a variety of other major and minor constituents that may have health concerns or can be used to define and trace the plumes. Almost all the common stable isotopes are used in studies of sewage contamination.

Ground Water Contamination from Sewage Treatment Plants

Sewage is treated through a variety of technologies. The degree of treatment can be classified as *primary*, *secondary*, and *tertiary (advanced)*. Primary treatment refers to the removal of solids by screens or settling tanks. Secondary treatment relies on microorganisms to remove the organic loading of the waste, typically measured by biochemical oxygen demand (BOD). Further treatment to remove specific chemical substances, such as nitrate or phosphate, is called tertiary, or advanced, treatment. After secondary treatment, the effluent can be discharged to natural waterways, allowed to infiltrate to the subsurface in infiltration beds, or used for irrigation of crops, golf courses, or other types of vegetation. Ground water impacts occur during these effluent disposal practices. The solids separated from the wastewater can be further treated and disposed of as sludges, either by burial in a landfill or by land spreading on crops to enhance soil fertility. Leaching of sludge components can also impact ground water.

One of the most common methods of providing secondary treatment for small communities in rural areas is the use of large ponds called *oxidation ponds*, or *waste stabilization lagoons*. These impoundments, which are commonly constructed in series so that wastewater is increasingly treated as it passes through a sequence of cells, utilize both aerobic and anaerobic processes to reduce the BOD in municipal wastewater. This method is less expensive than other treatment technologies and is particularly suited to areas in which land availability is not a constraint. Kehew et al. (1984) and Bulger et al. (1989) studied the effects on ground water of a system of this type in North Dakota. The impoundments were constructed in permeable glacial outwash, and the cells had to be lined in order to retain wastewater for the appropriate length of time for treatment. Only one cell of three at the site was actually lined, and when wastewater levels became too high for the lined cell, wastewater drained into an unlined cell, where it rapidly infiltrated into a shallow unconfined aquifer. Ground water downgradient from the second cell was elevated in dissolved solids, dissolved organic carbon, ammonium, iron, and other constituents (Fig. 10–9). Field-measured pe values were very low close to the cell and increased with distance from the impoundment, much like any organic-rich contaminant plume. One interesting aspect of this site was that a nitrate plume from an upgradient aerobic landfill appeared to mix with the reducing plume from the waste stabilization lagoon with the result that nitrate was reduced to ammonium (Bulger et al., 1989).

The treatment plant at Otis Air Force Base in Massachusetts is a well documented case of ground water contamination caused by infiltration of secondary-treated wastes from infiltration beds (LeBlanc, 1984; Barber, 1992). Secondary effluents at the treatment plant, which has been in operation since 1936, are discharged into 24, $\frac{1}{2}$-acre-size infiltration beds. A contaminant plume 4,000 m long, 1,000 m wide, and 30 m deep has developed downgradient from the infiltration beds. The plume can be delineated with numerous parameters, of which boron (Fig. 10–10) is particularly useful because it travels

FIGURE 10–9 Distribution of dissolved organic carbon at the McVille, North Dakota, waste stabilization lagoon site (from Bulger et al., 1989, reprinted by permission of the National Ground Water Association).

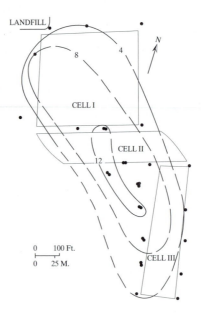

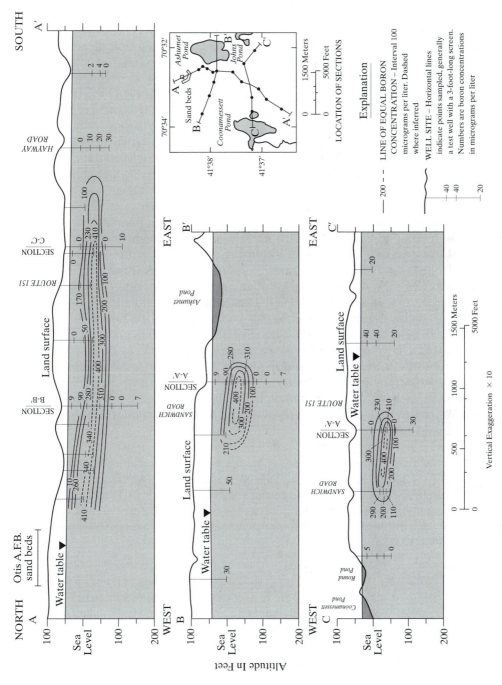

FIGURE 10–10 Vertical distribution of boron in ground water May, 1978, through May, 1979, at the Otis Air Force Base sewage plume, Cape Cod, Massachusetts (from LeBlanc, 1984).

329

conservatively and is not present in background ground water. Boron occurs in the plume because of the use of sodium perborate as bleach in detergent powders. In ground water, boron is present as orthoboric acid, $B(OH)_3$, which remains in the neutral acid molecular form because the pH of the plume is well below the pK_a of the acid. The plume can also be defined by specific conductance, chloride, and other parameters. Nitrogen is present mainly in the reduced, ammonium form in the plume. DOC remaining in secondary effluent is greatly reduced from raw sewage. Still, biodegradation of the greater-than-background DOC (2–5 mg/l) is high enough to create anoxic (denitrifying) conditions in the plume. Downgradient, mixing of the plume with oxic recharge water causes some nitrification of the ammonia, although nitrate-N concentrations are generally less than 5 mg/l. Sewage treatment effluents have relatively high concentrations of phosphorus, mainly in the form of orthophosphate (PO_4^{3-}). In ground water, phosphate is strongly adsorbed or precipitated as iron or aluminum solids of low solubility. As a result, phosphate is strongly retarded in the contaminant plume.

An interesting aspect of the Otis Air Force Base contaminant plume is the presence of detergent compounds derived from household use. Because of the test used to measure these substances, detergents are reported as MBAS (methylene blue active substances). These compounds consist of anionic surfactants that are very mobile in ground water. Detergent use in the United States dates from around 1946; in 1953 their use exceeded that of soaps. Prior to 1964, the most common surfactant used in detergents was alkyl benzene sulfonate (ABS). ABS is essentially non-biodegradable; and in 1964, it was replaced by the more biodegradable surfactant, linear alkyl sulfonate (LAS). The distribution of MBAS in the contaminant plume retains this history of detergent use. The highest concentrations of MBAS are present near the leading edge of the plume (Fig. 10–11). These higher concentrations reflect the presence of ABS, whereas the lower concentrations closer to the source indicate that LAS has been removed from the plume by biodegradation.

A wide range of synthetic volatile and semivolatile compounds were also measured in the plume. The most abundant among these were trichloroethene and tetrachloroethene (Barber, 1992). These and the other synthetic organics detected are derived from household cleaners and other products of various types. TCE and PCE concentrations are high enough in the plume to exceed common regulatory limits.

Septic Systems

Septic systems are the waste disposal method of choice for most areas in North America that lack sewers. It is estimated that one-third of the sewage effluent in the United States is disposed of in septic systems. In these systems, wastewater passes through a tank where solids are separated from liquid effluent by settling. The liquid effluent is discharged to perforated drain tiles that release wastewater to leach beds, where it infiltrates into the soil. Alternatively, perforated culverts called dry wells, installed vertically below the surface soils, are used instead of leach beds. The principle upon which the septic system is based is that percolation through the soil will remove contaminants from the wastewater. Unfortunately, it is abundantly clear that septic systems produce contaminant plumes in shallow, unconfined aquifers that can impact adjacent wells or surface water bodies.

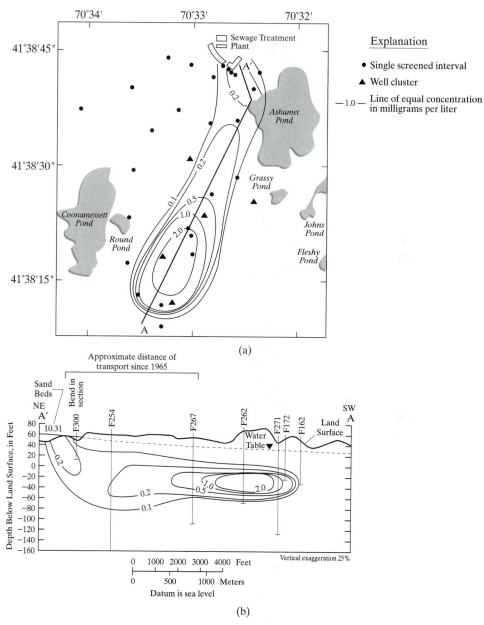

(a)

(b)

FIGURE 10–11   Plan view (a) and cross section (b) showing the distribution of anionic surfactants measured as methylene-blue-active substances (mg/l) in 1983. Also shown is the estimated transport distance since 1965 for the trailing edge of branched-chain alkyl benzene sulfonates (ABS) (reprinted from Barber, Hierarchial analytical approach to evaluating the transport and biogeochemical fate of organic compounds in sewage-contaminated ground water, Cape Cod, Massachusetts, in, Lesage, S. and Jackson, R. E., eds., *Ground Water Contamination and Analysis at Hazardous Waste Sites,* 1992, by Courtesy of Marcel Dekker, Inc.).

Detailed studies in recent years have focused on the hydrogeochemistry of septic system contaminant plumes (Harman et al., 1996; Robertson et al., 1991; Robertson et al., 1998; Tinker, 1991; Aravena and Roberston, 1998; Robertson, 1995; Robertson and Cherry, 1995). The compounds of greatest concern are nitrate and phosphate. Nitrate can cause the sometimes fatal disease methemoglobinemia in infants, which is caused by a decrease in the ability of the blood to carry oxygen. Nitrate is also a nutrient that contributes to eutrophication of water bodies into which the ground water may discharge. Phosphate, although less mobile than nitrate, is also a major cause of eutrophication. Transport of pathogenic organisms is also a concern in permeable aquifers.

Harman et al. (1996) studied a school septic system in Ontario located over a shallow, unconfined sand aquifer. In the septic tank, the wastewater effluent is a highly reducing solution with a high DOC and nitrogen mostly in the form of ammonium. The effluent is significantly altered in the vadose zone as it moves from the leach field to the water table. Oxidation, leading to a 90% decline in DOC and complete conversion of ammonium to nitrate, is the primary cause. Nitrate concentrations in the plume are shown in Fig. 10–12. Oxidation of the organic carbon produces CO_2, which would lower pH in the absence of carbonate minerals in the aquifer solids. Carbonate minerals, when present, dissolve to buffer the pH and increase alkalinity, calcium, and magnesium concentrations in the plume.

The mobility of phosphate was compared in ten septic system plumes in Ontario in variable hydrochemical environments (Robertson et al., 1998). Average PO_4^{2-}—P concentrations in the proximal plumes ranged from 0.03 to 4.9 mg/l, and the plume lengths ranged from 1 to 70 m. These results contradict the common assumption that phosphate is strongly adsorbed on aquifer solids and therefore does not present a problem in ground water. The results demonstrate that ground water transport of phosphate can be a major problem, particularly around small lakes that are fringed with houses with individual septic systems.

Robertson et al. (1998) concluded that phosphate is attenuated in the vadose zone by precipitation in minerals such as vivianite $[Fe_3(PO_4)_2 \cdot 8H_2O]$, strengite $[FePO_4 \cdot 2H_2O]$, and variscite $[AlPO_4 \cdot 2H_2O]$. The equilibrium phosphate concentrations are highly pH-dependent. At low pH values, typical of noncalcareous aquifers,

FIGURE 10–12   Nitrate concentrations shown in cross section along centerline of plume originating from school septic system in Ontario (from Harman et al., 1996, reprinted by permission of the National Ground Water Association).

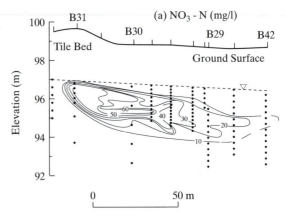

phosphate is maintained at very low levels by mineral solubility constraints. At moderate pH levels, however, which are attained when the aquifer contains carbonate minerals, phosphate concentrations can be much higher. Once the effluent reaches the water table, phosphate attenuation is controlled by adsorption to aquifer solids, especially metal oxides with positively charged surface sites (Chapter 4). Phosphate transport is retarded relative to the ground water flow velocity by a factor of about 20. Stable isotopes of nitrogen, carbon, oxgyen, and sulfur are useful in tracking septic system plumes and studying chemical transformations along the flow path (Aravena et al., 1993; Aravena and Robertson, 1998).

Relatively few studies have evaluated the threat of contamination by pathogenic bacteria and viruses from septic systems (Bitton and Gerba, 1984; Bales et al., 1995; Canter and Knox, 1985; Yates, 1985). Assays for the wide range of possible microorganisms are technically difficult to perform and expensive. Research has focused on determining the mobility of indicator organisms that might suggest the potential mobility of their pathogenic counterparts. Coliform bacteria have been used as bacterial indicator species, and human enteroviruses and coliphage, viruses that infect intestinal coliform bacteria, are used as proxy indicators of viral transport.

Deborde et al. (1998) addressed the fate and transport of viruses in a study of a high school septic system in Montana. The study involved monitoring of the septic tank and contaminant plume for human enteroviruses and coliphage, as well as injection of coliphage into the aquifer. Although enteroviruses were only rarely detected in either the septic tank or the aquifer, coliphage was consistently measured in monitor wells. Despite strong adsorption in the aquifer, detections of the virus were made in monitor wells more than 30 m from the injection wells. These results show that virus transport can potentially exceed minimum setback distances established for septic leach fields and water supply wells. The tremendous variability of aquifer properties does not yet permit accurate predictions of the transport of pathogenic organisms in all hydrogeologic settings.

Land Application of Sewage Effluents

Effluents and sludge from sewage treatment plants are commonly applied to the soil as a disposal method and to irrigate and/or fertilize crops. The effects on ground water chemistry are similar to those of septic systems except that much larger volumes of aquifer are impacted. The contaminant of greatest concern commonly is nitrate. Organic nitrogen or ammonia in the waste will be oxidized to nitrate if the site is underlain by an aerobic vadose zone. In the saturated zone, nitrate will travel rapidly as long as redox conditions remain oxic. Spalding et al. (1993) instrumented a site in Nebraska in which sewage sludge was applied to an irrigated cornfield. The sludge injection created a large nitrate plume downgradient from the fields (Fig. 10–13). NO_3—N concentrations above 10 mg/l extend to a depth of approximately 15 m, although a lens of fine-grained sediment may impede greater depth penetration. Nitrogen isotope analyses proved that the source of the nitrogen was animal waste.

Other changes in ground water chemistry result from the DOC derived from the waste. If significant DOC reaches the water table, oxygen consumption takes place. Anoxification was observed in a 30-m-deep aquifer in Israel below lands irrigated with

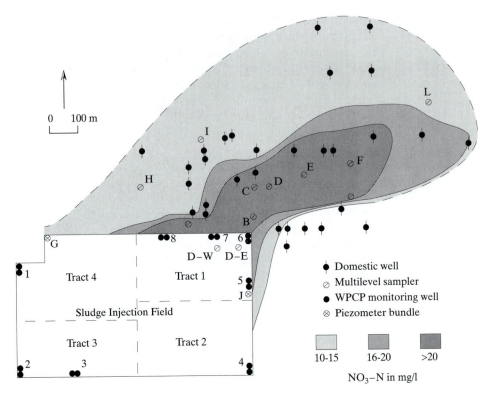

FIGURE 10–13 Plume of nitrate in ground water downgradient from a sludge injection field in Nebraska (reprinted from *Jour. Hydrol.*, v. 142, Spalding, R. F., Exner, M. E., Martin, G. E., and Snow, D. D., Effects of sludge disposal on ground water nitrate concentrations, pp. 213–228, Copyright 1993, with permission from Elsevier Science).

sewage effluent (Ronen et al., 1987). Transport of the organic carbon through the vadose zone takes as much as 15 years in this setting. At the Nebraska site described previously, DOC caused denitrification in deeper parts of the aquifer, where oxygen was not present (Spalding et al., 1993). Increases in other major ions in ground water generally accompany the increases in nitrate and DOC. Metals are concentrated in sewage sludges, but sorption and precipitation generally limit their mobility in the subsurface.

CONTAMINATION OF GROUND WATER BY AGRICULTURE AND OTHER NONPOINT SOURCES

The vast areas of the continents used for agricultural production constitute by far the largest source of nonpoint source contamination for shallow aquifers. Variability of agricultural practices leads to a wide range of ground water impacts. In general, these can be divided into contamination by nutrients, primarily nitrate, pesticides and herbicides, and, less commonly, pathogenic microorganisms, trace elements, and dissolved solids (salinization).

Nitrate Contamination

Contamination by nitrate derived from synthetic and organic fertilizers is the most widespread agricultural effect on ground water. Hallberg and Keeney (1993) summarize the scope of this problem in the United States, where row-crop agriculture is the major contributor to ground water nitrate contamination. U.S. crop production, including corn, cotton, soybeans, and wheat, occupies 70 to 80 million ha of land. The factors conducive to nitrate contamination include (1) the high nitrogen requirements of these crops, with the exception of soybeans, (2) the production of only one crop per year in many areas, leaving the soil susceptible to mineralization and leaching of nitrogen when plants are not available to take up the nitrogen, and (3) tillage activities that enhance mineralization of soil N. In oxic vadose zones, mineralization of organic N produces ammonium, which is subsequently nitrified to nitrate. Anionic nitrate is readily leached to the water table in permeable, oxic soils. The dominance of corn in the United States, about 25% of total U.S. cropland, elevates rural ground water nitrate contamination to a major national problem. In the midwestern U.S. corn belt, where domestic water supplies rely on individual wells, many of which are very shallow, elevated nitrate in domestic and municipal water supplies is the most serious ground water problem.

Nitrogen applications are divided between synthetic fertilizers and manure. $\delta^{15}N$ contents (Chapter 8) of nitrate in ground water are effective in distinguishing these two sources along with natural soil nitrogen. In addition, nitrogen isotopes are also used to determine if denitrification has been significant. Komor and Andersen (1993) utilized nitrogen isotopes for source identification in agricultural areas in Minnesota (Fig. 10–14). This technique worked reasonably well, although mixing of water from different sources and denitrification can cause some uncertainty in source identification. The magnitude and spatial distribution of nitrate in ground water in agricultural watersheds is a function of numerous hydrogeologic and crop-management factors. Crop type is linked directly to ground water impacts because of differing nitrogen-management practices. For example, Stephany et al. (1998) documented a rise in nitrate concentration from less than 5 mg/l NO_3-N to more than 20 mg/l at the water table beneath a field converted from alfalfa to manure-fertilized corn in a period of three months. The water table depth was about 15 m. Rapid movement through the vadose zone, even in clay-rich soils, is facilitated by the presence of macropores, (Iqbal and Krothe, 1995). Thus the vadose zone can be divided into slower matrix flow systems and rapid, macropore flow systems. Because of the chemical stability range of nitrate (Chapter 5), oxic conditions in the vadose and shallow saturated zones are required for the persistence of nitrate in ground water. Starr and Gilham (1993) showed that the depth to the water table is critical in maintaining oxic conditions and nitrate stability. In their study area in Ontario, where the depth to the water table was about 1 m, transport of the soil organic matter produced anoxic conditions at the water table. Nitrate is denitrified under these conditions. When the water table is deeper, soil organic matter is consumed by biodegradation in the vadose zone and denitrification at the water table is prevented.

Agricultural drainage in areas of high water table can short-circuit transport to the water table. Subsurface tile drains are installed at a depth of between 1 and 2 m so that crops can be planted earlier in the spring to maximize the growing season. This practice imposes a greater threat to surface water in drained watersheds because field

FIGURE 10–14 Histograms of $\delta^{15}N_{NO_3}$ for ground water samples from sand plain aquifers in central Minnesota. Ranges of $\delta^{15}N$ for possible sources are shown at top of figure (from Komor and Anderson, 1993, reprinted by permission of the National Ground Water Association).

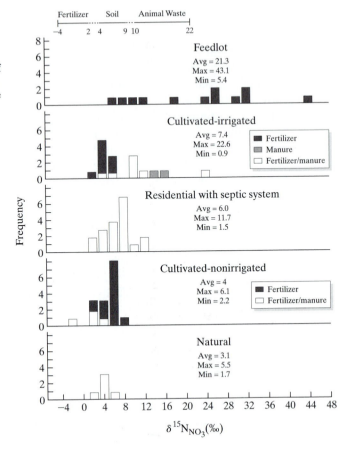

tiles discharge to surface drains and ultimately to streams. Thus, much of the nitrogen that would normally move slowly through the ground water flow system and perhaps be reduced by denitrification, is discharged directly to surface water bodies (Fenelon and Moore, 1998).

Irrigation may increase nitrogen losses to the water table. In humid, temperate areas, ground water recharge is limited to fall and spring, when precipitation is high and evapotranspiration is minimal. Crop irrigation maintains wetter soil conditions during the growing season. If the crop is over irrigated, or if heavy rainfall occurs when the soil is already wet from irrigation, nitrogen can be flushed below the root zone during a time when natural recharge would be rare.

Transport of nitrate in the aquifer, both laterally and vertically, depends upon maintenance of oxic conditions. In most aquifers, nitrate concentrations are restricted to shallow levels. For example, Hallberg (1986) found a strong inverse relationship between well depth and nitrate concentration (Fig. 10–15). Higher iron concentrations with depth in many aquifers indicate that nitrate would not be stable in that geochemical environment. Komor and Anderson (1993) demonstrated that $\delta^{15}N$ commonly

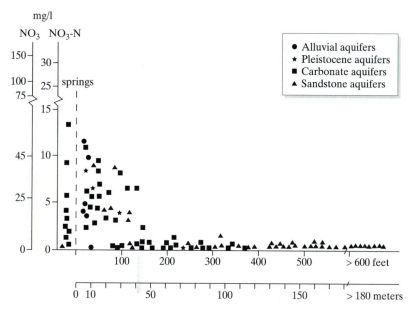

FIGURE 10–15 Nitrate concentration vs. well depth for wells in Iowa (from Hallberg, 1986, reprinted by permission of the National Ground Water Association).

increases with depth at a site, probably indicating progressive dentrification. Kehew et al. (1996) studied flow and chemistry in a thick aquifer in glacial drift in southern Michigan. Although there was no physical barrier to downward migration of nitrate in most areas, nitrate was limited to shallow flow cells that interact with lakes and wetlands. Surface water bodies in southern Michigan commonly have a flow-through relationship with the water table (Kehew et al., 1998). Nitrate is absent from long, narrow zones downgradient from surface water bodies, even when nitrate loading is heavy at the land surface, because of plumes of reducing recharge from the lakes or wetlands traveling downgradient in the upper part of the aquifer.

Pesticides

Pesticides encompass a wide range of compounds applied to plants or soil to control unwanted organisms. More specifically, pesticides can be grouped into insecticides, herbicides, rodenticides, and fungicides. The most common synthetic organic pesticides can be classified as *organochlorines, organophosphates, and carbamates*. Many of the important herbicides are halogenated compounds.

The worldwide concern with the environmental effects of pesticides initially focused on organochlorine compounds such as DDT (dichlorodiphenyltrichloroethane), which were applied to eradicate insect-borne diseases such as malaria. DDT, which is not particularly toxic to humans, is persistent and accumulates in the fatty tissue of organisms. It also tends to bioconcentrate, which means its concentrations increase in organisms higher in the food chain. Its interference in the reproductive systems of birds led to its ban several decades ago. Organochlorines also produce cancer and birth defects in mice and hamsters. Organophosphates such as parathion, malathion, and diazinon have largely replaced the organochlorines. Although the organophosphates are much less persistent in the environment, they are more toxic than the organochlorines, causing damage to the nervous system of humans upon acute exposure. Carbamate pesticides such as aldicarb also produce severe heath effects at high exposure levels.

In the late 1970s, when ground water was analyzed for a wide variety of compounds under CERCLA, RCRA, and other programs, pesticides and herbicides such as aldicarb, DBCP (dibromochloropropane), and EDB (ethylene dibromide) were discovered to be present. Extensive monitoring programs have since been implemented to assess the distribution of pesticides and their metabolites in ground water. In the USEPA National Pesticide Survey, for example, 1,300 community and domestic water wells were sampled for 126 pesticides and pesticide metabolites (Rao and Alley, 1993). This study and others like it have generally concluded that roughly 10% of wells in areas of pesticide application are likely to have detectable levels of 1 or more compounds. Exceedences of health advisory levels (HALs) or maximum contaminant levels (MCLs) are commonly less than one percent of wells sampled.

Leaching of pesticide compounds to ground water is basically a function of the sorptive properties of the compound in the soil in which it is applied and the rate of degradation of the compound. Sorptive properties are assessed by K_{oc} values of the compound in combination with the fraction of organic carbon in the soil. A pesticide that is highly sorbed will be more likely to be degraded prior to reaching ground water. K_{oc} values vary by several orders of magnitude. For example, atrazine, the most common pre-emergent herbicide applied in the midwestern corn belt of the United States, degrades by both abiotic and biotic processes. Biodegradation by dealkylation is a common reaction in which deethylatrazine and deiospropylatrazine are the metabolites (Fig. 10–16). Dechlorination and hydroxylation reactions are also possible (Chapelle, 1993). Degradation rates, by both biotic and abiotic processes, are also highly variable. Half-lives are calculated based on the assumption of pseudo first-order degradation reaction mechanisms.

FIGURE 10–16 Structures of atrazine and its major metabolites.

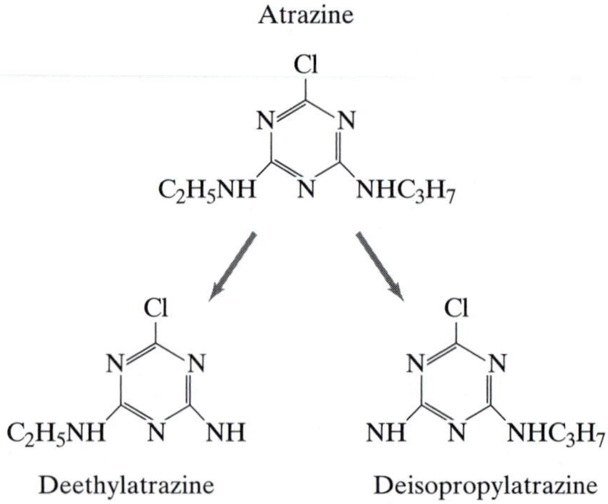

Atrazine

Deethylatrazine

Deisopropylatrazine

With reasonable assumptions of soil properties and half-lives, the vulnerability of aquifers to pesticide contamination can be assessed. One common problem with models of pesticide mobility is that preferential flow through macropores provides a short circuit to the aquifer through the vadose zone.

Concentrations of pesticides are typically higher in surface water than in ground water in agricultural watersheds as a result of runoff from fields carrying the compounds sorbed to soil particles or in dissolved form. During periods of heavy application or runoff, stream concentrations can be significantly elevated in pesticides. Although most streams in temperate areas are ground water discharge areas, construction of well fields adjacent to rivers creates conditions conducive to induced recharge of surface water. Pesticides can be transported through alluvial aquifers to municipal wells under the strong gradients induced by pumpage of large-capacity wells. In Lincoln, Nebraska, peaks of elevated concentrations of atrazine in wells correlated with atrazine peaks in the river (Duncan et al., 1991; Blum et al., 1993), although the ground water peaks were delayed relative to the river peaks (Fig. 10–17). Ground water concentrations peaked later because induced recharge reached its maximum levels in the summer, several months after the river peaks recorded during the early part of the growing season.

Although pesticide concentrations in wells are commonly very low, high concentrations have been detected in some areas. A good example of pesticide mobility is provided by a study of ethylene dibromide (EDB) in Florida (Katz, 1993). One of the two study sites, the Lake Pierce area, is shown in Figs. 10–18 and 10–19. EDB, which has the IUPAC name 1, 2-dibromoethane, was applied as a soil fumigant to control the burrowing nematode in commercial citrus groves. In one method of application, the push-and-treat method, infested trees were bulldozed into a concentrated area and burned. EDB was then applied to the plot in troughs. The soil was then tamped down over the trenches. EDB was also applied in buffer strips around groves in order to prevent the intrusion of the nematode.

Once in the soil, EDB is subject to various physical, chemical, and biochemical processes (Fig. 10–20). Because of the relatively high solubility (4,250 mg/l) and low Henry's constant (0.033) of EDB, it strongly partitions into the aqueous phase. Sorption is limited in the study area by the low organic carbon content of the aquifer materials. Half-times for chemical hydrolysis of EDB range from 1.5 to 15 years. Biodegradation half-times, 35 to 350 days, are significantly shorter. These properties are conducive to significant migration in ground water. In the Lake Pierce study area, 103 wells out of 290 had detections of EDB above the analytical detection limit of 0.02 μg/l. The detections had a mean of 2.74 μg/l, a median of 0.09 μg/l, and a maximum of 73 μg/l. EDB, a potent carcinogen and mutagen, has a maximum contaminant level (MCL) of 0.02 μg/l in the state of Florida. All agricultural uses of EDB were banned in 1983 by the U.S. EPA.

Irrigation

Intensive irrigation of agricultural lands, particularly in arid areas, can cause degradation of shallow ground water quality by evaporative concentration of dissolved solids. If the shallow ground water discharges to surface water, the contaminated water can negatively impact the water chemistry of surface water bodies. A case in point is the

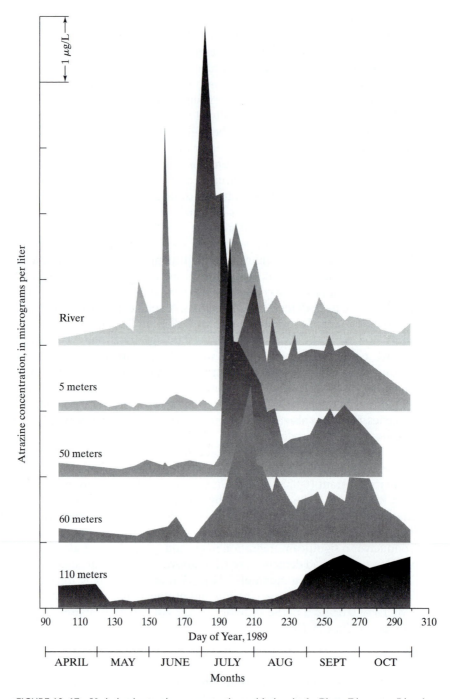

FIGURE 10–17 Variation in atrazine concentrations with time in the Platte River near Lincoln, Nebraska, and in wells in a transect from the river (from Duncan, 1991, reprinted by permission of the National Ground Water Association).

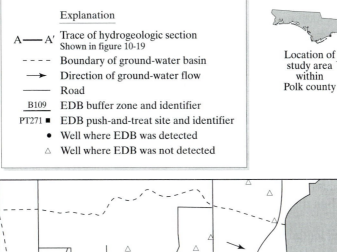

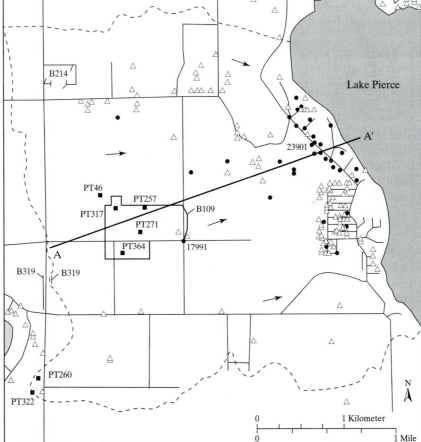

FIGURE 10–18   Lake Pierce study site showing buffer zones, push-and-treat sites, and wells where EDB was detected and not detected. Cross section A–A′ shown in Fig. 8–19 (from Katz, 1993).

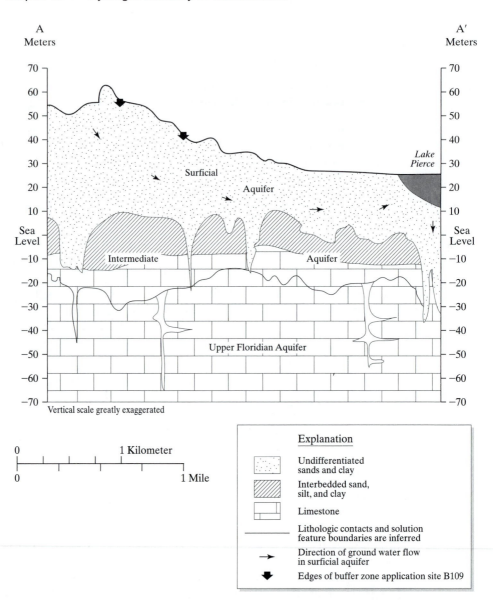

FIGURE 10–19   Generalized hydrogeologic cross section of the Lake Pierce study area (from Katz, 1993).

central San Joaquin Valley of California (Fig. 10–21). Dubrovsky et al. (1993) summarized the problems that have resulted from irrigation in this area over the past several decades. Large-scale irrigation in the valley began in the 1950s. The water table was naturally high near the center of the valley, which is bounded by the Coast Ranges on the west and the Sierra Nevada on the east. Because of the arid climate, direct evaporation from the water table in areas of high water table concentrated salts from ground water in the soils in these ground water discharge areas. With the advent of irrigation, water tables

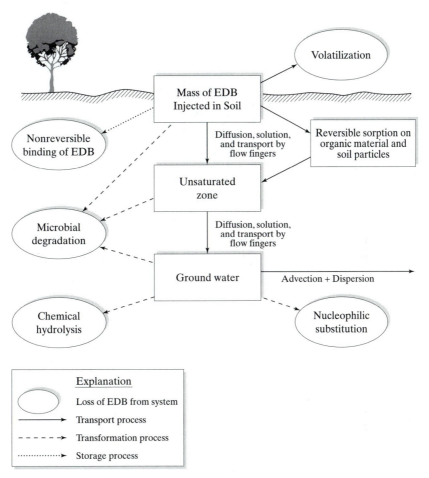

FIGURE 10–20   Processes controlling the fate and transport of EDB in the subsurface (from Katz, 1993).

rose and evaporative concentration of dissolved solids increased. In addition, the increased recharge from irrigation flushed the salts downward to the water table. The higher water table necessitated the installation of drains to prevent waterlogging of the soils and allowed the continued use of the land for irrigation. Agricultural drains include open ditches excavated down to the elevation of the water table or buried perforated pipes in a network beneath the fields that drain to natural streams or drainage ditches. Some of the drainage water was eventually routed to reservoirs in the center of the valley.

The main problem with the preceding scenario in the San Joaquin Valley was the abundance of selenium in sediments deposited in alluvial fans adjacent to the Coast Ranges on the western side of the valley. Selenium, along with other dissolved solids, was concentrated by evaporation to levels far above natural concentrations. Concentrations in drain tile effluent exceeded 100 μg/l. The drinking water standard for selenium is 10 μg/l, and the four-day-average criterion for aquatic life is 5 μg/l. The high levels of

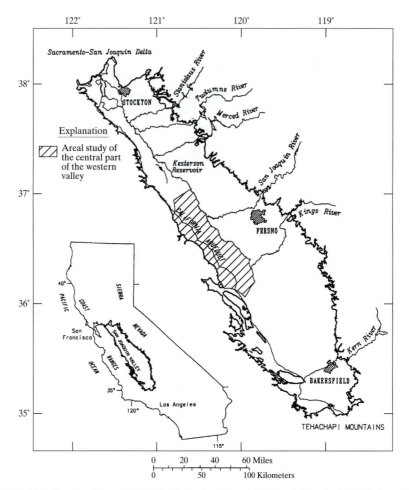

FIGURE 10–21 Location of study sites in the San Joaquin Valley, California (from Dubrovsky et al., 1993. Multi scale approach to regional ground water quality assessment: selenium in the San Joaquin Valley, California, in W. M. Alley, ed., *Regional Ground Water Quality*. Van Nostrand Reinhold. Copyright 1993 by John Wiley & Sons, Inc. Reprinted by permission of John Wiley & Sons, Inc.).

selenium in drainage water proved to be extremely toxic to birds that drank water from Kesterson reservoir, one of the surface water bodies that received drainage from irrigated farm lands. The ecological disaster at Kesterson Reservoir attracted national attention.

Subsequent ground water studies indicated elevated concentrations of selenium in shallow ground water near the eastern boundary of the alluvial fan sediments (Fig. 10–22). Oxidizing conditions, in which selenium exists in the form of the selenate (SeO_4^{2-}) or selenite (SeO_3^{2-}) ions, are necessary for mobility in ground water, although the mobility of the selenite species is restricted by adsorption. With increasing depth below the water table, reducing conditions prevail and selenium exists as elemental selenium (Se^0) or selenide (Se^{2-}). Selenium is immobile under these conditions because of solubility constraints.

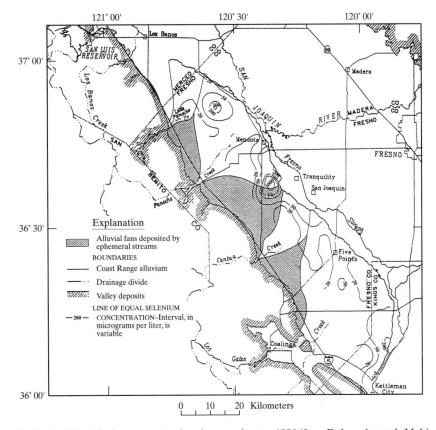

FIGURE 10–22   Selenium concentrations in ground water, 1986 (from Dubrovsky et al., Multi scale approach to regional ground water quality assessment: selenium in the San Joaquin Valley, California, in W. M. Alley, ed., *Regional Ground Water Quality*, Van Nostrand Reinhold. Copyright 1993 by John Wiley & Sons, Inc. Reprinted by permission of John Wiley & Sons, Inc.).

Other Nonpoint Problems

Other land use activities impact ground water quality in many areas. One such activity is road-salt application in areas with cold winters. The solutions used include sodium and calcium chloride. Runoff from road surfaces provides a pathway to surface waters and for infiltration to ground water. In high-population areas in which ground water is used for water supply, chloride concentrations gradually increase in municipal wells that are poorly protected from surface contamination (Fig. 10–23). In extreme cases, high chloride could necessitate well abandonment. Discharge of the impacted ground water to streams and lakes leads to degradation of those water bodies.

In areas with multiple aquifers, identification of the sources of high-chloride water may be difficult. In southern Ontario, Howard and Beck (1993) used trace element concentrations for source identification. For example, a plot of chloride vs. iodide (Fig. 10–24) shows separation of bedrock formation waters with high iodide and other potential sources from road salt sources.

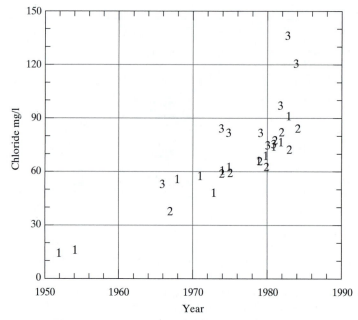

FIGURE 10–23   Changes in chloride concentrations in three municipal wells in Kalamazoo, Michigan. Road salt applications increased significantly during this period. (Kehew, unpublished data.)

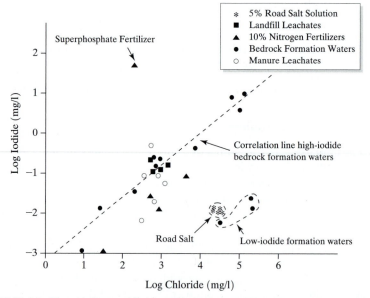

FIGURE 10–24   Plot of iodine vs. chloride concentrations to differentiate road-salt waters from other water types (reprinted from *Jour. Contam. Hydrol.*, v. 12, Howard, K. W. F., and Beck, P. J., Hydrogeochemical implications of ground water contamination by road salt de-icing chemicals, pp. 245–268, Copyright 1993, with permission of Elsevier Science).

PROBLEMS

1. What is the "dry-tomb" approach, and what alternative can be used for landfill design?

2. Describe the changes in leachate chemistry associated with the three stages of leachate generation.

3. What is the composition of alkalinity in landfill leachate? Do alkalinity values mean the same as in uncontaminated waters?

4. Discuss the redox zonation that develops downgradient from landfills in ground water contaminant plumes.

5. What possible factors control the mobility of organic compounds used in wood treatment in ground water flow systems?

6. In Henderson's (1994) study, how was the partitioning of Cr^{+6} and Cr^{+3} species determined? What did the study suggest about the persistence of chromium in ground water at the site?

7. What are the differences between primary, secondary, and tertiary treatment of waste water?

8. What compounds and processes characterize the occurrence and distribution of detergents in the Otis Air Force Base plume?

9. Discuss the hydrochemistry of septic system contaminant plumes from vadose zone source to the leading edge.

10. What are the sources of nitrate contaminants in ground water resulting from agricultural systems? How can these sources be differentiated?

11. What physical, chemical, and biochemical properties of pesticides and herbicides control their mobility in ground water? Give examples.

REFERENCES

Alexander, M., *Biodegradation and Bioremediation*. 1994. San Diego, Academic Press.

Amundson, R.G., Chadwick, O.A., and Sowers, J.M. 1989. A comparison of soil climate and biological activity along an elevation gradient in the eastern Mojave Desert. *Oceologic* 80, 395–400.

Amrhein C. and Suarez, D.L. 1988. The use of a surface complexation model to describe the kinetics of ligand-promoted dissolution of anorthite. *Geochim. Cosmochim. Acta* 52, 2785–2793.

Antweiler, R.C. and Drever, J.I. 1983. The weathering of a late Tertiary volcanic ash: Importance of organic solutes. *Geochim. Cosmochim. Acta* 47, 623–629.

Appelo, C.A.J. and Postma, D., 1993. *Geochemistry, groundwater and pollution*. A.A. Balkema, Rotterdam, 536.

April, R., Newton, R., and Coles, L.T. 1986. Chemical weathering in two Adirondack watersheds: Past and present-day rates. *Geol. Soc. Amer. Bull.* 97, 1232–1238.

Aravena, R. and Robertson, W.D. 1998. Use of multiple isotope tracers to evaluate denitrification in ground water: Study of nitrate from a large-flux septic system plume. *Ground Water* 36, 975–982.

Arndt, J.L. and Richardson, J.L. 1993. Temporal variations in the salinity of shallow groundwater from the periphery of some North Dakota wetlands (USA). *Jour. Hydrol.* 141, 75–105.

Back, W. and Hanshaw, B.B. 1970. Comparison of chemical hydrogeology of the carbonate peninsulas of Florida and Yucatan. *Jour. Hydrol.* 10, 330–368.

Baedecker, M.J. and Apgar, M.A. 1984. Hydrogeological studies at a landfill in Delaware, in National Research Council, *Groundwater Contamination*. Washington, DC: National Academy Press, 127–138.

Baedecker, M.J. and Back, W. 1979a. Hydrogeological processes and chemical reactions at a landfill. *Ground Water* 17, 429–437.

Baedecker, M.J. and Back, W. 1979b. Modern marine sediments as a natural analog to the chemically stressed environment of a landfill. *Jour. Hydrol.* 43, 393–414.

Baedecker, M.J., Cozzarelli, I.M., Eganhouse, R.P., Siegel, D.I., and Bennett, P.C. 1993. Crude oil in a shallow sand and gravel aquifer—III. Biogeochemical reactions and mass balance modeling in anoxic groundwater. *Appl. Geochem.* 8, 569–586.

Bales, R.C., Li, S., Maguire, K.M., Yahya, M.T., Gerba, C.P., and Harvey, R.W. 1995. Virus and bacteria transport in a sandy aquifer, Cape Cod, Massachusetts. *Ground Water* 33, 653–661.

Barber, L.B. II. 1992. Hierarchical analytical approach to evaluating the transport and biogeochemical fate of organic compounds in sewage-contaminated groundwater, Cape Cod, Massachusetts, in Lesage, S. and Jackson, R.E. (eds.), *Groundwater Contamination and Analysis at Hazardous Waste Sites*. New York: Marcel Dekker, Inc., 73–120.

Barcelona, M.J. and Holm, T.R. 1991. Oxidation–reduction capabilities of aquifer solids. *Environ. Sci. Technol.* 25, 1565–1572.

Barcelona, M.J., Holm, T.R., Schock, M.R., and George, G.K. 1989. Spatial and temporal gradients in aquifer oxidation–reduction conditions. *Water Resour. Res.* 25, 991–1003.

Barker, J.F., Patrick, G.C., and Major, D. 1987. Natural attenuation of aromatic hydrocarbons in a shallow sand aquifer. *Ground Water Monit. Rev.* 7, 64–71.

Bennett, P.C. and Casey, W., 1993. Chemistry and mechanisms of low-temperature dissolution of silicates by organic acids. In, E.D. Pittman and M.D. Lewan (eds.), *Organic Acids in Geological Processes.* Springer-Verlag, Berlin, 162–200.

Bennett, P.C., Melcer, M.E., Siegel, D.I., and Hassett, J.P. 1988. The dissolution of quartz in dilute aqueous solutions of organic acids at 250C. *Geochim. Cosmochim. Acta* 52, 1521–1530.

Bennett, P.C. and Siegel, D.I. 1987. Increased solubility of quartz in water due to complexation by dissolved organic compounds. *Nature* 326, 684–687.

Berner, R.A. 1975. The role of magnesium in the crystal growth of calcite and aragonite from sea water. *Geochim. Cosmochim. Acta* 39, 489–504.

Berner, R.A. and Morse, J.W. 1974. Dissolution kinetics of calcium carbonate in sea water: IV. Theory of calcite dissolution. *Am. J. Sci.* 274, 108–134.

Bishop, P.K., Lerner, D.N., Jakobsen, R., Gosk, D., Burston, M.W., and Chen, T. 1993. Investigation of a solvent polluted sequence in the UK. Part 2. Contaminant sources, distributions, transport and retardation. *Jour. Hydrol.* 149, 231–256.

Bitton, B. and Gerba, C.P., eds. 1984. *Groundwater Pollution Microbiology.* New York: Wiley.

Blum, D.A., Carr, J.D., Davis, R.K., and Peterson, D.T. 1993. Atrazine in a stream-aquifer system: Transport of atrazine and its environmental impact near Ashland, Nebraska. *Ground Water Mon. Rev.* 13, 125–133.

Borden, R.C., Daniel, R.A., LeBrun, L.E., IV, and Davis, C.W. 1997a. Intrinsic biodegradation of MTBE and BTEX in a gasoline contaminated aquifer. *Water Resour. Res.* 33, 1105–1115.

Borden, R.C., Gomez, C.A., and Becker, M.T. 1995. Geochemical indicators of intrinsic bioremediation. *Ground Water* 33, 180–188.

Borden, R.C., Hunt, M.J., Shafer, M.B., and Barlaz, M.A. 1997b. Anaerobic biodegradation of BTEX in aquifer material. U.S. EPA Environmental Research Brief. EPA/600/S-97/003.

Bouwer, E.J. 1994. Bioremediation of chlorinated solvents using alternate electron acceptors, in Norris, R.D., et al. *Handbook of Bioremediation.* Boca Raton, FL: CRC Press, Inc., 149–176.

Bradbury, K.R. 1984. Major ion and isotope geochemistry of ground water in clayey till, northwestern Wisconsin, USA. In, *Proceedings of the First Canadian/American Conference on Hydrogeology,* 284–289. Banff, Alta. Alberta Res. Council and Nat. Ground Water Assoc.

Bradbury, K.R. 1991. Tritium as an indicator of ground water age in central Wisconsin. *Ground Water* 29, 398–404.

Bradley, P.M. and Chapelle, F.H. 1996. Anaerobic mineralization of vinyl chloride in Fe(III)-reducing, aquifer sediments. *Environ. Sci. Technol.* 30, 2084–2086.

Brown, W.H., 1982. *Introduction to Organic Chemistry,* 3d ed. Boston: Willard Grant Press, 500 p.

Bulger, P.R., Kehew, A.E., and Nelson, R.A. 1989. Dissimilatory nitrate reduction in a waste-water contaminated aquifer. *Ground Water* 27, 664–671.

Burchett, C.R., Rettman, P.L. and Boning, C.W. 1986. *The Edwards Aquifer: Extremely Productive, but....* San Antonio, TX; The Edwards Underground Water District.

Canter, L.W. and Knox, R.C. 1985. *Septic System Effects on Ground Water Quality.* Chelsea, Michigan: Lewis Publishers, Inc.

Chapelle, F.H. 1993. *Ground-Water Microbiology & Geochemistry*. New York: John Wiley & Sons, Inc., 424 p.

Chapelle, F.H., McMahon, P.B., Dubrovsky, N.M., Fujii, R.F., Oaksford, E.T., and Vroblesky, D.A. 1995. Deducing the distribution of terminal electron-accepting processes in hydrologically diverse groundwater systems. *Water Resour. Res.* 31, 359–371.

Chebotarev, I.I. 1955. Metamorphism of natural waters in the crust of weathering. *Geochim. Cosmochim. Acta* 8., 22–48, 137–170, 198–212.

Chian, E.S.K. and DeWalle, F.B. 1976. Sanitary landfill leachates and their treatment. *Jour. Envr. Engr. Div., ASCE* 103 (EE2), 411–431.

Chian, E.S.K. and DeWalle, F.B. 1977. *Evaluation of Leachate Treatment, V. 1, Characterization of Leachate*. EPA-600/2–77–186a, Cincinnati: U.S. EPA.

Chiou, C.T., Schmedding, D.W. and Manes, M. 1982. Partitioning of organic compounds in octanol-water systems: *Environ. Sci. Technol.* 16, 4–10.

Clark, I.D. and Fritz, P. 1997. *Environmental Isotopes in Hydrogeology*. Boca Raton, FL: CRC Press.

Cozzarelli, I.M., Baedecker, M.J., Eganhouse, R.P., and Goerlitz, D.F. 1994. The geochemical evolution of low-molecular weight organic acids derived from the degradation of petroleum contaminants in groundwater. *Geochim. Cosmochim. Acta* 58, 863–877.

Cozzarelli, I.M., Eganhouse, R.P., and Baedecker, M.J. 1990. Transformation of monoaromatic hydrocarbons to organic acids in anoxic groundwater environment. *Environ. Geol Water Sci.* 16, 135–141.

Craig, H. 1961. Isotopic variations in meteoric waters. *Science* 133, 1702–1703.

Criddle, C.S., DeWill, J.T., and McCarty, P.L. 1990. Transformation of carbon tetrachloride by *Pseudomonas* sp. strain KC under denitrification conditions. *Applied and Environ. Microbiol.* 56, 1966–2000.

Dance, J.T. and Reardon, E.J. 1983. Migration of contaminants in groundwater at a landfill: A case study; 5: Cation migration in the dispersion test. *Jour. Hydrol.* 63, 109–130.

Daniels, D.P., Fritz, S.J., and Leap, D.I. 1991. Estimating recharge rates through unsaturated glacial till by tritium tracing. *Ground Water* 29, 26–34.

Davis, A., Kempton, J.H., Nicholson, A., and Yare, B. 1994. Groundwater transport of arsenic and chromium at a historical tannery, Woburn, Massachusetts, U.S.A. *App. Geochem.* 9, 569–582.

Davis, E.L. 1997. How heat can enhance in-situ soil and aquifer remediation: Important chemical properties and guidance on choosing the appropriate technique. U.S. EPA Ground Water Issue, EPA/540/S-97/502 18.

Davis, J.W., Klier, N.J., and Carpenter, C.L. 1994. Natural biological attenuation of benzene in ground water beneath a manufacturing facility. *Ground Water* 32, 215–226.

Davis, S.N. and De Wiest, R.J.M. 1966. *Hydrogeology*. New York: John Wiley & Sons.

DeBorde, D.C., Woessner, W.W., Lauerman, B., and Ball, P.N. 1998. Virus occurrence and transport in a school septic system and unconfined aquifer. *Ground Water* 36, 825–834.

De Jong, E. and Schappert, H.J.V. 1972. Calculation of soil respiration and activity from CO_2 profiles in the soil. *Soil Science* 113, 328–333.

Delcore, M.R. and Larson, G.J. 1987. Application of the tritium interface method for determining recharge rates to unconfined drift aquifers. II. Non-homogeneous case. *Jour. Hydrol.* 91, 73–81.

Desaulniers, D.E., Cherry, J.A., and Fritz, P. 1981. Origin, age and movement of porewater in argillaceous Quaternary deposits at four sites in southwestern Ontario. *Jour. Hydrol.* 50, 231–257.

Domenico, P.A. and Schwartz, F.W. 1998. *Physical and Chemical Hydrogeology*, 2d ed. New York: John Wiley & Sons.

Donovan, J.J. and Rose, A.W. 1994. Geochemical evolution of lacustrine brines from variable-scale groundwater circulation. *Jour. Hydrol.* 154, 35–62.

Dragun, J. 1988. *The Soil Chemistry of Hazardous Materials*. Silver Spring, MD: Hazardous Materials Control Research Institute, 458 p.

Drever, J.I. 1997. *The Geochemistry of Natural Waters*, 3d ed. Upper Saddle River, NJ: Prentice-Hall, Inc., 436 p.

Dubrovsky, N.M., Deverel, S.J., and Gilliom, R.J. 1993. Multiscale approach to regional ground-water-quality assessment: Selenium in the San Joaquin Valley, California, in W.M. Alley, ed., *Regional Ground-Water Quality*. New York: Van Nostrand Reinhold, 537–562.

Duncan, D., Peterson, D.T., Shepard, T.R., and Carr, J.D. 1991. Atrazine used as a tracer of induced recharge. *Ground Water Mon. Rev.* 11, 144–150.

Eganhouse, R.P., Baedecker, M.J., Cozzarelli, I.M., Aiken, G.R., Thorn, K.A., and Dorsey, T.F. 1993. Crude oil in a shallow sand and gravel aquifer—II. Organic geochemistry. *Appl. Geochem.* 8, 551–567.

Egboka, B.C.E., Cherry, J.A., Farvolden, R.N., and Frind, E.O. 1983. Migration of contaminants in groundwater at a landfill: A case study. 3. Tritium as an indicator of dispersion and recharge. *Jour. Hydrol.* 63, 51–80.

Feenstra, S., Mackay, D.M., and Cherry, J.A. 1991. A method for assessing residual NAPL based on organic chemical concentrations in soil samples. *Ground Water Monit. Rev.* 11, 128–136.

Fetter, C.W., 1993. *Contaminant Hydrogeology*. New York: Macmillan, 458.

Flyvbjerg, J., Arvin, E., Jensen, B.K., and Olsen, S.K. 1993. Microbial degradation of phenols and aromatic hydrocarbons in creosote-contaminated groundwater under nitrate-reducing conditions. *Jour. Contam. Hydrol.* v. 12, 133–150.

Foster, M.D. 1950. The origin of high sodium bicarbonate waters in the Atlantic and gulf coastal plains. *Geochim. Cosmochim. Acta* 1, 33–48.

Freeze, R.A. and Cherry, J.A. 1979. *Groundwater*. Englewood Cliffs, NJ: Prentice-Hall, Inc., 604.

Freeze, R.A. and Witherspoon, P.A. 1966. Theoretical analysis of regional groundwater flow: 1. Analytical and numberical solutions to the mathematical model. *Water Resour. Res.* 3, 641–656.

Freeze, R.A. and Witherspoon, P.A. 1967. Theoretical analysis of regional groundwater flow: 2. Effect of water table configuration and subsurface permeability variation. *Water Resour. Res.* 3, 623–634.

Freeze, R.A. and Witherspoon, P.A. 1966. Theoretical analysis of regional groundwater flow: 3. Quantitative interpretations. *Water Resour. Res.* 4, 581–590.

Fritz, P., Cherry, J.A., Weyer, K.U., and Sklash, M. 1976. Storm runoff analyses using environmental isotopes and major ions, in: *Interpretation of Envrional Isotope and Hydrochemical Data in Groundwater Hydrology*. Workshop proceedings, Vienna: IAEA, 111–130.

Fritz, S.J. 1994. A survey of charge-balance errors on published analyses of potable ground and surface waters. *Ground Water* 32, 539–546.

Garrels, R.M. 1967. Genesis of some ground waters from igneous rocks, in Abelson, P.H., ed., *Researches in Geochemistry* 2, 405–420. New York: John Wiley & Sons.

Garrels, R.M. and Mackenzie, F.T. 1967. Origin of the chemical compositions of some springs and lakes, in Stumm, W., ed., *Equilibrium concepts in natural water systems*. Adv. in Chem. Series 67, Am. Chem. Soc., 222–242.

Gerla, P.J. 1992. Pathline and geochemical evolution of ground water in a regional discharge area, Red River Valley, North Dakota. *Ground Water* 30, 743–754.

Godsy, E.M., Georlitz, D.F., and Grbić-Galić, D., 1992. Methanogenic biodegradation of creosote contaminants in natural and simulated ground-water ecosystems. *Ground Water* 30, 232–242.

Goerlitz, D.F. 1992. A review of studies of contaminated groundwater conducted by the U.S. Geological Survey Organics Project, Menlo Park, California, 1961–1990, in Lesage, S. and Jackson, R.E., eds., *Groundwater Contamination and Analysis at Hazardous Waste Sites*. New York: Marcel Dekker, Inc., 295–355.

Goldich, S.S. 1938. A study in rock weathering. *J. Geol.* 46, 17–58.

Gonfiantini, R., Dincer, T., and Derekoy, A.M. 1974. Environmental isotope hydrology in the Hodna region, Algeria, in *Isotope Techniques in Groundwater Hydrology,* V. 1 Vienna: IAEA, 293–314.

Gonfiantini, R., Gallo, G., Payne, B.R., and Taylor, C.B. 1976. Environmental Isotopes and Hydrogeochemistry in Groundwater of Gran Canaria, in *Interpretation of Environmental Isotopes and Hydrochemical Data in Ground Water Hydrology,* Vienna: IAEA, 159–170.

Griffin, R.A, Cartwright, K., Shimp, N.F., Steele, J.D., Ruch, R.R, White, W.A., Hughes, G.M., and Gilkeson, R.H. 1976. *Attenuation of Pollutants in Municipal Landfill Leachate by Clay Minerals: Part I—Column Leaching and Field Verification.* Illinois State Geological Survey, Environmental Geology Notes No. 78, 34.

Grim, R.E. 1968. *Clay Mineralogy*, 2d ed. New York: McGraw-Hill.

Haag, F.M., Reinhard, M., and McCarty, P.L. 1991. Degradation of toluene and *p*-xylene in an anaerobic microcosm: Evidence for sulfate as a terminal electron acceptor. *Environ. and Toxicolog. Chem.* 10, 1379–1389.

Hackley, K.C., Liu, C.L., and Coleman, D.D. 1996. Environmental isotope characteristics of landfill leachates and gases. *Ground Water* 34, 827–836.

Hallberg, G.R. 1986. Overview of agricultural chemicals in ground water, in *Agricultural Impacts on Ground Water*, 1–63. Worthington, OH: National Water Well Association.

Hallberg, G.R. and Keeney, D.R. Nitrate, in W.M. Alley, ed. *Regional Ground-Water Quality*. New York: Van Nostrand Reinhold, 297–322.

Harman, J., Robertson, W.D., Cherry, J.A., and Zanini, L. Impacts on a sand aquifer from an old septic system: Nitrate and phosphate. *Ground Water* 34, 1105–1114.

Hayashi, M., van der Kamp, G., and Rudolph, D.L. 1998. Water and solute transfer between a prairie wetland and adjacent uplands: 2. Chloride cycle. *Jour. Hydrol.* 207, 56–67.

Hendry, J., 1989, Short course notes from "Applications of Environmental Isotopes to Practical Ground Water Solutions," 6: 76. Westerville, OH, National Ground Water Association.

Hendry, M.J. 1983. Groundwater recharge through a heavy-textured soil. *Jour. Hydrol.* 63, 201–209.

Hendry, M.J. 1988. Hydrogeology of clay till in a prairie region of Canada. *Ground Water* 26, 607–614.

Hendry, M.J. and Buckland, G.D. 1990. Causes of soil salinization: 1. A basin in southern Alberta, Canada. *Ground Water* 28, 385–393.

Hendry, M.J., McCready, R.G.L., and Gould, W.D. 1984. Distribution, source and evolution of nitrate in a glacial till of southern Alberta, Canada. *Jour. Hydrol.* 70, 177–178.

Hendry, M.J. and Schwartz, F.W. 1990. The chemical evolution of ground water in the Milk River aquifer. *Ground Water* 28, 253–261.

Hem, J.D. 1985. *Study and Interpretation of the Chemical Characteristics of Natural Water*, 3d ed., U.S. Geological Survey Water Supply Paper 2254.

Herman, J.S., 1982. The dissolution kinetics of calcite, dolomite, and dolomitic rocks in the CO_2–water system. The Pennsylvania State University, Ph.D. Dissertation, 214 p.

Heron, G. and Christensen, T.H. 1994. Impact of sediment-bound iron on redox buffering in a landfill leachate polluted aquifer (Vejen, Denmark). *Environ. Sci. Technol.* 29, 187–192.

Hess, J.W. and White, W.B. 1989. Chemical hydrology, in White, W.B. and White, E.L., eds., *Karst Hydrology: Concepts from the Mammoth Cave Area*. New York: Van Nostrand Reinhold, 145–174.

Henderson, T. 1985. *Geochemistry of Ground Water in Two Sandstone Aquifer Systems in the Northern Great Plains in Parts of Montana and Wyoming*. U.S. Geol. Surv. Prof. Paper 1402-C, 84 p.

Henderson, T. 1994. Geochemical reduction of hexavalent chromium in the Trinity Sand aquifer. *Ground Water* 32, 477–486.

Hillel, D. 1980. *Introduction to Soil Physics*. New York: Academic Press, Inc.

Hounslow, A.W. 1995. *Water Quality Data*. Boca Raton, FL: Lewis Publishers, 397 p.

Howard, K.W.F. and Beck, P.J. 1993. Hydrogeochemical implications of groundwater contamination by road salt de-icing chemicals. *Jour. Contam. Hydrol.* 12, 245–268.

Hutchins, S.R., Downs, W.C., Wilson, J.T., Smith, G.B., Kovacs, D.K., Fine, D.D., Douglass, R.H., and Hendrix, D.J. 1991. Effect of nitrate addition on biorestoration of fuel-contaminated aquifer field demonstration. *Ground Water* 29, 571–580.

Iqbal, M.Z. and Krothe, N.C. 1995. Infiltration mechanisms related to agricultural waste transport through the soil mantle to karst aquifers of southern Indiana, USA. *Jour. Hydrol.* 164, 171–192.

Jackson, R.E., Patterson, R.J., Graham, B.W., Bahr, J., Belanger, D., Lockwood, J., and Priddle, M. 1985. Contaminant hydrogeology of toxic organic chemicals at a disposal site, Gloucester, Ontario: 1. Chemical concepts and site assessment. Environment Canada; National Hydrology Research Institute Paper No. 23.

Jones, C.C. 1996. Natural attenuation: a remedial option for a dissolved phase plume at an MGP tar disposal site. *Proceedings, 2d Annual International Symposium on Intrinsic Bioremediation*, Southborough, MA: IBC.

Karickhoff, S.W., Brown, D.S., and Scott, T.A. 1979. Sorption of hydrophobic pollutants on natural sediments. *Water Res.* 13, 241–248.

Kasenow, M. 1995. *Hydrogeology and Hydrogeochemistry of a Northern Bog*. Highlands Ranch, CO: Water Resources Publications.

Katz, B.G. 1993. *Biogeochemical and Hydrological Processes Controlling the Transport and Fate of 1,2-Dibromoethane (EDB) in Soil and Ground Water, Central Florida*. U.S. Geological Survey Water-Supply Paper 2402.

Kehew, A.E. and Passero, R.N. 1990. pH and redox buffering mechanisms in a glacial drift aquifer contaminated by landfill leachate. *Ground Water* 28, 728–737.

Kehew, A.E., Passero, R.N., Krishnamurthy, R.V., Lovett, C.K., Betts, M.A. and Dayharsh, B.A. 1998. Hydrogeochemical interaction between a wetland and an unconfined glacial drift aquifer, southwestern Michigan. *Ground Water* 36, 849–856.

Kehew, A.E., Schwindt, F.J., and Brown, D.J. 1984. Hydrogeochemical interaction between a municipal waste stabilization lagoon and a shallow aquifer. *Ground Water* 22, 746–754.

Kehew, A.E., Straw, W.T., Steinmann, W.K., Barrese, P.G., Passarella, G., and Peng, W.S. 1996. Ground-water quality and flow in a shallow glaciofluvial aquifer impacted by agricultural contamination. *Ground Water* 34, 491–500.

Knobel, L.L. and Phillips, S.W. 1988. *Aqueous Geochemistry of the Magothy Aquifer, Maryland.* U.S. Geological Survey Water-Supply Paper 2323, 27.

Komor, S.C. and Anderson, H.W., Jr. 1993. Nitrogen isotopes as indicators of nitrate sources in Minnesota sand-plain aquifers. *Ground Water* 31, 260–270.

Krabbenhoft, D.P., Bowser, C.J., Anderson, M.P., and Valley, J.W. 1990. Estimating groundwater exchange with lakes: 1. The stable isotope mass balance method. *Water Resources Research* 26, 2445–2453.

Krauskopf, K.B. and Bird, D.K. 1995. *Introduction to Geochemistry*, 3d ed., New York: McGraw-Hill, Inc.

Kreitler, C.W. and Jones, D.C. 1975. Natural soil nitrate: The cause of the nitrate contamination of ground water in Runnels County, Texas. *Ground Water* 13, 53–61.

Langmuir, D. 1971. The geochemistry of some carbonate ground waters in central Pennsylvania. *Geochim. Cosmochim. Acta* 35, 1023–1045.

Langmuir, D. and Mahoney, J. 1984. Chemical Equilibrium and Kinetics of Geochemical Processes in Ground Water Studies, in Hitchon, B. and Wallick, E.I., eds., *Proceed. First Canadian/American Conf. on Hydrogeology*, Banff Alta. Alberta Res. Coun. and Nat. Water Well Assoc., 69–95.

Lasaga, A.C. 1984. Chemical kinetics of water-rock interaction. *J. Geophys. Res.* 89, 4009–4025.

LeBlanc, D.R. 1984. *Sewage Plume in a Sand and Gravel Aquifer, Cape Cod, Massachusetts.* U.S. Geological Survey Water-Supply Paper 2218.

Lee, G.F. and Jones, R.A. 1990. Managed fermentation and leaching: An alternative to MSW land fills. *BioCycle* 31(5), 78–80, 83.

Lindberg, R.D. and Runnells, D.D. 1984. Ground Water Redox Reactions: An Analysis of Equilibrium State Applied to Eh Measurements and Geochemical Modeling. *Science* 225, 925–927.

Lorah, M.M. and Herman, J.S. 1988. The chemical evolution of a travertine-depositing stream: Geochemical processes and mass transfer reactions. *Water Resour. Res.* 24, 1541–1552.

Lovley, D.R., Baedecker, M.J., Lonergan, D.J., Cozzarelli, I.M., Philipps, E.J.P., and Siegel, D.I. 1989. Oxidation of aromatic contaminants coupled to microbial iron reduction. *Nature* 339, 297–299.

Lovley, D.R. and Goodwin, S. 1988. Hydrogen concentrations as an indicator of the predominant terminal electron-accepting reactions in aquatic sediments. *Geochim. Cosmochim. Acta* 52, 2993–3003.

Lovley, D.R. and Lonergan, D. J. 1990. Anaerobic oxidation of toluene, phenol, and *p*-cresol by the dissimilatory iron-reducing organism, GS-15. *Appl. Environ. Microbiol.* 56, 1858–1864.

Lovley, D.R. and Phillips, E.J.P., 1988. Novel mode of microbial energy metabolism: Organic carbon oxidation coupled to dissimilatory reduction of iron or manganese. *Appl. Environ Microbiol.* 54(6), 1472–1480.

Lovley, S.R., Chapelle, F.H., and Phillips, E.J.P. 1990. Recovery of Fe(III)-reducing bacteria from deeply buried sediments of the Atlantic Coastal Plain. *Geology* 18, 954–957.

Lyngkilde, J. and Christensen, T.H. 1992. Redox zones of a landfill leachate plume (Vejen, Denmark). *Jour. Contam. Hydrol.* 10, 273–289.

Machavaram, M. and Krishnamurthy, R.V. 1994. Survey of factors controlling the stable isotope ratios in precipitation in the Great Lakes Region. *Israel Jour. of Earth Sci.* 43, 195–202.

Mackay, D.M. and Cherry, J.A. 1989. Groundwater contamination: Pump-and-treat remediation. *Environ. Sci. Technol.* v. 23, 630–636.

Major, D.W., Mayfield, C.I. and Barker, J.F. 1988. Biotransformation of benzene by denitrification in aquifer sand. *Ground Water* 26, 8–14.

Majoube, M. 1971. Fractionnement en oxygéne-18 et en deutérium entre l'eau et sa vapeur. *Jour. Chem. Physics,* 197, 1423–1436.

Manning, D.A.C. 1997. Acetate and propionate in landfill leachates: Implications for the recognition of microbiological influences on the composition of waters in sedimentary systems. *Geology* 25, 279–281.

Mayo, A.L., Nielsen, P.J., Loucks, M., and Brimhall, W.H. 1992. The use of solute and isotopic chemistry to identify flow patterns and factors which limit acid mine drainage in the Wasatch Range, Utah. *Ground Water* 30, 243–249.

McCarty, P.L. and Semprini, L. 1994. Ground-water treatment for cholorinated solvents, in Norris, R.D., et al., *Handbook of Bioremediation.* Boca Raton, FL: CRC Press, Inc., 87–116.

McCarty, P.L. and Wilson, J.T. 1992. Natural anaerobic treatment of a TCE plume St. Joseph, Michigan, NPL site, in *Bioremediation of Hazardous Wastes.* EPA/600/R-92/126, 47–50.

McMahon, P.B., Williams, D.F., and Morris, J.T. 1990. Production and carbon isotope composition of bacterial CO_2 in deep coastal plain sediments of South Carolina. *Ground Water* 28, 693–702.

Montgomery, J. H. and Welkom, L. M. 1989. *Groundwater Chemical Desk Reference.* Boca Raton, Fl. Lewis Publishers.

Morse, J.W., Millero, F.J., Cornwell, J.C., and Rickard, D. 1987. The chemistry of the hydrogen sulfide and iron sulfide systems in natural waters. *Earth Sci. Rev.* 24, 1–42.

Nebergall, W.H., Schmidt, F.C., and Holtzclaw, H.F., Jr. 1972. *College Chemistry*, 4th ed., Lexington, MA: D.C. Heath and Company, 1009.

Nicholson, J.A., Cherry, J.A., and Reardon, E.J. 1983. Migration of contaminants in groundwater at a landfill: A case study: 6. Hydrogeochemistry. *Jour. Hydrol.* 63, 131–176.

Norris, R.D., Hinchee, R.E., Brown, R., McCarty, P.L., Semprini, L., Wilson, J.T., Kampbell, D.H., Reinhard, M., Bouwer, E.J., Borden, R.C., Vogel, T.M., Thomas, J.M., and Ward, C.H. 1994. *Handbook of Bioremediation.* Boca Raton, FL: CRC Press, Inc.

O'Neill, P. 1985. *Environmental Chemistry.* London: George Allen & Unwin.

Owen, J.A. and Manning, D.A.C. 1997. Silica in landfill leachates: Implications for clay mineral stabilities. *Appl. Geochem.* 12, 267–280.

Palmer, A.N. 1990. Groundwater processes in karst terraines, in C.G. Higgins and D.R. Coates, eds., *Groundwater Geomorphology: The role of Subsurface Water in Earth-Surface Processes and Landforms.* Geological Society of America, Special Paper 252, 177–210.

Palmer, C.D. and Cherry, J.A. 1984. Geochemical evolution of groundwater in sequences of sedimentary rocks. *Jour. Hydrol.* 75, 27–65.

Parkin, T.B. and Simpkins, W.W. 1995. Contemporary groundwater methane production from Pleistocene carbon. *Jour. of Environ. Qual.* 24, 367–372.

Pearson, F.J. and White, D.E. 1967. Carbon 14 ages and flow rates of water in Carrizo Sand, Atascosa County, Texas. *Water Resour. Res.*, 3, 23–33.

Perlmutter, N.M and Lieber, M. 1970. Disposal of plating wastes and sewage contamination in ground water and surface water, South Farmingdale–Massapequa area, Nassau County, New York. U.S. Geological Survey Water Supply Paper 1879-G.

Phillips, P.J. and Shedlock, R.J. 1993. Hydrology and chemistry of groundwater and seasonal ponds in the Atlantic coastal plain in Deleware, USA. *Jour. Hydrol.* 141, 157–178.

Plumb, R.H., Jr. 1992. The importance of volatile organic compounds as a disposal site monitoring parameter, in Lesage, S. and Jackson, R.E., eds., *Groundwater Contamination and Analysis at Hazardous Waste Sites.* New York: Marcel Dekker, Inc., 173–198.

Plummer, L.N., Busby, J.F., Lee, R.W., and Hanshaw, B.B. 1990. Geochemical modeling of the Madison aquifer in parts of Montana, Wyoming, and South Dakota. *Water Resour. Res.* 26, 1981–2014.

Plummer, L.N. and Busenberg, E. 1982. The solubilities of calcite, aragonite, and vaterite in CO_2—H_2O solutions between 0 and 90°C and an evaluation of the aqueous model for the system $CaCO_3$—CO_2—H_2O. *Geochim. Cosmochim. Acta* 46, 1011–1040.

Plummer, L.N., Jone, B.F., and Truesdell, A.H. 1976. WATEQF-a FORTRAN IV version of WATEQ, a computer program for calculating chemical equilibrium in natural waters. *U.S. Geological Survey Water Resour. Invest.* 76–13, 61.

Plummer, L.N. and Mackenzie, F.T. 1974. Predicting mineral solubility from rate data: Application to the dissolution of magnesian calcites. *Am. J. Sci.* 274, 61–83.

Qasim, S.R. and Chiang, W. 1994. *Sanitary Landfill Leachate: Generation, Control and Treatment.* Lancaster, PA: Technomic Pub. Co.

Rao, P.S.C. and W.M. Alley. 1993. Pesticides, in W.M. Alley, ed. *Regional Ground-Water Quality.* New York: Van Nostrand Reinhold, 345–382.

Reardon, E.J., Allison, G.B., and Fritz, P. 1979. Seasonal chemical and isotopic variations of soil CO_2 at Trout Creek, Ontario. *Jour. Hydrol.* 43, 355–371.

Remenda, V.H., Cherry, J.A., and Edwards, T.W.D. 1994. Isotopic composition of old groundwater from Lake Agassiz: Implications for late Pleistocence climate. *Science* 266, 1975–1978.

Rightmire, C.T., Pearson, F.J., Jr., Back, W., Rye, R.O., and Hanshaw, B.B. 1974. Distribution of sulphur isotopes of sulphates in groundwaters from the principal artesian aquifer of Florida and the Edwards aqufier of Texas, United States of America, in *Isotope Techniques in Groundwater Hydrology.* v. 2, Vienna: IAEA, 191–207.

Ring, J.J. 1995. The Variability of Carbon Dioxide in the Vadose Zone. M.S. thesis, Western Michigan Univ., Kalamazoo, MI, 129.

Robertson, W.D. 1995. Development of steady-state phosphate concentrations in septic system plumes. *Jour. Contam. Hydrol.* 19, 289–305.

Robertson, W.D. and Cherry, J.A. 1995. In situ denitrification of septic-system nitrate using reactive porous media barriers: Field trials. *Ground Water* 33, 99–111.

Robertson, W.D., Schiff, S.L., and Ptacek, C.J. 1998. Review of phosphate mobility and persistence in 10 septic system plumes. *Ground Water* 36, 1000–1010.

Robie, R.A. and Hemingway, B.S. 1995. *Thermodynamic Properties of Minerals and Related Substances at 298.15K and 1 Bar (10^5 Pascals) Pressure and at Higher Temperatures.* U.S. Geological Survey Bulletin 2131.

Rodgers, R.J. 1989. Geochemical comparison of ground water in areas of New England, New York, and Pennsylvania. *Ground Water* 27, 690–712.

Ronen, D., Margaritz, M., Almon, E., and Amiel, A.J. 1987. Anthropogenic anoxification ("eutrophication") of the water table region of a deep phreatic aquifer. *Water Resour. Res.* 23, 1554–1559.

Rozanski, K., Araguás-Araguás, L., and Gonfiantini, R. 1993. Isotopic patterns in modern global precipitation, in *Continental Isotope Indicators of Climate*, Washington, D.C., American Geophysical Union Monograph.

Ryan, M. and Meiman, J. 1996. An examination of short-term variations in water quality at a karst spring in Kentucky. *Ground Water* 34, 23–30.

Sawyer, C.N., McCarty, P.L., and Parkin G.F. 1994. *Chemistry for Environmental Engineering*, 4th ed. New York: McGraw-Hill, 658.

Scanlon, B.R. 1989. Physical controls on hydrochemical variability in the Inner Bluegrass karst region of central Kentucky. *Ground Water* 27, 639–646.

Semprini, L., Kitanidis, P.K., Kampbell, D.H., and Wilson, J.T. 1995. Anaerobic transformation of chlorinated aliphatic hydrocarbons in a sand aquifer based on spatial chemical distributions. *Water Resour. Res.* 31, 1051–1062.

Semprini, L., and McCarty, P.L. 1991. Comparison between model simulations and field results for in-situ biorestoration of chlorinated aliphatics: Part I. Biostimulation of methanotrophic bacteria. *Ground Water* 30, 239–250.

Senger, R.K. and Kreitler, C.W. 1984. *Hydrogeology of the Edwards Aquifer, Austin Area, Central Texas.* Bureau of Economic Geology, University of Texas at Austin, Report of Investigations No. 141.

Senstius, M. 1958. Climax forms of chemical rock-weathering. *Am. Scientist* 46, 355–367.

Shedlock, R.J., Wilcox, D.A., Thompson, T.A. and Cohen, D.A. 1993. Interactions between ground water and wetlands, southern shore of Lake Michgain, USA. *Jour. Hydrol.* 141, 127–155.

Shuster, E.T. and White, W.B. 1971. Seasonal fluctuations in the chemistry of limestone springs: A possible means for characterizing carbonate aquifers. *Jour. Hydrol.* 14, 93–128.

Shuster, E.T. and White, W.B. 1972. Source areas and climatic effects in carbonate groundwaters determined by saturation indices and carbon dioxide pressures. *Water Resour. Res.* 8, 1067–1073.

Siegel, D.I. 1988. The recharge–discharge function of wetlands near Juneau, Alaska: Part II. geochemical investigations. *Ground Water* 26, 580–586.

Siegel, D.I. 1990. Sulfur isotope evidence for regional recharge of saline water during continental glaciation, north-central United States. *Geology* 18, 1054–1056.

Siegel, D.I. and Pfannkuch, H.O. 1984. Silicate dissolution influence on Filson Creek chemistry, northeastern Minnesota. *Geol. Soc. Amer. Bull.* 95, 1446–1453.

Simpkins, W.W. and Parkin, T.B. 1993. Hydrogeology and redox geochemistry of CH_4 in a late Wisconsinan till and loess sequence in central Iowa. *Water Resour. Res.* 29, 3643–3657.

Smart, P.L. and Hobbs, S.L. 1986. Characterization of carbonate aquifers: A conceptual base, in *Proceedings of the Environmental Problems in Karst Terranes and Their Solutions Conference.* Dublin, OH: National Water Well Association, 1–14.

Snoeyink, V.L. and Jenkins, D. 1980. *Water Chemistry.* New York: John Wiley & Sons, 463.

Solomon, D.K., Poreda, R.J., Cook, P.G., and Hunt, A. 1995. Site characterization using $^3H/^3He$ ground-water ages, Cape Cod, MA. *Ground Water* 33, 988–996.

Spalding, R.F., Exner, M.E., Martin, G.E., and Snow, D.D. 1993. Effects of sludge disposal on groundwater nitrate concentrations. *Jour. Hydrol.* 142, 213–228.

Sposito, G. 1989. *The Chemistry of Soils.* New York: Oxford University Press, 277 p.

Sprinkle, C.L. 1989. *Geochemistry of the Floridan Aquifer System in Florida and in Parts of Georgia, South Carolina, and Alabama.* U.S. Geological Survey Professional Paper 1403-I.

Starr, R.C. 1988. An investigation of the role of labile organic carbon in denitrification in shallow sandy aquifers. Ph.D. thesis, Univ. Waterloo, 148 p.

Starr, R.C. and Gilham, R.W. 1993. Denitrification and organic carbon availability in two aquifers. *Ground Water* 31, 934–947.

Stephany, C.L., Kirby, M.J., and Kehew, A.E. 1998. Spatial and temporal variability in the distribution of nitrate in an agriculturally-impacted aquifer, Cass County, Michigan. *Proceedings, Animal Production Systems and the Environment,* Ames, IA: Iowa State University, S13–S18.

Stumm, W. 1992. *Chemistry of the Solid-Water Interface.* New York: John Wiley & Sons, 428.

Stumm, W. and Morgan, J.J. 1996. *Aquatic Chemistry,* 3d ed. New York: John Wiley & Sons.

Stute, M., Deák, J., Révész, K., Bohlke, J.K., Deseö, É, Weppernig, R., and Schlosser, P., 1997. Tritium/^{3}He dating of river infiltration: An example from the Danube in the Szigetköz area, Hungary. *Ground Water* 35(5), 905–911.

Tardy, Y. 1971. Characterization of the principal weathering types by the geochemistry of waters from some European and African crystalline massifs. *Chem. Geol.* 7, 253–271.

Tinker, J.G., Jr. 1991. An analysis of nitrate–nitrogen in ground water beneath unsewered subdivisions. *Ground Water Mon. Rev.* 11, 141–150.

Thorpe, R.K., Isherwood, W.F., Dresen, M.D., and Webster-Scholten, C.P. 1990. *CERCLA Remedial Investigations Report of the LLNL Livermore Site.* Livermore, CA: Lawrence Livermore National Laboratory.

Thorstenson, D.C., Fisher, D.W., and Croft, M.G. 1979. The geochemistry of the Fox Hills–Basal Hell Creek Aquifer in Southwestern North Dakota and Northwestern South Dakota. *Water Resour. Res.* 15, 1479–1498.

Tóth, J. 1963. A theortical analysis of groundwater flow in small drainage basins. *J. Geophys. Res.* 68, 4795–4812.

Tóth, J. 1980. Cross-formational gravity-flow of ground water: A mechanism of the transport and accumulation of petroleum (The generalized hydraulic theory of petroleum migration), in *AAPG Studies in Geology, No. 10: Problems of Petroleum Migration*, 1585–1597.

Tóth, J. 1984. The role of regional gravity flow in the chemical and thermal evolution of ground water, in B. Hitchon and E.I. Wallick, eds., *First Canadian/American Conference on Hydrogeology, Practical Applications of Ground Water Geochemistry.* Dublin, OH: National Ground Water Association.

Trainer, F.W. and Heath, R.C. 1976. Bicarbonate content of groundwater in carbonate rock in eastern North America. *Jour. Hydrol.* 31, 37–55.

U.S. Environmental Protection Agency. 1988. *Summary of Data on Municipal Solid Waste Landfill Leachate Characteristics Criteria for Municipal Solid Waste Landfills (40 CFR Part 258).* EPA/530-SW-88-038, Office of Solid Wastes, Washington, DC.

Vogel, J.C. 1967. Investigation of ground water flow with radiocarbon, in *Isotopes in Hydrology.* Vienna: IAEA, 355–368.

Vogel, T.M. 1994. Natural bioremediation of chlorinated solvents, in R.D. Norris, et al., *Handbook of Bioremediation.* Boca Raton, FL: CRC Press, 201–225.

Warzyn, Inc. 1991. Sturgis Well Field Remedial Investigation/Feasibility Study. Sturgis, MI: Remedial Investigation Report 12686.

Wassenaar, L.I. 1995. Evaluation of the origin and fate of nitrate in the Abbotsford aquifer using the isotopes of ^{15}N and ^{18}O in NO_3^-. *Appl. Geochem.* 10, 391–405.

Welch, S.A. and Ullman, W.J. 1993. The effect of organic acids on plagioclase dissolution rates and stoichiometry. *Geochim. Cosmochim. Acta* 57, 2725–2736.

Wentz, D.A., Rose, W.J., and Webster, K.E. 1995. Long-term hydrologic and biogeochemical responses of a soft water seepage lake in north central Wisconsin. *Water Resour. Res.* 31, 199–212.

Westjohn, D.B. and Godsy, E.M. 1998. Microbial origin of methane in ground water and glacial deposits at Kingsford, Michigan. U.S. Geological Survey. Administrative report prepared for the U.S. EPA.

White, W.B. and White, E.L. (eds.) 1989. *Karst Hydrology: Concepts from the Mammoth Cave Area.* New York: Van Nostrand Reinhold.

Wiedemeier, T.H., Wilson, J.T., Miller, R.N., and Kampbell, D.H. 1995. United States Air Force guidelines for successfully supporting intrinsic remediation with an example from Hill Air Force

Base. *Proceedings, 1994 Petroleum Hydrocarbon Conference, National Ground Water Association*, 317–334.

Wildung, R.E., Garland, T.R., and Buschbom, R.L. 1975. The interdependent effects of soil temperature and water content on soil respiration rate and plant root decomposition in arid grassland soils. *Soil Biol. and Biochem.* 7, 373–378.

Wilson, J.T., Kampbell, D.H., and Armstrong. 1994. Natural bioreclamation of alkylbenzenes (BTEX) from a gasoline spill in methanogenic groundwater, in R.E. Hinchee, B.C. Alleman, Hoeppel, and R.N. Miller, eds., *Hydrocarbon Bioremediation*. Boca Raton, FL: CRC Press, 201–218.

Wilson, J.T. and Wilson, B.H. 1985. Biotransformation of trichloroethylene in soil. *App. Environ. Microbiol.* 49, 242–243.

Wood, W.W. 1976. Guidelines for collection and field analysis of ground-water samples for selected unstable contituents. *U.S. Geological Survey Techniques of Water Resources Investigations.* Book 1, Chapter D-2, 24.

Wood, W.W. and Low, W.H. 1986. Aqueous geochemistry and diagenesis in the eastern Snake River Plain aquifer system, Idaho. *Geol. Soc. Amer. Bull.* 97, 1456–1466.

Yates, M.V. 1985. Septic tank density and ground water contamination. *Ground Water* 23, 586–591.

Yates, M.V. and Yates, S.R. 1993. Pathogens, in W.M. Alley, ed. *Regional Ground-Water Quality*. New York: Van Nostrand Reinhold, 383–404.

Zack, A. and Roberts, I. 1988. *The Geochemical Evolution of Aqueous Sodium in the Black Creek Aquifer, Horry and Georgetown Counties, South Carolina*. U.S. Geological Survey Water-Supply Paper 2324, 15.

APPENDIX

Species	Name (form)	S° (J·mol⁻¹·K⁻¹)	ΔH_f^0 (kJ/mol)	ΔG_f^0 (kJ/mol)	Source
Aluminum					
Al^{3+}	aq	−332	−538.4	−489.4	1
$Al(OH)^{2+}$	aq	−204	−778	−696.54	3
$Al(OH)_2^+$	aq	−16	−1000	−901.7	3
$Al(OH)_3^0$	aq	108	−1230	−1100.6	3
$Al(OH)_4^-$	aq	160	−1487	−1305.8	3
$Al(OH)_3$	gibbsite	68.4	−1293.1	−1154.9	1
$Al_2Si_2O_5(OH)_4$	kaolinite	200.4	−4119.0	−3797.5	1
$Al_2Si_2O_5(OH)_4$	halloysite	203.0	−4101.5	−3780.7	1
$Al_2Si_4O_{10}(OH)_2$	pyrophyllite	239.4	−5640.0	−5266.2	1
Barium					
Ba^{2+}	aq	8.40	−532.5	−555.4	1
$BaSO_4$	barite	132.2	−1473.6	−1362.5	1
$BaCO_3$	witherite	112.1	−1210.9	−1132.2	1
Calcium					
Ca^{2+}	aq	−56.2	−543.0	−553.6	1
$CaCO_3$	calcite	91.7	−1207.4	−1128.5	1
$CaCO_3$	aragonite	88.0	−1207.4	−1127.4	1
$CaMg(CO_3)_2$	dolomite	155.2	−2324.5	−2161.3	1
$CaSO_4$	anhydrite	107.4	−1434.4	−1321.8	1
$CaSO_4 \cdot H_2O$	gypsum	193.8	−2023.0	−1797.0	1
$Ca_5(PO_4)_3OH$	hydroxyapatite	390.4	−6738.5	−6337.1	1
$CaAl_2Si_2O_8$	anorthite	199.3	−4234.0	−4007.9	1
Carbon					
CH_4	g	186.26	−74.8	−50.7	1
CO	g	197.3	−110.5	−137.1	1
CO_2	g	213.7	−393.5	−394.4	2
H_2CO_3	aq	184.7	−699.7	−623.2	1
HCO_3^-	aq	98.4	−689.9	−586.8	1
CO_3^{2-}	aq	−50.0	−675.2	−527.0	1
CN^-	aq	94.1	150.6	172.4	2

Appendix 1.   *(continued)*

Species	Name (form)	S° ($J \cdot mol^{-1} \cdot K^{-1}$)	ΔH_f^0 (kJ/mol)	ΔG_f^0 (kJ/mol)	Source
Chlorine					
Cl^-	aq	56.60	−167.1	−131.2	1
Chromium					
$FeCr_2O_4$	chromite	146.0	−1444.7	−1343.8	2
CrO_4^{2-}	aq	50.2	−881.2	−727.8	2
$Cr_2O_7^{2-}$	aq	261.9	−1490.3	−1301.1	2
Copper					
Cu^{2+}	aq	−98.0	64.9	65.1	1
Cu_2S	chalcocite	116.2	−83.9	−89.2	1
CuS	covelite	67.4	−54.6	−55.3	1
$Cu_2(OH)_2CO_3$	malachite	166.3	−1054.0	−890.2	1
Fluorine					
F^-	aq	−13.8	−335.4	−281.5	1
Hydrogen					
H_2O	aq	70.0	−285.8	−237.1	1
H_2O	g	188.8	−241.8	−228.6	1
Iron					
Fe^{2+}	aq	−107.1	−91.1	−90.0	1
Fe^{3+}	aq	−280.0	−49.9	−16.7	1
Fe_3O_4	magnetite	146.1	−1115.7	−1012.7	1
Fe_2O_3	hematite	87.4	−826.2	−744.4	1
$Fe(OH)_2$	s (amorphous)	88.	−569.0	−486.5	3
$Fe(OH)_3$	s (amorphous)	106.7	−823.0	−696.5	2
$FeCO_3$	siderite	95.5	−755.9	−682.8	1
FeS	troilite	60.3	−101.0	−101.3	1
FeS	mackinawite	−	−	−93.0	3
FeS_2	pyrite	52.9	−171.5	−160.1	1
Lead					
Pb^{2+}	aq	18.5	0.9	−24.2	1
PbS	galena	91.7	−98.3	−96.8	1
$PbSO_4$	anglesite	148.5	−920.0	−813.1	1
$PbCO_3$	cerrusite	131.0	−699.2	−625.5	1
Magnesium					
Mg^{2+}	aq	−137.0	−467.0	−455.4	1
$Mg(OH)_2$	brucite	63.2	−924.5	−833.5	1
$MgCO_3$	magnesite	65.1	−1113.3	−1029.5	1
Mg_2SiO_4	forsterite	94.1	−2173.0	−2053.6	1
$MgSiO_3$	enstatite	66.3	−1545.6	−1458.3	1
$Mg_3Si_2O_5(OH)_4$	chrysotile	221.3	−4360.0	−4032.4	1
$Mg_3Si_4O_{10}(OH)_2$	talc	260.8	−5900.0	−5520.2	1
$Mg_5Al_2Si_3O_{10}(OH)_8$	chlorite	465.3	−8857.4	−8207.8	3
$Mg_4Si_6O_{15}(OH)_2 \cdot 6H_2O$	sepiolite	613.4	−10,116.9	−9251.6	3

Appendix 1. *(continued)*

Species	Name (form)	S° (J·mol⁻¹·K⁻¹)	ΔH_f^0 (kJ/mol)	ΔG_f^0 (kJ/mol)	Source
Manganese					
Mn^{2+}	aq	−73.60	−220.8	−228.1	1
MnO	manganosite	59.7	−385.2	−362.9	1
$Mn(OH)_2$	pyrochroite	–	–	−616.5	3
MnO(OH)	manganite	–	–	−133.3	3
Mn_3O_4	hausmannite	164.1	−1384.5	−1282.5	1
Mn_2O_3	bixbyite	113.7	−959.0	−882.1	1
MnO_2	birnessite	–	–	−453.1	3
MnS	alabandite	80.3	−213.9	−218.7	1
$MnSiO_3$	rhodonite	100.5	−1321.6	−1244.7	1
Nitrogen					
N_2	g	191.6	0	0	2
NH_3	g	192.77	−45.9	−16.4	1
NH_3	aq	111.3	−80.29	−26.5	3
NH_4^+	aq	111.17	−133.3	−79.4	1
NO_3^-	aq	146.7	−206.9	−110.8	1
Oxygen					
O_2	g	205.15	0.0	0.0	1
OH^-	aq	−10.7	−230.0	−157.3	1
Potassium					
K^+	aq	101.20	−252.1	−282.5	1
KCl	sylvite	82.6	−436.5	−408.6	1
$KAlSi_3O_8$	microcline	214.2	−3974.6	−3749.3	1
$KAl_3Si_3O_{10}(OH)_2$	muscovite	287.7	−5990.0	−5608.4	1
Silicon					
SiO_2	quartz	41.5	−910.7	−856.3	1
SiO_2	amorphous		−899.7	−849.1	3
H_4SiO_4	aq	180.0	−1460.0	−1307.8	1
Sodium					
Na^+	aq	58.45	−240.3	−261.5	1
NaCl	halite	72.1	−411.3	−384.2	1
$NaHCO_3$	nahcolite	102.1	−949.0	−851.2	1
$NaHCO_3 \cdot Na_2CO_3 \cdot 2H_2O$ trona		–	−2682.1	–	1
Na_2SO_4	thenardite	149.6	−1387.8	−1269.8	1
$Na_2SO_4 \cdot 10H_2O$	mirabilite	592.0	−4327.1	−3646.4	3
$NaAlSi_3O_8$	albite	207.4	−3935.0	−3711.6	1
$NaAlSi_2O_6 \cdot H_2O$	analcime	227.7	−3310.1	−3090.0	1
Strontium					
Sr^{2+}	aq	−31.5	−550.9	−563.8	1
$SrCO_3$	strontianite	97.1	−1218.7	−1137.6	1
$SrSO_4$	celestite	117.0	−1453.2	−1339.6	1

Appendix 1.   *(continued)*

Species	Name (form)	$S°$ $(J \cdot mol^{-1} \cdot K^{-1})$	ΔH_f^0 (kJ/mol)	ΔG_f^0 (kJ/mol)	Source
Sulfur					
S	orthorhombic	32.05	0.0	0.0	1
H_2S	g	205.8	−20.6	−33.4	1
H_2S	aq	126	−38.6	−27.7	3
SO_2	g	248.2	−296.8	−300.1	1
HS^-	aq	67.0	16.3	44.8	1
S^{2-}	aq	−14.6	33.1	85.8	1
HSO_4^-	aq	131.7	−886.9	−755.3	3
SO_4^{2-}	aq	18.5	−909.3	−744.0	1
Uranium					
UO_2	uraninite	77.0	−1084.9	−1031.7	1
U^{4+}	aq	−410.	−591.2	−531.0	2
UO_2^{2+}	aq	−97.5	−1019.6	−953.5	2
Zinc					
ZnO	zincite	43.2	−350.5	−320.4	1
ZnS	sphalerite	58.7	−204.1	−199.9	1
$ZnCO_3$	smithsonite	81.2	−817.0	−735.3	1
Zn_2SiO_4	willemite	131.4	−1636.7	−1523.2	2
Zn^{2+}	aq	−109.8	−153.4	−147.3	1
$Zn(OH)_4^{2-}$	aq	–	–	−858.5	2

Sources:
1 Robie and Hemingway, 1985.
2 Krauskopf, and Bird, 1995.
3 Drever, 1997.

INDEX

A

Abiotic transformations of organic compounds, 220
Acclimation period, 229, 237–241
Acid-base reactions, 40
Activity coefficient, 20
Activity diagrams, 93–95
Activity product, 23
Activity, 20
Aerobic respiration, 232–233
Agricultural contamination, 334–344
Alkalinity, 49–53
Alkanes, 176–180
Alkenes, 180–182
Alkynes, 180–182
Ammonification, 152
Anaerobic respiration, 233–235
Analysis of organic compounds, 223
Anion, 1
Atom, 1
Atomic weight, 1
Atomic number, 1
Autotrophic microorganisms, 228

B

Bad water line, 299–300
Bemidji, Minnesota pipeline rupture, 90, 186–190, 215, 247
Biodegradation of chlorinated compounds, 248–254
Biodegradation of petroleum compounds, 241–248
Bioremediation, 242, 254

Bjerrum plot, 43
Borden landfill, 283
BTEX, 182–183, 242–248

C

Calcite, equilibria, 54–59
Carbon tetrachloride, biodegradation, 251
Carbon dioxide, 41
Carbonate mineral equilibria, 53
Carbonic acid, 41
Cation, 1
Cation exchange capacity, 115
Charge balance error, 10
Chebotarev sequence, 289
Chemical potential 19
Chemical equilibrium, 16
Chemolithotrophic microorganisms, 227
Chlorofluorocarbons, 193
Clay mineral structures, 82–86
Cometabolism, 36, 241
Common ion effect, 59
Complexes, 31
Conduit flow systems, 71
Covalent bond, 3

D

Debye-Huckel equation, 29
Denitrification, 152, 270–272
Density, organic compounds, 168
Detergents in ground water, 330
Diffuse double layer, 109
Diffuse flow systems, 71
Dioctahedral minerals, 83